RFID
Implementation

Dennis E. Brown

Mc Graw Hill

New York   Chicago   San Francisco   Lisbon
London   Madrid   Mexico City   Milan   New Delhi
San Juan   Seoul   Singapore   Sydney   Toronto

The **McGraw·Hill** Companies

McGraw-Hill books are available at special quantity discounts to use as premiums and sales promotions, or for use in corporate training programs. For more information, please write to the Director of Special Sales, Professional Publishing, McGraw-Hill, Two Penn Plaza, New York, NY 10121-2298. Or contact your local bookstore.

RFID Implementation

1234567890 DOC DOC 019876

ISBN-13: 978-0-07-226324-4

ISBN-10: 0-07-226324-5

Sponsoring Editor
Jane K. Brownlow

Editorial Supervisor
Janet Walden

Project Editor
LeeAnn Pickrell

Acquisitions Coordinator
Jennifer Housh

Technical Editor
Joseph A. Yacura

Copy Editor
Sally Engelfried

Proofreader
Susie Elkind

Indexer
Karin Arrigoni

Production Supervisor
George Anderson

Composition
International Typesetting
and Composition

Illustration
International Typesetting
and Composition

Art Director, Cover
Brian D. Boucher

Cover Designer
12E

DEDICATION

This book is dedicated to Resa, my wife and my life for nearly 30 years.

CONTENTS AT A GLANCE

CONTENTS

ABOUT THE AUTHOR

Dennis Brown is president of RFID Runner, a company specializing in selection, configuration, sale, and support of RFID systems. The company is distinguished by its focus on data modeling, software integration, and business value beyond the simple requirements of customer mandates. Dennis has been a technical consultant, entrepreneur, and executive for nearly 30 years, helping companies discover and adopt new technologies. A senior executive at Ashton-Tate, he helped guide one of the most successful software companies of all time. He has spent the past ten years selling and installing asset management systems. His clients include Warner Bros., Boeing, Sony Pictures, the National Imagery and Mapping Agency (NIMA), and the National Security Agency (NSA). As a software author, Brown wrote the Financial Planning Language, an early bestseller in the microcomputer arena. He founded Information Access, a software publisher, and KRM/Analytics, a distributor of analytical software. Brown has a BS from Georgetown University and a masters degree from the University of Southern California (USC). He taught business information systems at the university level for six years.

About the Technical Editor

Joseph A. Yacura is co-founder and chief strategist of Supply Chain Management, LLC, a supply chain and cost-management consulting firm. The company provides consulting services related to supply chain management, accounts payable, finance, change management, quality management, and information systems. Joseph has spent over 30 years assisting companies in utilizing processes, quality, financial controls, and technology to enhance their competitive posture. During his early career he spent 10 years with IBM in various management positions. As a senior vice president and/or chief procurement officer at Pacific Bell, American Express, Bank of America, and InterContinental Hotels Group, he gained insight into several different vertical industries. Joseph received his undergraduate degree from the State University of New York at Oswego. He received an MBA and MS degree from Binghamton University and an MQM degree from Loyola University. He has also completed the Senior Executive Program at Stanford University.

ACKNOWLEDGMENTS

I would like to acknowledge and thank Mr. Gaylon Morris of the MetLab Company. When I needed exactly the right fact, Gaylon was always there.

I would also like to thank Mr. Patrick Sweeney of Odin Corporation for his encouragement and support.

Thanks are due to the folks at *RFID Journal*, RFID Update, Lowry Computer Products, and Paxar Corporation. That having been said, responsibility for any errors is strictly my own.

INTRODUCTION

RFID stands for *Radio-Frequency Identification*. It is the name given to systems that put "tags" on objects (items bought and sold commercially, documents, people, animals, vehicles, containers, and so on) so they can be identified, tracked, and managed automatically utilizing radio frequency equipment and supporting computer systems. Tags are very small radio transmitter/responders (also called "transponders") that store and broadcast data. There are a variety of types of tags, and they come in various shapes, sizes, and capabilities. The data in the tag may be a simple identification number that identifies the object, or it may fully describe it. It may give its history or contain relevant information such as warnings or instructions. The tag most often contains just a number, similar to a license plate on a car. The number serves as a key to a record in a database on a "host computer," which stores the actual data. The format of the number is an important consideration, as it must be read by a variety of systems, not all of which may be known at the time the system is designed. When the tags are in the vicinity of a reader, they broadcast their contents. The reader captures this information and sends it to a host computer. The host uses the data in an application program, such as a warehouse management system, inventory system, database or Enterprise Resource Planning (ERP) system. The data may also be stored in a data warehouse where analysts can use it to study, evaluate, and improve the movement of goods; reduce the time involved; and uncover various hidden dependencies and correlations.

RFID is one of many tools that perform automatic collection of data. Similar tools you may be familiar with are bar codes and credit cards with magnetic stripes on them. RFID differs from these in that it is more automatic, and it is capable of higher speeds of operation. Thus, RFID can automatically gather data that might not be collected otherwise. Having collected the data, you can then use it to improve your operations and solve several very complex and time-sensitive problems. In 2004, Wal-Mart and the Department of Defense decided they wanted to solve those problems. They announced a schedule by which their suppliers needed to attach RFID tags on the pallets and cases of goods they shipped to particular distribution centers. Best Buy, Target, Albertson's, and Metro Group in Europe all quickly made similar demands. These demands have been called "mandates" in the press. As people began to understand the scope of the mandates, RFID went from obscurity to celebrity very quickly.

The excitement around RFID centers on its "new" applicability to supply chain management, but RFID is in widespread use in other applications as well. As Chapter 5 describes, RFID technology has applicability for baggage handling, security, asset management, and a host of other uses. These applications, less dramatic in many cases than those in the supply chain, are nonetheless profitable uses of the technology. They benefit from the falling costs of tags and readers and the increased number of persons familiar with the techniques that arise from the supply chain uptake.

Background

RFID's basic technology dates from World War II, where it was used to identify friendly ships and airplanes. Today, low-cost tags and equipment and sophisticated applications support a variety of new uses for the technology beyond aircraft identification.

In June 2003, it was apparent that the following developments were either fully available or would be available soon:

- Protocols were becoming standardized.
- The price of tags was coming down.
- Higher-frequency technology was becoming available.
- The ability to send and receive data at faster data rates was being developed.
- The ability to read multiple tags in the same read zone at the same time was demonstrated.
- Software was becoming available that would utilize the RFID-generated data.

In addition to these developments, the rise of the Internet as a critical part of most companies' infrastructure means that connectivity is always on and always available. Tags anyplace in the world can communicate with a nearby reader and thus signal their location and telemetry (temperature, moisture, pedigree, radiation) to listening applications.

The United States Department of Defense (DoD) and Wal-Mart and other retailers felt they could justify requiring their vendors to tag pallets and cases being shipped to them. In July 2005, tags containing integrated circuit chips were commercially available for about 25–30 cents each in large quantities. By September, one firm announced they had reduced their tag prices to 15 cents each. It is widely believed that when tags reach a price of 5 cents or less each, retailers will require that they be used on individual low-cost items, not just the cases, pallets, and containers. Experts now project that this will happen in the 2007–2008 timeframe.

RFID systems generate data that organizations can use to improve operations. RFID's proponents claim that these systems:

- Improve management of facilities, assets, and resources
- Reduce theft or misplacement of goods, tools, equipment, files, prisoners, or small children
- Keep detailed records of the history of each item in trade
- Increase the speed and accuracy of nearly all business transactions
- Enable suppliers to comply with customer mandates

Accomplishing these goals requires that you keep data that is current, complete, and error free. This perfecting of the data in databases is an often-overlooked prerequisite to utilizing the collected data for decision making. But RFID is unique because other currently available methods of data collection are costly, cumbersome and, in some cases, unreliable and too slow to provide high quality, synchronized data. The payoff for an RFID project comes from the utilization of the data to drive down errors, theft, cost, and inefficiency and to increase the speed of a company's activities and transactions.

Major Vendor Support for RFID

RFID has attracted the attention of the most sophisticated companies in the technology sector. IBM, Accenture, Sun, HP, Microsoft, and Intel have all announced major investments in RFID products and services. Oracle and SAP have RFID initiatives. And companies such as Symbol and Intermec, RFID industry leaders, are seeing substantial growth. Many large consumer goods companies are conducting pilots, and several have begun large-scale deployments.

RFID today delivers value for companies, with the ability to provide real-time error-free information that they can use to solve complex business problems. As the cost of tags continues to come down, companies will find new and innovative uses for the technology. As new global standards are announced and embodied in available products, the barriers to adoption are reduced. With leading companies demanding RFID tags on products and shipping containers, the technology will continue its rapid adoption throughout the total supply chain. Consumers will become better educated about the technology to allay their privacy fears. As with other disruptive technologies, companies that master RFID early will enjoy competitive advantages as compared with those that wait. Companies around the world are starting pilot projects and learning how to work with this exciting technology.

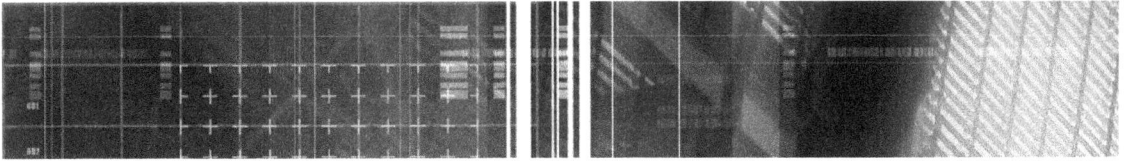

PART I

The Basics

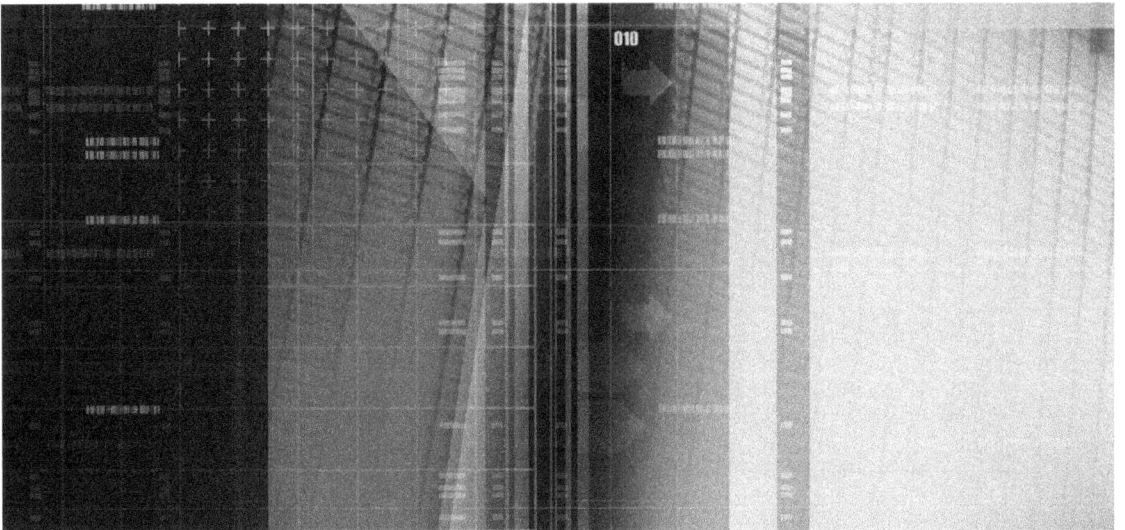

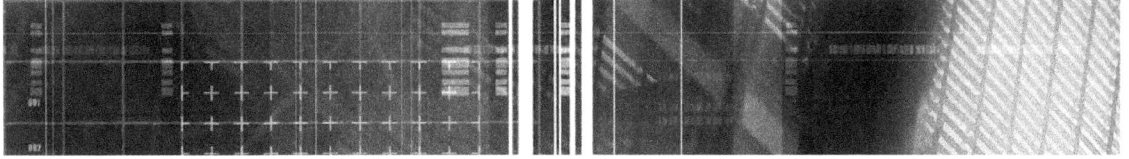

CHAPTER 1

RFID Physics, Standards, and Regulations

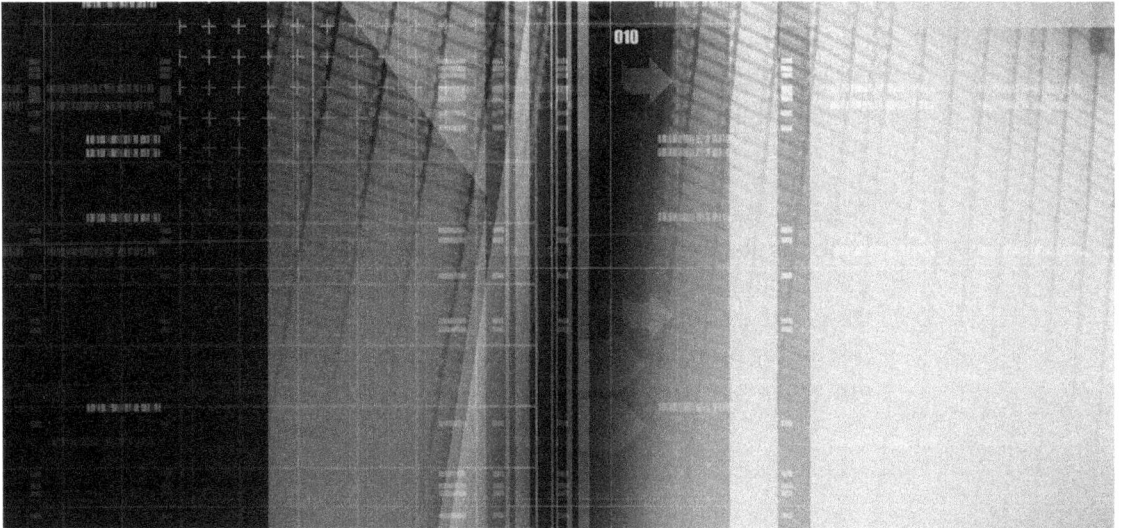

Too many RFID systems are installed based on "one-size-fits-all" starter kits and trial-and-error project methodologies. These two approaches ignore the physics and distinctive characteristics of the radio waves that govern how well the systems will work in practice. The result is predictable: tag-read rates are low and the systems perform ineffectually. This chapter describes the science behind your RFID system, the underlying mechanics on which all RFID systems are built. It then goes on to explain the standards and regulations that enable systems built by different manufacturers to work together.

1.1 RFID Physics

A wave is a disturbance that carries energy from one place to another. Radio waves are created when electrons are passed through a conductor, like an electrical wire. The current creates a magnetic field. Fluctuations in the current produce changes in the magnetic field, creating *waves* of electromagnetic energy. These are called electromagnetic (EM) waves.

There are many types of EM waves you encounter other than radio waves, such as microwaves, gamma waves, x-rays, and light. EM waves oscillate, or vibrate. *Radio waves* are low-frequency electromagnetic waves, which means they oscillate more slowly, and their wavelengths are longer than other types of EM waves. Radio waves share many characteristics with other EM waves, but they differ from light and microwaves in important ways.

Radio waves and light waves both pass readily through air. But unlike light, radio waves can pass through many other materials such as plastic, cardboard, wood, cloth, and so on. Radio waves are less efficient at passing through metal, graphite, sodium, and liquids. We call these *opaque materials*. See Section 1.1.1 for a discussion of this important topic. Different wave frequencies differ in their ability to penetrate opaque materials. We will return to this point often in this book, as it impacts many aspects of implementing your system. Frequency and wavelength are discussed in detail in Section 1.1.2.

You probably have experience with radio waves every day, as you make use of radio and television entertainment and cordless telephones. All of these use radio waves to deliver sound or pictures to your car or house (excluding alternate delivery mechanisms such as cable television). Every day, you make use of the ability of radio waves to carry data such as voice, music, television programming, or information. The method of encoding such data is called *modulation*, which means the wave is mixed with the information in a systematic fashion when the wave is transmitted. The receiver extracts the information when it receives the wave. RFID systems use radio waves to transmit data from the reader to the tag and from the tag to the reader.

The engineering processes that enable us to work with radio waves in communication are very well known to communications engineers, since they underlie our television and radio entertainment, as well as many of our communications systems. However, most Information Technology (IT) people have not had deep training or experience with radio communications technology, and they may find its intricacies challenging. This chapter will help.

1.1.1 Opaque Materials

Unlike light waves, radio waves readily penetrate most materials. They pass easily through paper, cardboard, wood, cloth, plastic, leather, and corn flakes,. This feature underlies a key advantage RFID tags have over bar codes. The RFID tags can be read without having to be within direct view of the reader. A bar code must be scanned individually by a reader, usually involving a person paid to perform the scan. An RFID tag in a well-designed system can be read automatically as it passes by the reader, without the necessity of paying for human intervention.

But opaque materials are difficult for some radio waves to penetrate, as shown in Figure 1-1. Low-frequency and high-frequency waves can penetrate them or go around them, but low-frequency and high-frequency tags are too large and too expensive to support the supply chain initiatives. Ultra-high-frequency tags are much less expensive to produce, so they have become the standard for supply-chain applications, where tag quantities are high and tag costs are a limiting factor. However, high-frequency radio waves are absorbed by water, sodium, and graphite, and they are reflected by metals. If your application involves any of these materials, you will need to engineer the reader/tag relationship to mitigate these effects.

Radio waves have three characteristics that will concern you: frequency, wavelength, and power. As you will see, frequency and wavelength are two sides of the same coin, and they are the most important physical characteristics of the system you install.

Figure 1-1
Radio waves and
opaque materials

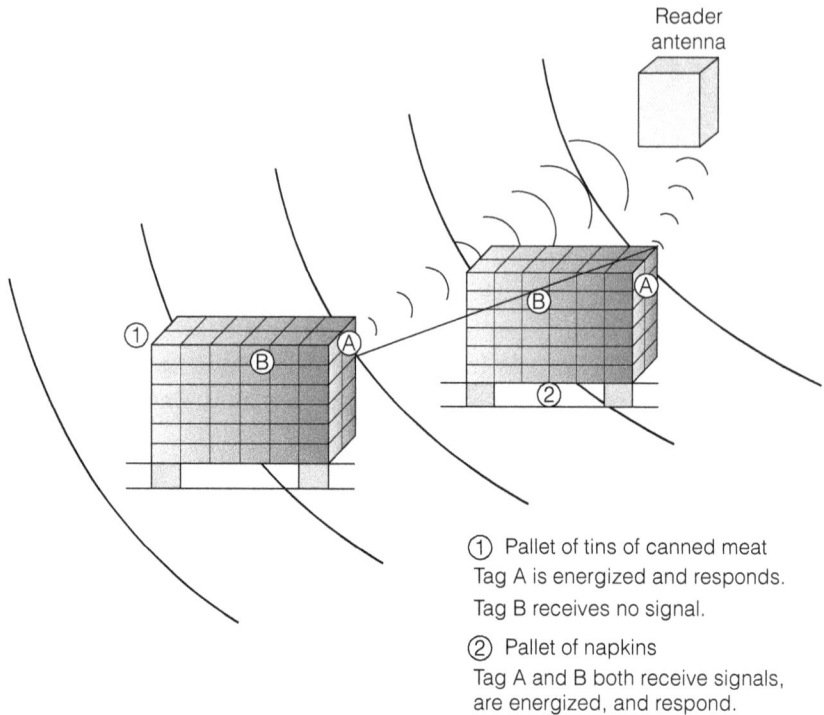

Reader
antenna

① Pallet of tins of canned meat
Tag A is energized and responds.
Tag B receives no signal.

② Pallet of napkins
Tag A and B both receive signals,
are energized, and respond.

1.1.2 Frequency and Wavelength

Radio waves *oscillate*. This means that they repeatedly rise in intensity to a peak, fade to a minimum, and then rise to their peak level. The complete path from one peak to the next is called a *cycle*. The term *frequency* refers to the wave's rate of oscillation, the number of cycles that take place in one second.

Frequency is measured in hertz, named after Heinrich Hertz, an early radio pioneer. In fact, radio waves were originally called hertzian waves. One hertz (Hz) is one complete cycle per second. Thus a radio wave of 960 megahertz (MHz) oscillates 960 million cycles per second.

As shown in Figure 1-2, the distance between two successive peaks is called the energy's *wavelength*. Wavelength is measured, like other distances, in meters, centimeters, and millimeters. A particular frequency is uniquely associated with a particular wavelength; a higher frequency corresponds to a shorter wavelength. Every frequency has a unique corresponding wavelength. In many books and articles, these two terms,

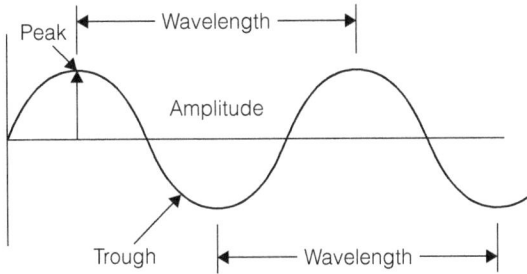

Figure 1-2
Parts of a radio wave

Table 1-1

Wavelengths
for Selected
Frequencies

Frequency Range (band)	Wavelength (approx.)
Low frequency (9–135 KHz)	2300 meters
High frequency (13.553–15.567 MHz)	22 meters
Amateur radio band (430–440 MHz)	69 cm
Ultra-high frequency (860–930 MHz)	33 cm
Microwave frequency (2.4–2.4835 and 5.8 GHz)	12 cm

frequency and *wavelength*, are used interchangeably to designate a particular set of radio wave characteristics. So, for example, a wave 33 centimeters long has a corresponding frequency of 915 MHz.

Table 1-1 shows the wavelengths associated with the five frequency ranges commonly used by RFID systems. You can see that low-frequency waves are much longer than the others, and microwaves are very short, by comparison. These wavelengths explain the behavior of systems that utilize the various frequencies, and why different frequencies are used by different applications. Longer waves can go around obstacles that would stop shorter ones, but they take more power to traverse the same distance. Opaque materials will obstruct shorter waves, but if they are unimpeded, they can go significant distances using very low energy resources. The frequency of a radio wave is determined by the antenna and its associated electronics. A particular antenna has permanent physical characteristics, such as its length and shape. The length of the antenna limits it to a range of frequencies. The antenna is then fine-tuned by the associated electronics to a particular frequency within that range. You use this process when you choose a specific channel on your television set. The physical characteristics of the antenna limit it to what is called the television band, the range of frequencies for VHF television is 54 to 88 MHz, and 174 to 216 MHz.

Each channel in the television band is 6 MHz wide. A complete discussion of antennas is contained in Section 3.3.

NOTE   *For example, your television's VHF Channel 2 uses the range 54–60 MHz. The video is broadcast at 55.25 MHz and the audio is broadcast at 59.75 MHz. A signal is also broadcast on 58.825 MHz to help your television set decode the color signals.*

When you turn the knob or press the button to select a channel, you are actually using the set's electronics to tune your antenna to a particular channel within the television frequency band, the range of frequencies on which television stations broadcast. The antenna will ignore all the other channels. That is why broadcasters on different frequencies in your city do not interfere with one another. Your antenna is tuned to some particular frequency, and it ignores all others. Similarly, RFID tags, and other radio-wave devices, have antennas designed to operate within a particular range, and then the associated electronics fine-tune them to particular frequencies. Systems operating on different frequencies do not interfere with one another. But when there are numerous tags on the same frequency in the reader's read zone at the same time, the system must employ sophisticated anticollision protocols to keep them from interfering with each other. When there are numerous devices operating on the same or nearby frequencies, the likelihood of interference is high.

As listed in Table 1-1, five frequency ranges are utilized in most of the industrial world for RFID applications. Systems operating in each frequency range have different abilities to receive signals across distances and to pass through opaque materials such as metals, sodium, graphite, and liquids.

There are technical differences in the systems associated with each frequency range. Antenna size, for example, is proportional to the wavelength, so each of the various frequency ranges is associated with an antenna of a particular length. Tags for each frequency range also have different cost characteristics and distance restrictions. The governing body for RFID worldwide, EPCglobal, has published standards only for ultra-high-frequency (UHF) RFID systems. EPCglobal recognizes three regions around the world (North America, Europe, and the Far East), and each region has a different UHF RFID frequency allocation. In the Far East, different countries such as Singapore and Korea use different frequencies. Japan has made an allocation for experimental purposes only, not production, and China has not yet made any allocation at all.

This complicates RFID installations for companies with worldwide operations or supply chains.

These technical and regulatory differences make your selection of operating frequency (wavelength) the most important single decision you will make about your system. The characteristics and ultimate cost and success of your installation will flow largely from this single decision.

1.1.2.1 Low-Frequency RFID Systems

The low-frequency (LF) range is 9–135 kHz. This range is characterized by long waves, and it is used by long range radio communications. Typical radio services in this frequency range are aeronautical and marine navigational services (such as LORAN, DECCA), time signal services, and military radio services. To prevent interference with these services, European authorities have defined a protected zone between 70 and 119 KHz where RFID systems are expressly prohibited.

Low-frequency RFID systems use the frequency range 125–134 KHz. This corresponds to radio waves about 2300 meters long. Low-frequency communications are governed by ISO specification 18000-2, and they are uniformly standard around the world. As of this writing, there is no EPCglobal low-frequency specification published, although the organization plans to publish one. ISO 18000 Part 2 defines some standards for low-frequency RFID systems.

In RFID, these systems are mostly used for animal tagging, access control, and vehicle immobilizers. They are not very susceptible to opaque materials. The low-frequency systems are the oldest RFID systems in existence, and thus they have had, until very recently, the largest installed base. The tag-read rates are relatively slow, which is not a problem in their current application set, but it makes them inappropriate for supply chain or RTLS applications. The antennas needed to work with such long waves are relatively large and relatively expensive. Low-frequency tags can be read to distances of about 20 inches. Low-frequency tags are available commercially that can store up to about 60 characters. They can readily penetrate opaque materials. Low-frequency systems are ideal when the environment contains dirt, snow, or mud. Some vendors offer active LF tags (explained in Section 3.2.5.3), but most systems use passive tags (explained in 3.2.1).

1.1.2.2 High-Frequency RFID Systems

The high-frequency (HF) range is located in the middle of the so-called short wave (length) range. This range is suitable for transcontinental connections, so it is used by press agencies and telecommunications.

High-frequency (HF) RFID systems around the world use the 13.553–15.567 MHz frequency. 13.56 MHz waves are about 22 meters long. HF communications are governed by three publications: ISO/IEC specification 18000-3, ISO/IEC specification 15693, and ISO/IEC specification 14443, Parts A and B. These tags are less expensive than the low-frequency tags, and the technology is very stable and mature.

HF communication for RFID actually takes place using magnetic coupling rather than exchange of radio-waves. A complete discussion of magnetic coupling is beyond the scope of this book. High-frequency systems are widely used in smart cards, access control, luggage control, biometric identification systems, libraries, apparel management systems, and laundries. There have been successful HF trials for item-level tagging (for example, by Tesco), but these companies have delayed implementation pending expected improvements in UHF. The shipping company DHL has announced plans to tag all of the one billion parcels they ship every year. They have successfully carried out item-level tagging using 13.56 MHz Identec tags.

In the United States, HF systems are permitted to broadcast up to 30 microvolts effective radiated power (ERP) measured at 30 meters.

1.1.2.3 Amateur Radio Band Systems

The frequency range 430–440 MHz is allocated to amateur radio usage around the world. "Ham" radio operators use it to communicate voice and data. The waves are readily obstructed by buildings or other obstacles. The ISM band 433.05–434.790 is located near the middle of the amateur radio band and is heavily utilized by baby intercoms, wireless thermometers, cordless telephones, walkie-talkies, keyless entry systems, and a host of other users. Mutual interference is not uncommon.

The amateur radio band has emerged as an RFID channel in a number of applications. The frequency range has been called the "optimal frequency for global use of Active RFID." There are several factors cited for this. First, the amateur band has a wavelength of about a meter, so it can propagate around obstacles such as vehicles, containers, and other large obstructions. Second, the power requirement in an amateur band active system is only 1 milliwatt for 100 meter communication, whereas a UHF system would require 100 milliwatts or more. Last, proponents of the amateur band systems note that the frequency is available in most industrial countries. The U.S. Department of Defense (DoD) has used amateur band RFID systems for equipment tagging for over ten years.

1.1.2.4 Ultra-High-Frequency RFID Systems

UHF systems use the frequencies from 860–930 MHz. This corresponds to wavelengths about 33 centimeters. This range is part of the Industrial-Scientific-Medical (ISM) allocation. The other members of ISM are 2.45 GHz and 5.8 GHz. UHF systems are governed by the ISO/IEC specification 18000-6 and the new Gen-2 specification from EPCglobal. Tags and readers manufactured under the older EPCglobal specifications for Class 0, Class 1, and Class 2 also remain in use. UHF tags are by far the least expensive to produce, and these are required by the mandates from retailers and DoD. The tags are used extensively in supply chain applications for pallet and box tagging. They are also used for electronic toll collections, and some companies are using them for asset management applications. Both passive and active UHF tags (see Section 3.2.1) are available, and UHF tags are being produced that will store up to about 8000 characters. UHF tags have a read range of about 4–5 meters.

One problem area for UHF tags is the inconsistent frequency applications around the world. Table 1-2 describes the frequency allocations and bandwidth used for RFID in various countries around the world.

Table 1-2

UHF Frequency Allocations in Selected Regions

Country	Frequency Allocation	Bandwidth	Power Allocation
North America	902–928 MHz	26 MHz	4W EIRP
Japan*	952–954	1 MHz	4W EIRP
Korea	910–914	4 MHz	4W EIRP
Europe	865–868	3 MHz	2W EIRP
Singapore	923–925	2 MHz	2W ERP
Hong Kong	920–925	5 MHz	
Australia	918–926	8 MHz	4W EIRP
Argentina, Brazil, Peru	902–928	26 MHz	4W EIRP
New Zealand	864–929 (spotty)		.5 –1-1 4W EIRP
China	none allocated		
India	865–867 MHz	2 MHz	

*Experimental use only

The EPCglobal Gen-2 standard calls for readers that can read tags across the entire UHF spectrum, which mitigates the inconsistencies in frequency allocations.

Read rates for multiple tags are different in the different countries because of the interaction between bandwidth and the methods utilized to communicate between tags and readers. The wider the bandwidth is, the faster a large group of tags in a read zone can be read. So in the United States, UHF tag read rates are rated at about 1600 tags per second; in Europe, they are rated about 600 tags per second. Older Class 0 and Class 1 tags were rated in the United States around 60 tags per second.

Regulations in each jurisdiction divide the frequency allocation into *channels*. Each channel is 200 KHz wide. In Europe, there are 15 channels available; in North America, the broader bandwidth makes 200 channels available for the readers to use, significantly reducing the opportunities for channel conflict. The net result is that readers will operate faster in North America than in Europe.

1.1.2.5 Microwave RFID Systems

Microwave RFID systems use the 2.45 GHz frequency. They are governed by ISO/IEC specification 18000-4. Note that this is the same frequency as many cordless telephones, and some medical equipment also uses this frequency. Also, 2.45 GHz is the frequency at which water boils, so it is the frequency used by microwave ovens. This means there is a potential for interference between these devices and your RFID system. Read rates for microwave tags are even faster than those for UHF, and they are even more susceptible than UHF tags to degradations from opaque materials.

Microwave frequencies are often used with active tags, and storage capacities up to 16,000 characters are available. The read range is up to about 10 meters for passive tags and 100 meters for active tags. Microwave tags are often used in electronic toll collection applications and for real-time location of assets. Table 1-3 summarizes the characteristics of the various frequencies.

1.1.2.6 How to Choose a Frequency for Your Applications

If you are implementing an RFID system based on a customer's mandate, it is likely that the customer will specify your frequency range for you. If you are implementing a closed-loop system, the nature of the application

	LF	HF	Amateur Band	UHF	Microwave
Frequency	9–135 kHz	13.553–15.567 kHz	430–440 MHz	860–930 MHz	2.4—2.4835 GHz and 5.8 GHz
Typical applications	Aeronautical and marine communication	Smart cards, personal identification	Baby monitors, amateur radio		Microwave ovens, cordless telephones, 802.11 wireless computer networks
Standards	ISO/IEC 18000 Part 2	ISO/IEC 18000 Part 3, ISO 15693, ISO 14443 Parts A and B	ISO/IEC 18000 Part 7	ISO/IEC 18000-6, EPCglobal Gen-1 and Gen-2 standards	ISO/IEC 18000-4
RFID applications	Animal tagging, access control, vehicle immobilizers	Access control, payment ID, item level tagging, luggage control, biometrics, library books, laundries, apparel, pharmaceuticals	Active tags identifying containers, vehicles, other RTLS applications	Supply chain (case and pallet level), asset management, and access control	Security, access control, work tracking for factory automation, RTLS
Opaque materials	Not susceptible	Somewhat susceptible	Somewhat susceptible	Very susceptible	Very susceptible
Read rates	Slow		Fast	Fast	Very fast
Read range	20 inches	1 meter	30 meters	4-5 meters	10 meters

Table 1-3 Frequency Characteristics' Summary

will drive the frequency selection. This may take place because of tag and equipment availability. For example, the systems built for tagging and management of cattle will be low-frequency systems. Systems built for tracking of luggage will be high-frequency systems, although there is some discussion of using UHF. Toll systems will probably be microwave systems, as will real-time location systems (RTLS). In some cases, you will have a choice. For example, RTLS systems are emerging as of this writing that utilize UHF frequencies, although most RTLS systems use microwave frequencies. These additional choices offer differing profiles of read ranges, tag-read speeds, and tag and equipment costs.

1.1.3 Power

The FCC regulates the level of transmission power permitted in RFID systems operating within the United States. The relevant document for UHF systems is Part 15, Rule 247. This rule governs frequencies 902–928 MHz, 2400–2483.5 MHz, and 5725–5850 MHz. Systems are permitted to transmit up to 1 watt of transmitted power with an omnidirectional antenna (an antenna that propagates in all directions, as opposed to a directional antenna that focuses its propagation in one direction). Such systems are permitted up to 4 watts with a directional antenna, if they hop across a minimum of 50 channels. Other countries have their own regulatory bodies with power limitations of their own, as listed previously in Table 1-1.

1.1.4 Radio Communication Risks

Radio transmission is subject to a set of risks that you must manage. Your system must be designed, engineered, tested, installed, and operated so as to avoid external influences that can cause transmission failures, false positives, and false negatives. Even when signals are successfully transmitted and received, adverse external influences may cause degradation of tag read rates below acceptable levels. Good network design and operation will mitigate these risks. Your success in this mitigation will largely determine the success of your project. This section describes the influences that may impact your installation. The three influences we will discuss are

- Interference
- Attenuation
- Multipath distortion

1.1.4.1 Interference

Interference arises when a radio signal encounters other radio transmissions on the same or nearby frequencies from electrical equipment such as other readers, fluorescent lights, bug zappers, malfunctioning or poorly shielded equipment, spark plugs, or cordless telephones. Most cordless telephones today work at 900 MHz, which can cause interference with UHF RFID readers. Others operate at 2.45 GHz, which can interfere with the microwave RFID readers.

1.1.4.2 Attenuation

Attenuation is the weakening of a radio transmission over the distance it must travel. In general, a radio signal weakens as the distance from the source increases. Opaque objects and materials located between source and receiver will weaken the signal further, thus increasing attenuation.

1.1.4.3 Multipath distortion

Multipath distortion results when a radio wave reflects off nearby objects such as walls, furniture, appliances, coins in your pocket, or forklifts. Several copies of the same signal can reach the receiver at slightly different times, creating reception problems. In a related scenario, two or more tags may send signals on the same frequency to a reader, which can create delays. This occurs because neither signal is received correctly and both must be re-sent.

As previously mentioned, these sources can be managed by careful design. Placement and tuning of antennas, analysis of ambient electronic noise, location of antennas and readers, management of traffic around the read zones, careful setting of power levels, and of course, selection of equipment and tag performance all play a role. These design issues are discussed extensively in Chapter 12.

1.2 Deciding on Your Frequency

The first important decision you will make about your implementation is to select your frequency of operation. This section gives important information to help you make this decision. These are the factors to consider:

- Is there a mandate that specifies the frequency you should use?
- Are you purchasing a hardware-software application package for your entire application?
- Is there an international standard that specifies a frequency for your application?
- Is there a typical frequency selection for your type of application? (See Appendix A for more information.)
- What are the salient characteristics of your application?

Let's look at each one of these questions in detail.

1.2.1 Mandates

The following list describes the frequency selection of major mandates around the world:

- The retailer mandates in the United States and Europe have specified a system conforming to EPCglobal's standards. This requires working with UHF passive tags in a specified way. See Chapter 2 for details.

- The United States Department of Defense has specified an EPCglobal system for item level tagging and nonstandard active tags for pallets and cases. See Chapter 6 for details.

- Boeing has issued a mandate for its suppliers, but no commercially available tags meet their requirements. Boeing has announced that Intelleflex will provide the tags they have specified.

- The International Air Transport Association (IATA) has published a specification for baggage handling (RPC 1740c) using UHF tags and ISO-18000-6-C standards for air interface, and ISO-15691 and ISO-15692 to describe data compression and the tag's command set.

- The United States Food and Drug Administration (FDA) has called for RFID use to combat counterfeiting of drugs. Solutions have been proposed using UHF and HF systems. See Chapter 7.

- Editeur, the Book Industry Standards Development Organization, has called for upgrade of the ISBN to conform with EAN13 number and bar code, which can work with EPCglobal UHF RFID systems.

1.2.2 Application Packages

As the industry moves forward, more and more hardware-software application packages become available. The vendors of these packages have determined the optimal frequencies for their application, and the choice is not yours to make. In most cases, their choice is an informed one, and only rarely will you need to change it. Nonetheless, it is good practice for you to review the frequency characteristics listed previously (see Table 1-3) to make sure the frequency selection works for you.

1.2.3 International Standards

The International Organization for Standardization (ISO) has standards, which include frequency choices, for many RFID applications. These standards usually represent a good choice for your application and only rarely will you want to deviate.

1.2.4 Typical Frequency Selection

Appendix A describes the customary frequency choices for major application types.

1.2.5 Application Characteristics

In the unlikely event that none of the preceding circumstances enables you to select a frequency, this section breaks down the application characteristics that vary the most by frequency, so you can profile your application. Frequencies vary in the way they deal with the following characteristics:

- Read range
- Read speed
- Ability to penetrate opaque substances
- Cost

Let's examine each characteristic in detail.

1.2.5.1 Read Range

Some applications require that tags be read at great distances. Active tags operating at UHF or microwave frequencies can be read up to about 300 feet with current technology. Some applications will find the ability to read tags at longer distances to be a disadvantage. For example, consider an electronic payment transaction. A long read range opens the possibility that the person behind you in line might end up paying for your groceries! Similarly, pharmaceutical identification systems require a limited read range. So give thought to what the optimal read range for your application should be, and review Table 1-3.

1.2.5.2 Read Speed

Hundreds of items rushing by on a conveyor belt require a very high read speed. So do applications reading tags on cars on a freeway or applications that count currency. Applications that admit people through a secure doorway can work with much slower speeds. Determine what your read speed needs to be and select a system that delivers it.

1.2.5.3 Opaque Materials

As described in Section 1.1, UHF and microwave systems are seriously impacted by opaque materials. LF systems are impacted much less, and HF systems are not impacted greatly by water. The HF systems cannot penetrate metal, but their wavelength is often long enough to go around a metal obstacle.

1.2.5.4 Cost

UHF passive tags are seeing the greatest decline in unit prices in the marketplace. To some extent, this is the result of manufacturers trying to jump start an early market, but it also represents the real economies of large volumes being manufactured.

Readers can be purchased as stand-alone units or as modules to be inserted in other devices. Very low-cost reader modules can be added to handheld, desktop, and laptop computers. Low-frequency readers feature about a 50 percent cost advantage over UHF readers. High-frequency readers' costs are in between the two.

1.3 RFID Standards Bodies and Organizations

In any emerging technology marketplace there is tension between the forces of standardization and the companies selling proprietary solutions. Proprietary solutions are quicker for a vendor to implement. This is particularly true in the early days of a technology, when positions are just being established by regulatory or governing bodies and by early vendors of the technology. Companies with proprietary systems can build unique features and value propositions and use them as leverage to make sales. They can build value by taking out patents on associated processes, as well as their new unique components. Standards tend to evolve more slowly, as they involve negotiating compromises with competitors and other stakeholders.

But standards benefit the consumer. They level the playing field among competitors. They force competition on issues of price, performance, and quality, rather than on inconsistent lists of features. They tend to force prices down, turning even the most advanced technology into a less costly commodity.

When you install systems that are designed to meet accepted standards, it is more likely that your systems will work with those of other companies, both now and in the future. Your choices of equipment will be broader, your costs of ownership will be lower, components of your system will be more interchangeable, you will likely have an easier time finding knowledgeable people, and you will be able to more easily adopt the inevitable improvements in performance and cost structures.

Some companies view this trade-off differently. They see the competitive advantages of adopting disruptive new technologies without waiting for standards to develop. They accept the costs and risks of adopting new technologies as the price of developing expertise early, being able to influence development of standards, and gaining advantage against their competitors. The RFID marketplace has seen this process play out in a uniquely public drama. Major retailers and the Department of Defense are known for using information technology as a major source of economies, flexibility, and competitive advantage. Among the suppliers required by mandates to implement RFID solutions are several companies who traditionally prefer to await a more developed, orderly, and standardized marketplace before adopting new technologies. Several technology aggressor companies have taken the RFID mandates as an opportunity to review work processes and address heretofore intractable problems of speed-of-decision, time, cost, theft, inaccuracy, and forecasting accuracy. The technology laggards have preferred to do as little as possible, adopting minimally engineered slap-and-ship or tag-and-ship solutions. These installations are the least costly way to comply with the mandates, but they offer no return on the investment other than compliance with the idiosyncrasies of a large customer.

RFID standards are published and managed worldwide by several organizations. Foremost is the EPCglobal organization, which is described in the next section. Applicable standards are also administered by ISO, the International Organization for Standardization. In addition, the Internet Engineering Task Force (IETF) has begun work on a standard called

Simple Lightweight Remote RFID-Reader Protocol (SLRRP). SLRRP is in the early stages of development, and a detailed discussion is outside the scope of this book.

1.3.1 EPCglobal

The most influential organization in RFID is EPCglobal. With the publication of its landmark Generation-2 standard, EPCglobal created a single marketplace in UHF RFID systems, opening the world to compatible interoperable readers, tags, and encoder printers.

1.3.1.1 EPCglobal Origins

We say that RFID had its origins in World War II, and this is correct. But the World War II system for identifying friendly airplanes is very different from RFID today. The World War II system really had little in common with today's commercial tags.

The first U.S. patent for an active RFID tag with rewritable memory was awarded to Mario Cardullo in 1973. That same year, a California entrepreneur named Charles Walton patented a passive transponder used to unlock a door without a key. In the 1970s, a system was designed at Los Alamos National Laboratory that would secure nuclear materials by putting a transponder in a truck and installing readers at the gates of secure facilities. Los Alamos also developed a passive RFID tag to track cows. When cows were ill, they would be given hormones and medicine, but it was hard to make sure the right cow got the right dosage and was not accidentally medicated twice. RFID tags were introduced in order to uniquely identify the particular cow being medicated. The system used 125 KHz radio waves. In the early 1990s, IBM engineers patented a UHF system for RFID. The company never commercialized the technology, and they sold the patents to Intermec, a bar code systems vendor, in the mid-1990s.

As all this was going on, Dr. Sanjay Sarma, a robotics researcher at the Massachusetts Institute of Technology (MIT) was researching the possibility of putting low-cost tags on all products and tracking them through the supply chain. He and his colleague David Brock recognized that the high cost of tags was keeping adoption rates low, and they were frustrated by the cycle the industry was in: low adoption kept costs high, and high costs kept adoption low. Their ideas began to receive attention from other researchers and from executives in the consumer packaged-goods industry as well. They were joined by Kevin Ashton of Procter & Gamble. Ashton was a brand manager for P&G's Oil of Olay product line, and he was struggling with out-of-stocks. He found that fast-selling items are out-of-stock on store shelves about 10 percent of the time. This group of interested parties was joined by Alan Haberman, who is frequently called the "father of the bar code." Haberman, now associated with the Uniform Code Council (the UCC), was looking for a research organization to find a successor technology to replace the bar code. Sarma and Brock agreed to launch the Auto-ID Center at MIT with funding from the UCC, Gillette, and Procter & Gamble. The Center was launched in September 1999. Its first priority was to reduce the prohibitively high cost of RFID tags. The Center proposed increasingly simpler and smaller chips and simplified communication protocols.

The breakthrough idea was the marriage of low-cost chips and ubiquitous networks. The chips would only need to store a relatively small "license plate" that would serve as a pointer to the data stored on servers accessible via the network.

Partner companies Alien Technologies, which had a technique for making inexpensive chips, and Rafsec Corporation, which had a technique for making inexpensive antennas, teamed up and announced that low-cost tags were possible in volume.

Having resolved the hardware problem, the Center turned to the issue of reducing the flood of data produced by the tags. They came up with the Savant, middleware that would decentralize the decision process, making local data available quickly and aggregating, sorting, smoothing, and filtering it up to higher-level Savants. They started working with a small consulting company called Oat Systems to develop the first Savants.

From 1999 to 2003, the Auto-ID Center grew from one lab to six and from three sponsors to more than one hundred. By January 2002, it was outgrowing the universities. The research was becoming less theoretical, more applied, and the commercial demands were growing. The Uniform Code Council (UCC), along with its sister organization, the European Article Number (EAN), agreed to take over the work of the Auto-ID Center and run the EPC. They agreed to set up a not-for-profit organization to do so and to license the technology from MIT. The licensing agreement has a clause that the technology will not be used to tag human beings except in two scenarios: hospitals and defense. The new organization would be called EPCglobal.

EPCglobal was chartered in October 2003 to develop and administer standards for RFID technology. The organization was created to commercialize the work of the Auto-ID Center at MIT. Auto-ID Center Inc. has closed, and their research now goes on under a new organization, the Auto-ID Labs. Auto-ID Labs partially funds and coordinates the work of universities in Cambridge, England; Adelaide, Australia; Keio, Japan; Fudan, China; St. Gallen, Switzerland; and most recently, ICU, Korea. They have transferred their technology and the tasks of creating and publishing standards to EPCglobal. EPCglobal is a subsidiary of GS-1, which is the joint venture of the Uniform Code Council (UCC) in the United States and the EAN organization in Europe. UCC has become GS-1 U.S., and there are national GS-1 bodies in most major countries.

The Auto-ID Center created the original concepts for implementing an "Internet of Things" to parallel the "Internet of Information" we have now. They developed four core concepts, which are the foundation for today's RFID industry:

- Electronic Product Code (EPC)
- ID system
- EPC middleware (Savant)
- EPC Information System (EPCIS)

Each of these concepts is described fully in Chapter 2.

1.3.1.2 EPCglobal First Generation Specification

EPCglobal initially classified tags in a system of tiers. This classification system, and the tags and readers manufactured to its specification, are now referred to as its first-generation effort. Now that the second generation (Gen-2) standard has been released, this older classification standard is called *Gen-1*.

The Gen-1 classification scheme differentiates tag classes by their data-handling capacity. This scheme is still widely used today and is described in Table 1-4.

1.3.1.3 EPC Information System (EPCIS)

The EPCIS is the mechanism to acquire, secure, and deliver real-time data about individual items as they move through the supply chain. It is conceived as the globally recognized and universally available mechanism to provide a pedigree of product identification and movements accessible to authorized users but secured behind firewalls, encoding, and other security measures.

Table 1-4

EPCglobal Gen-1
Tag Classes

Class of Tag	Description
Class 0	Read-only passive tags, UHF. The number is programmed into the tag during manufacture. Class 0 tags were added after the other classes were published.
Class 1	Passive tags, UHF or HF, field programmable Write Once-Read -Many (WORM). The newer standard is UHF Generation 2 as of December 2004. WORM tags are written once and then read many times, as required.
Class 2	Passive read-write. The most flexible type of passive tag. Data can be added to the tag by a qualified reader at any point in the supply chain.
Class 3	Read-write with on-board sensors. Class 3 tags can record parameters like temperature, pressure, and motion, which are recorded by writing into the tag's memory. Class 3 tags are either semipassive or active.
Class 4	Read-write active tags with integrated transmitters. Class 4 tags are like miniature radio devices that can communicate with other tags and readers.
Class 5	Class 4 tags with added capability. Class 5 tags can provide power to other tags and can communicate with devices other than readers. A Class 5 tag is, fundamentally, a reader.

1.3.1.4 EPC Road Map to the Future

In December 2004, EPCglobal approved and released its Second Generation Tag and Reader specification. The Gen-2 specification is the most significant event in EPCglobal's history. The organization claims that the standard opens the door for a number of manufacturers to make interoperable, compatible products quickly. Gen-2 allows for global interoperability of EPC systems and creates a single converged standard. End users won't have to worry about using equipment specified for separate standards such as Class 0, Class 1, Class 2, ISO, or 18000-6 UHF standards.

The EPCglobal Network is the name given to the entire range of EPCglobal specifications: the EPC data format, the air interface specification, and EPCIS. The Gen-2 specification covers a very wide range of functionality. It has been embraced by nearly all of the mandating organizations, so its importance cannot be overstated. As a result, we devote the entire next chapter to describing the EPCglobal Network.

1.3.2 International Organization for Standardization (ISO)

The ISO is a worldwide body that publishes standards for a wide variety of technical specifications. When the Gen-2 standard was ratified by its board of directors, EPCglobal immediately submitted it to the ISO for ratification. The ISO specification 18000 defines the air interface standards, the protocols by which readers and tags communicate, for various frequencies used around the world. Unlike EPCglobal's Gen-2, it does not define data characteristics or the format or content of the data being transmitted and only describes how readers and tags communicate with one another. ISO publication 15693 defines the physical standards and communications protocols for contactless smart vicinity cards. ISO publication 14443 describes physical standards and communications protocols for smart proximity cards. Philips and Texas Instruments have adopted these standards for their RFID offerings, where appropriate.

1.3.2.1 ISO Standard 18000

The ISO 18000 standard specifies the communication protocol: the air interface standard for low-frequency, high-frequency, ultra-high-frequency, and microwave communications around the world. The document is divided into six parts, summarized in this section. The entire document may be purchased from the ISO at www.iso.org.

As the Auto-ID Center and later EPCglobal were working on their specification, many industry leaders questioned whether an additional air interface protocol was needed. They pointed out that the 18000 family was flexible and sufficient and indeed could support all the requirements of a modern RFID system. An advantage, they said, was that the ISO development process required that all participants make their intellectual property available to anyone wishing to develop conforming products on a reasonable and equitable basis. This argument did not prevail, but it underlines the flexibility of the ISO 18000 standard.

Part 1 describes generic parameters for all RFID frequencies; it defines the parameters to be determined in any standardized air interface. It limits its scope to transactions and data exchanges; the means of generating and managing such transactions, which EPCglobal specifies, are outside its scope, as is the definition or specification of any supporting hardware, firmware, software, or equipment. The only requirement is a specified transactional performance.

Part 2 describes parameters for low-frequency (LF) RFID systems. This part of the standard is aimed at devices used in item management application. The specification defines the forward and return link parameters' technical attributes, including operating frequency, operating channel accuracy, occupied channel bandwidth, spurious emissions, modulation, duty cycle, data coding, bit rate, bit rate accuracy, and bit transmission order. The specification distinguishes between two types of tags: Type A tags are full duplex and Type B are half duplex tags. Full duplex Type A tags are powered by the reader both while receiving and transmitting content and operate at 125 kHz. Half duplex Type B tags are powered by the reader during reception, but not during transmission. They operate at 134.2 kHz. An optional anticollision mechanism is also described.

Part 3 describes parameters for high frequency (HF) RFID systems and provides physical layer, collision management, and protocol values for systems operating at 13.56 MHz. The standard provides for two modes of operation that are not interoperable but do not interfere with one another.

Part 4 describes parameters for microwave RFID systems operating at 2.45 GHz. This part of the standard defines the air interface for RFID devices operating in the Industrial-Scientific-Medical (ISM) band used in item management application and contains two modes. The first describes a passive tag operating as responding only to reader inquiries, a modality called *reader-talks-first*. The second is a battery-operated tag operating as tag-talks-first. The only common feature between the two modes is that they share the same frequency. This is markedly different from other 18000-n standards, where there is some overlap in the technologies specified.

Part 5 describes parameters for microwave RFID systems operating at 5.8 GHz.

Part 6 describes parameters for UHF RFID systems and describes one mode with two types. The two types are both reader-talks-first protocols, and both use the same return link but differ in the forward link. Type A uses Pulse Interval Encoding (PIE) in the forward link and an adaptive ALOHA collision arbitration algorithm. Type B uses Manchester encoding in the forward link and an adaptive binary tree collision arbitration algorithm. It is expected that there will be a Type C soon, which will reflect new technological developments. Compliant readers are expected to read all three types.

The EPCglobal specification, a much broader effort (see Chapter 2), incorporates and extends these standards. EPCglobal's specification is being incorporated into a forthcoming ISO standard.

Part 7 describes parameters for RFID systems operating in the amateur band.

1.3.2.2 ISO Standard 15693 Vicinity Card

The ISO 15693 specification is actually a smart card standard. Shortly after it was announced, however, both TI and Philips announced their intention to use it as a guideline for their upcoming high-frequency (13.56 MHz) RFID tags. ISO 15693 defines protocols for transmissions between readers and tags operating within a range of 7–15 centimeters, operating at 13.56 MHz. The standard is broken into three main sections. Section 1 describes the physical characteristics and how contactless cards containing integrated circuits communicate through inductive coupling. It specifies the environment the devices must tolerate, which includes physical bending and twisting, X-ray and ultra-violet susceptibility, electric and magnetic field exposure, and operating temperature ranges. Other concerns are the surface quality of the card for printing purposes and the location of where a slot can be cut without damaging the internal circuits.

Section 2 describes the signal interface, that is, the radio transmission parameters. It describes the power levels, radio carrier frequencies, type and percentage of modulation, data rates, data encoding algorithms, and system timing. These parameters form the communication protocol operating between a card (or tag) and its reader. This section defines the air interface for the contactless connection.

Section 3 describes the transmission protocol, which details the anticollision strategy; it describes protocols necessary to allow multiple tags located together in a reader's field to communicate with the reader.

The transmission protocol section also defines the complete communication interchange for requests and responses. The 15693 command set defines reads, writes, and locking of data for single and multiple blocks. A CRC (cyclic redundancy check) block is included to assure the integrity of the data received.

1.3.2.3 ISO Standard 14443 Proximity Card

This standard consists of four parts. It defines the functioning of a Proximity card used for identification purposes. Philips' Mifare and I-Code cards comply with ISO 14443, as does TI's Tag-It system. The three systems were at first incompatible, but the two companies agreed to share some of their intellectual property and created the ISO 15693 and ISO 14443 standards. Cards adhering to this standard broadcast at 13.56 MHz.

The general application of 14443 cards is limited to about 4 inches, although this is not stated in the specification. The transmission speed is 106 kilobits per second. The standard consists of four parts and describes Type A and Type B cards. Both cards use the same high-level protocol, but they differ in modulation method, coding scheme, and protocol initialization.

Part 1 describes the size of the card (the size of a credit card), and several environmental stresses that the card must withstand. These are tested at the card level, and they also influence antenna design. The card must be functional at all temperatures between 0 and 50 degrees centigrade.

Part 2 describes the RF power and signal interface. Communication schemes are half duplex, with a 106 Kbps (kilobits per second) data rate in each direction. Data transmitted by the card is load modulated with a 847.5 kHz subcarrier. The card must be powered by the radio frequency field; no battery is permitted.

Part 3 describes the initialization and anti-collision protocols. 14443 is designed to permit construction of multiprotocol readers that can communicate with Type A and Type B cards. Both cards awaken upon receipt of a polling command; the reader polls Type A cards, completes all transactions for all Type A cards, and then polls and reads all Type B cards.

Part 4 describes optional data protocols and is optional. The card reports to the reader whether or not it supports the Part 4 commands, as part of its response to the Part 3 polling command.

1.4 United States Federal Communications Commission (FCC)

The Federal Communications Commission (FCC) governs and regulates use of radio waves in the United States. The FCC controls and allocates commercial airwaves for radio, television, shortwave, microwave, and RFID communications. The importance of the FCC for implementers is twofold: first, it specifies which frequencies you are permitted to use without specific licensing; second, it places limitations on the power levels at which you are permitted to broadcast.

RFID devices operating at UHF frequencies in the Industrial-Scientific-Medical (ISM) bands in the United States are allowed to operate without a special license under conditions defined in FCC Rules, section 15.247.

Section 15.247 defines operations within bands 902–928 MHz, 2.4–2.4835 GHz, and 5.725–5.85 GHz. The 902–928 MHz band is used in the United States for supply chain applications.

Part 15–compliant UHF readers are permitted by the FCC to operate at a maximum transmitted power of 1 watt or up to 4 watts with a directional antenna if they hop across a minimum of 50 channels. Field strength limits are specified for a distance of 3 meters from the antenna.

1.5 Singapore

Singapore makes available only 923–925 MHz frequencies for UHF RFID.

1.6 European Telecommunications Standards Institute (ETSI)

The European Telecommunications Standards Institute (ETSI) regulates radio communications in most of the countries in Europe. The regulations governing RFID in Europe are rather more restrictive than those in the United States. The frequency allocation in Europe is established by ETSI regulation 302–208. It allocates the frequency range from 865.6 to 867.6, and divides it into three sub-bands. The sub-band between 865.0 and 865.6 is divided into three channels, and readers are limited to 0.1 watt of output. The sub-band between 865.6 and 867.6 is divided into ten channels. Within those channels, readers may broadcast up to 2 watts EIRP. The sub-band between 867.6 and 868.0 is divided into two channels, and readers may broadcast up to 0.5 watts ERP. The readers hop randomly from one channel to another as they are transmitting. This frequency hopping is part of the protocol to keep readers from interfering with one another. A reader is permitted to operate in a particular channel for up to 4 seconds. Then it must stop for 0.1 second to provide other devices with the opportunity to use the channel. The device may then either listen to the channel and resume transmitting if it is unused, or it may immediately hop to any other unoccupied channel and transmit for up to an additional four seconds. This logic is called Listen-Before-Talk (LBT), and it is effective but slower than the North American alternative.

1.7 Japan

The Japanese MIC (Ministry of Information and Communication) makes available the frequency band 952 to 954 MHz for experimental RFID use only. Antenna power is 10 milliwatts to 1 watt and antenna gain is 6 dBi (see Section 3.3.1.3). Low-power (10 milliwatts and below) systems require no license; high-power systems (10 milliwatts to 1 watt) require a license from the government. Note that Japan is constrained in its allocations because mobile phones in Japan use 800–1000 MHz and interference would be a problem.

1.8 China

Chinese activities are very important for the Wal-Mart and other retail initiatives. More than 60 percent of Wal-Mart's nonfood products are made in China, so any barriers China puts up to RFID adoption and usage have a meaningful impact. China's activities are also important in the rest of world trade. China's status as one of the top-three exporting nations in the world ensures that her decisions will have a worldwide impact.

The Standardization Administration of China (SAC) has announced that its RFID Tag Standard Working Group will develop China's national standards for UHF systems. China has yet to accept any of the three elements of the EPCglobal system: the Electronic Product Code, the tag/reader communication protocol, or the supporting information network.

China has its own numbering system called the National Product Code (NPC), which is not compatible with the EPCglobal system. The NPC has recently been expanded to allow item-level serial numbers, which suggests Chinese authorities intend to continue working with it. As described in Section 2.2.12.1, EPCglobal assigns manager numbers to companies and ensures their uniqueness and charges an annual fee for maintaining these numbers. Some Chinese officials have expressed reservations about requiring Chinese companies to join, pay EPCglobal, and incorporate an additional EPC numbering system.

EPCglobal has made its air protocol technology for communications between tags and readers available license-free and royalty-free to vendors. But United States company Intermec has successfully asserted ownership by patent of key elements of the technology. This gives China motivation and reason to develop and assert its own standards. China has historically been reluctant to pay royalties to Western companies. As of October 2005, however, Chinese authorities announced their intention to adopt the ISO Standard 18000 in its entirety.

As described in Chapter 2, EPCglobal recommends that companies share product-specific information by using a registry of companies with databases containing information about product specifications, manufacturing, or transport. These databases would be maintained by EPCglobal. Chinese officials contend that such a system, involving critical information about their production and logistics, would violate its state security. Chinese officials have made it clear that China intends to develop its own product-information registry. In exchange for fees, EPCglobal, and other organizations, will be able to access China's registry, but the data will reside in China.

China does recognize and allow low-frequency (125–134 KHz) high-frequency (13.56 MHz) and microwave (2.446–2.454 MHz) RFID systems compatible with those of other nations.

1.9 Chapter Summary

RFID networks operate using radio waves for communications. The physics of radio wave communication form a foundation on which RFID systems operate, and they present a host of conditions and constraints which the RFID implementer must deal with. The conditions vary according to the frequency and wavelength selected. The set of standards forms another part of the foundation for your system. Like physics, these standards present a host of very specific conditions and constraints, but they are inconsistent in important ways across different jurisdictions. As we move forward in this book, we will examine how the technology uses both physics and standards to encode and read the tags in the wide variety of circumstances required.

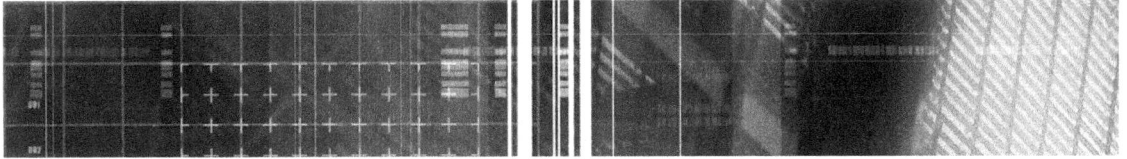

CHAPTER 2

The EPCglobal Network

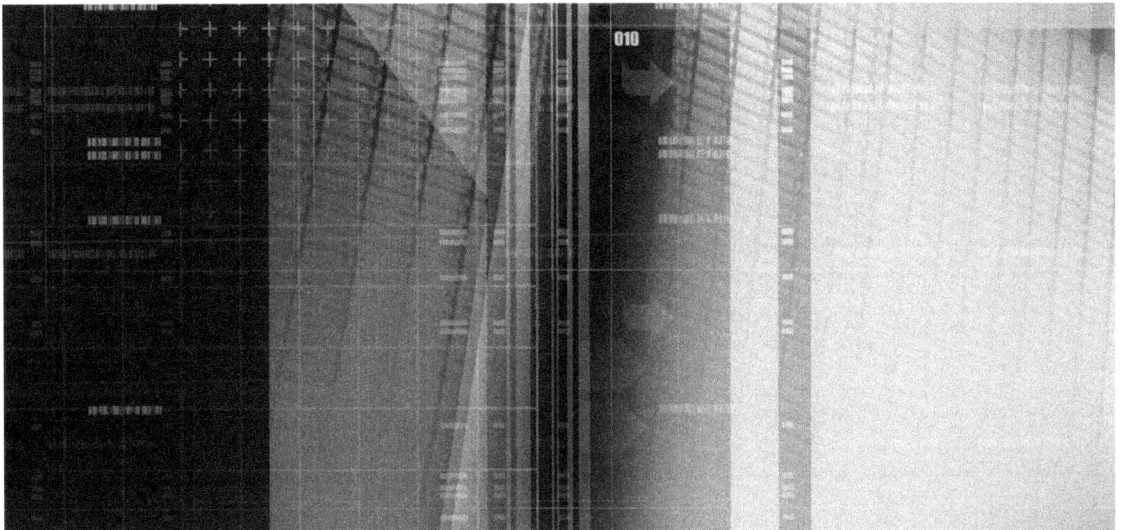

The scope of the projected EPCglobal Network is breathtaking. It is a platform to enable identification, tracking, and tracing of every object in global commerce. It will have the capabilities to uniquely identify trillions of items. It will be able to record every significant change in location, status, or ownership. As capabilities mature, it will record changes in moisture, temperature, security, radioactivity, and vibration as well. It will detect and record tampering events. The network, the conception of Dr. Sanjay Sarma and Dr. David Brock of MIT (see Section 1.3.1.1), will make baggage tracking in airlines efficient and safe; it will make pharmaceuticals and other products harder to counterfeit; food will be delivered fresher; airplanes will be safer; and automobiles will be repaired more efficiently. This technology will have a wide range of positive impacts on industry, commerce, and our personal lives. The mechanism by which this will occur is the EPCglobal Network; this network will reduce the time and costs associated with nearly every process in global trade.

The key components of the EPCglobal Network are an extension of the four core ideas developed by Dr. Sarma and his colleagues at MIT's Auto-ID Center back in 1999:

- The Electronic Product Code (EPC)
- The ID system
- EPC Middleware
- EPC information services

This chapter will explore each of these ideas and its subsequent realization as a part of a functioning system operating today. The system as a whole is represented in the diagram shown in Figure 2-1.

The system starts with the EPC stored on an RFID tag. The EPC is a number that uniquely identifies any item, and the scope of the network is illustrated by the scale of the EPC. At 96 bits, it can uniquely identify and track several thousand *trillion* items. Every one of these items will have a description and a history of its movements recorded on a database accessible to authorized users. Such a system requires a number of components: a standardized way to communicate with readers; a way to filter the relevant, actionable information from the flood of data; and a way to create, update, and manage the database. This chapter describes each of these components and how they work. It starts with the EPC.

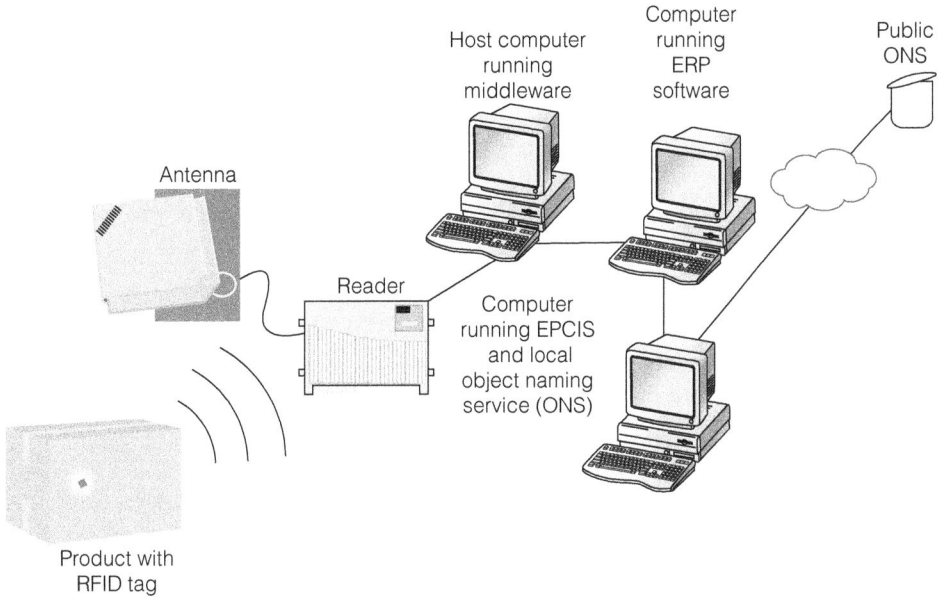

Figure 2-1
EPC network architecture

2.1 Electronic Product Code (EPC)

Commercial automatic identification is dominated today by two systems: the Universal Product Code (UPC) in the United States and the European Article Number (EAN) system in Europe. These numbers identify nearly all consumer goods and many industrial goods purchased around the world. They are typically rendered in bar codes, and they enable identification of the *type* of article being identified. For example, the bar code shown in Figure 2-2 uses UPC encoding to identify the object as a 5.1-ounce box of Jello instant pudding.

Just as the UPC is rendered in a bar code, the number can be rendered in a database or printed on a report. These are different renderings of the same number. In order to facilitate adoption of its Electronic Product Code and to enable organizations to adopt RFID technology more easily, EPCglobal has enabled companies to render their existing UPC and EAN

Figure 2-2
UCC bar
code example

numbers within its structure. The structure is called the Electronic Product Code (EPC).

Like the UCC and EAN, the EPC is simply a standardized number that identifies something. It is an identification number, like a license plate on a car or your social security number. No information beyond the number itself is conveyed in the EPC. The information associated with an EPC, such as company name, product information, history, shipment date, expiration date, and so on, is accessible only to authorized users behind security systems such as firewalls and encryption. Without access to that secure information, the EPC is said to be meaningless.

However, the EPC is more ambitious. First, it does not merely identify the *type* of object. It identifies the *individual* object. Thus, while every 5.1-ounce box of vanilla Jello in the world would have the same UPC number, in the EPC scheme, every single box would have a different number. It could thus be identified and tracked individually.

The EPC provides a flexible framework that supports multiple numbering schemes beyond just the UPCs and EANs, which means the EPC structure, described next, can express such pre-existing identification systems as the VIN (Vehicle Identification Number), CAGE (the DoD's Commercial and Governmental Entity codes), DODAAC (the DoD's ActivityAddress Code), and ISBN (International Standard Book Number). All these and more can be expressed within the EPC framework. Numerous industries besides the ones named have already evolved their own standardized data structures, and EPC provides a way to encompass them all within a single system. This means that EPC-compliant RFID readers will be useful across all but the most obscure application areas without forcing these industries to change their identification schemes.

The structure of the EPC is the key to its flexibility. The structure consists of a variable-length header followed by a series of value fields, as shown in Figure 2-3. The length, structure, and function of the value fields are completely determined by the header value.

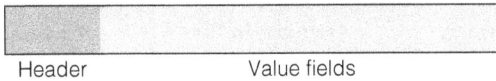

Figure 2-3
General structure
of EPC

Header   Value fields

2.1.1 The Header

When the EPC Gen-2 was approved by EPCglobal's board of directors in December 2004, the company acknowledged that one issue stood in the way of its being accepted by the International Organization for Standardization (ISO). ISO's code scheme began with a two-digit preface, which the organization called the Application Family Interface (AFI). EPC also began its code scheme with a two-digit preface, which it called the EPC header. EPC was concerned that ISO's use of the AFI would not leave enough numbers to accommodate all of the EPCglobal's source numbering needs. The resolution, announced the following month, was the addition of a single bit to precede the EPC. For example, if the bit is set to 0, the next eight bits are interpreted as an EPC header; if it is set to 1, the next eight bits are interpreted as an AFI. This means that an end user can buy tags and readers that will work across both numbering domains.

EPCglobal says of the header:

"The header tells the reader the overall length, identity type, and structure of the EPC tag encoding, including its filter value, if any. The header is of variable length. It uses a zero value in each tier, beginning at the left, to indicate that the header value is drawn from the next longer tier. For encoding defined in this specification (EPC Generation 1 Tag Data Standards Version 1.1 Rev. 1.27), headers are either 2 bits or 8 bits. Given that a zero value is reserved to indicate a header in the next longer tier, the 2-bit header can have three possible values (01, 10, and 11; not 00), and an 8-bit header can have 63 possible values (recognizing that the first 2 bits must be 00, and 0000 0000 is reserved to allow headers longer than 8 bits)."

The header tells the reader which scheme has been used to encode the tag. With the exception of 0011, the general identifier (see Section 2.2.10), all of the schemes thus far approved are derived from bar code formats.

Table 2-1 lists all currently approved EPC encoding schemes and gives their fixed header values.

Table 2-1

EPCglobal Standard
Encoding Schemes

Header in Binary	Length in Bits	Encoding Scheme
01	64	Reserved 64-bit scheme
10	64	SGTIN-64
1100 1110	64	DoD-64
0000 1000	64	SSCC-64
0000 1001	64	GLN-64
0000 1010	64	GRAI-64
0000 1011	64	GIAI-64
0001 1111	96	DoD-96
0011 0000	96	SGTIN-96
0011 0001	96	SSCC-96
0011 0010	96	GLN-96
0011 0011	96	GRAI-96
0011 0100	96	GRAI-96
0011 0101	96	GID-96

2.1.2 Accommodating 64-bit Tags

The vast majority of EPC today uses 96-bit tags. In the early days of
mandate compliance, only 64-bit tags were available, so many companies
purchased 64-bit equipment and tags. The standard makes an effort to
accommodate them.

However, the accommodation to the limits of 64-bit tags is considered
transitional. In the near future, when we no longer need to accommodate 64-bit components, the header format will be revised to enable 255
different encoding schemes. EPCglobal explains that the convoluted
tiered scheme simplifies the header processing required by the reader to
determine the tag data format and the location of the filter value, while
attempting to conserve bits for data values in the 64-bit tags. The accommodation included an external table to map the limited company code
space in the 64-bit tag to the full-sized company code.

2.2 EPC Encoding Schemes

In the EPC Generation 1 Tag Data Standards Version 1.1 Rev. 1.27 (Rev 1.27), released in May 2005, EPCglobal identified 13 encoding schemes for the EPC. One is the general identifier, and twelve identify specific items in global trade. There are schemes to identify such items as returnable assets, items in trade, locations, and shipping containers. Except for the general identifier, they are derived from existing bar code schemes, as listed in Table 2-1. The schemes' design enables companies to utilize the same upstream processes and systems with RFID as they use today with bar codes. Bar code standards are referred to as EAN.UCC notation in this chapter. This makes reference to the EAN and UCC, the two organizations that manage these standards (even though together they also own EPCglobal Inc.). The EPCglobal tags can replace the bar codes that have identified items in trade for the last 25 years. This section describes in detail the bar code encoding schemes currently in use and shows how each is rendered in the EPCglobal format.

Let's look now at the EAN.UCC data schemes.

2.2.1 EAN.UCC Data Formats

The EAN.UCC data formats that follow are the basis for commercial bar coding systems today. They provide the method most companies use to describe seven specific types of items used in commerce.

The EAN.UCC data formats may be preceded by an Application Identifier (AI) that qualifies the number. AIs have been defined for identification, traceability data, dates, quantities, measurements, locations, and many other types of information. For example, AI 00 is a serial shipping container code. Application Identifiers may be concatenated, so the following is valid under EAN.UCC encoding standards.

(00)345123451234567892(420)1000			
AI	SSCC	AI	Postal code

In this example, AI(00) designates the 18-digit number that follows it as an SSCC, and AI(420) designates the 4-digit number that follows it as a ship-to postal code location. 1000 is a postal code number. Note that the AI of (00) is included by common practice. In an EPC, the identity of this number as an SSCC can be inferred by the header so the (00) is unnecessary. EPCglobal permits, but does not require, the (00) for SSCCs.

The EAN.UCC data formats are

- **GTIN**   Global Trade Item Number
- **GLN**   Global Location Number
- **GRAI**   Global Returnable Asset Identifier
- **GIAI**   Global Individual Asset Identifier
- **GSRN**   Global Service Relation Number
- **SSCC**   Serial Shipping Container Code
- **UID**   Department of Defense Unique Identifier

2.2.2 Global Trade Item Number (GTIN)

The Global Trade Item Number (GTIN) is used to identify trade items sold, delivered, warehoused, and billed throughout all commercial distribution channels. GTIN is the number you most often see on products in stores. Its numbering system identifies the manufacturer (company code) and the SKU, or class of the item, but not the unique item.

EAN.UCC supports four data structures for GTIN. Each provides a unique number when it is right justified in a 14-digit database field. The four structures all utilize a company code and an item reference number. The company code can be as small as six digits or as large as nine. The item reference number's length varies to make *company + item reference number* equal to the desired length.

GTIN on a bar code can be rendered in any of four structures: UCC-12, EAN/UCC-13, EAN/UCC-14, or EAN/UCC-8.

- UCC-12 consists of
  - 6, 7, 8, or 9 digits for the EAN.UCC company code
  - 5, 4, 3, or 2 digits for the item reference number
  - 1 check digit (discarded when encoded into EPC)
- EAN/UCC-13 consists of
  - 12 digits for *company code + item reference number*
  - 1 check digit (discarded when encoded into EPC)
- EAN/UCC-14 consists of
  - 1 digit for package indicator to indicate packaging level
  - 12 digits for *company code + item reference number*
  - 1 check digit

The package indicator indicates package variants. For example, you might package the same product into two different cases, with one case containing 10 items and the other containing 12. You could assign package indicator to 1 for the first case and 2 for the second. There is no standard; the codes are up to you.

- EAN/UCC-8 consists of
  - 7 digits containing a *company code + item reference number*
  - 1 check digit

2.2.3 Serialized Global Trade Identification Number (SGTIN)

EPCglobal introduced the Serialized Global Trade Identification Number (SGTIN) as a new identity type based on the GTIN. The GTIN itself does not allow for item-level identification, so EPCglobal added a serial number. A SGTIN could be rendered in EAN.UCC format by concatenating AI(01), the individual item, AI(21), and the serial number. The specification for AI(21) allows 1 to 20 characters, and it permits alphabetic upper- and lowercase letters, as well as certain punctuation marks. However, it is unwise to use anything but numeric digits, since the specification for the EPCglobal serial number permits only digits.

The filter (see Table 2-2) is not part of the SGTIN pure identity but is additional data that is used for fast filtering and preselection of basic logistics types. The values for SGTIN filters are given in Table 2-3. Value 03 should be used when a single item is also the logistical unit, such as a couch, large-screen television, or bicycle.

Table 2-2 describes the precise bit pattern for SGTIN-96.

	Header	Filter Value	Partition	Company Code Index	Item Reference	Serial Number
SGTIN-96	8 bits	3 bits	3 bits	20–40 bits*	24–4 bits*	38 bits
	0011 0000 binary fixed value	See Table 2-3	See Table 2-4	999,999– 999,999,999,999 possible values	9,999,999–9 max decimal range	274,877,906,943 max decimal value

Company width + item reference width is 44 bits.

Table 2-2 SGTIN-96-bit Allocation, Header, and Maximum Decimal Values

Table 2-3

Filter Values for
SGTIN Encodings

00	All others
01	Retail consumer trade item
02	Standard trade item grouping
03	Single shipping/consumer trade item

The SGTIN-96 format includes a value called the partition, which tells the reader how to split the single-number *company code + item reference*. It designates how many of the bits or digits constitute the company code, and how many constitute the item reference. Table 2-4 describes how to interpret the partition values for SGTIN-96.

Table 2-4

SGTIN-96 Partition
Values

Partition Value	Company Code		Item Reference and Indicator Digit	
	Bits	Digits	Bits	Digits
0	40	12	4	1
1	37	11	7	2
2	34	10	10	3
3	30	9	14	4
4	27	8	17	5
5	24	7	20	6
6	20	6	24	7

Example 2-1

If the partition value is 6, the company code is the leftmost 20 bits (6 digits), and the item reference concatenated with the indicator digit constitutes the rightmost 24 bits (7 digits).

2.2.4 Global Location Number (GLN)

The Global Location Number (GLN) provides a standard means to identify legal entities, trading parties, and locations. GLNs for EAN.UCC may
be functional entities such as the purchasing department within a legal
entity, a returns department, or a nursing station. They may be physical
entities such as a particular room in a building, a warehouse, a gate, a
loading dock, or a cabinet shelf. GLNs can apply to legal entities such as
trading partners, buyers, sellers, and companies, as well as subsidiaries
or divisions. The EPCglobal concept of a GLN is slightly different: it is
meant to apply only to the physical location. EPCglobal has articulated a
concept of a serialized GLN but has reserved it pending a decision by the
EAN.UCC community as to whether and how to use it.

EAN.UCC gives the following Application Identifiers to be used
with GLNs.

Ship To – Deliver To <GLN>	AI 410
Bill To – Invoice To <GLN>	AI 411
Purchased From <GLN>	AI 412
Ship For – Deliver For <GLN>	AI 414
<GLN> Invoicing Party	AI 415

EPCglobal's encoding scheme for GLN permits the direct embedding
of the EAN.UCC System Standard GLN on EPC tags. The serial number field is not used. The check digit is not encoded. The SGLN-96 is
described here.

The filter is not part of the SGLN pure identity but is additional data
that is used for fast filtering and preselection of basic logistics types. No
values have yet been designated for SGLN filters.

Table 2-5 describes the encoding for SGLN-96 within the EPC
structure.

The SGLN-96 format includes the partition, which designates how
the single-number *company code + item reference* is to be split. Table 2-6
describes how to interpret the partition values for SGLN-96.

	Header	Filter Value	Partition	Company Code Index	Location Reference	Serial Number
SGLN-96	8 bits	3 bits	3 bits	20–40 bits*	21–1 bits*	41 bits
	0011 0000 fixed value	No values assigned	See Table 2-6	999,999– 999,999,999,999 possible values	9,999,999–0 max decimal range	2,199,023,255,551 max decimal value

Company width + location reference width is 41 bits.

Table 2-5   SGLN-96-bit Allocation, Header, and Maximum Decimal Values

Table 2-6

SGLN-96
Partition Values

Partition Value	Company Code		Location Reference	
	Bits	Digits	Bits	Digits
0	40	12	1	0
1	37	11	4	1
2	34	10	7	2
3	30	9	11	3
4	27	8	14	4
5	24	7	17	5
6	20	6	21	6

Example 2-2

If the partition value is 6, the company code is the leftmost 20 bits (6 digits), and the location reference constitutes the rightmost 21 bits.

2.2.5 Global Returnable Asset Identifier (GRAI)

Certain assets can be used in commerce and returned and reused by their owners as they conduct their business. Generally, a fee is charged for use of the asset. Examples of returnable assets are totes, beer kegs, gas cylinders, rail cars, trailers, fruit containers, and all rental equipment. The Global Returnable Asset Identifier (GRAI) is a tool to unambiguously identify and track returnable assets. In the following discussion, pay attention to the EAN.UCC rendition of GRAI, as compared with the EPCglobal rendition.

The EAN.UCC GRAI contains a fixed-value Application Identifier (8003), plus the EAN.UCC company code, asset type, check digit, and an optional serial number.

Unlike the GTIN, the GRAI is already intended for assignment to individual objects and therefore does not require any additional fields to serve as an EPC item-level identifier.

The EPC encoding scheme for GRAI permits the direct embedding of an EAN.UCC standard GRAI on EPC tags. The header takes the place of the AI, so the (8003) is not translated, nor is the check digit encoded.

The GRAI-96 encoding scheme is described in Table 2-7.

The filter is not part of the GRAI pure identity but is additional data used for fast filtering and preselection of basic logistics types. At this point in time, no values have been assigned for filters. This specification anticipates that useful filter values will be determined once there has been time to consider the possible use cases.

The partition designates how the single-number *company code + asset type* is to be split. Table 2-8 describes how to interpret the partition values.

	Header	Filter Value	Partition	Company Code Index	Asset Type	Serial Number
GRAI-96	8 bits	3 bits	3 bits	20–40 bits*	24–1 bits*	38 bits
	0011 0011 binary fixed value	No values assigned	See Table 2-8	999,999– 999,999,999,999 possible values	9,999,999–9 max decimal range	274,877,906,943 max decimal value

Company code + asset type is 44 bits long.

Table 2-7 GRAI-96-bit Allocation, Header, and Maximum Decimal Values

Table 2-8

GRAI-96
Partition Values

Partition Value	Company Code		Item Reference and Indicator Digit	
	Bits	Digits	Bits	Digits
0	40	12	4	0
1	37	11	7	1
2	34	10	10	2
3	30	9	14	3
4	27	8	17	4
5	24	7	20	5
6	20	6	24	6

Example 2-3

If the partition value is 6, the company code is the leftmost 20 bits (6 digits), and the item reference concatenated with the indicator digit constitutes the rightmost 24 bits.

2.2.6 Global Individual Asset Identifier (GIAI)

The Global Individual Asset Identifier (GIAI) is the license plate for individual fixed assets. Fixed assets are defined as "any property used in carrying on the operation of a business, which will not be consumed through use or converted into cash during the current fiscal period."

The GIAI consists of an Application Identifier (8004), the EAN.UCC company code, and an individual asset reference.

The EPCglobal encoding scheme for GIAI can encode the EAN.UCC standard GIAI codes on EPC tags. Two encoding schemes are specified: GIAI-64 and GIAI-96.

In the 64-bit EPC, the space is too short to fully encode the company code, so an intermediate table is used, as described in previous sections.

The header value 0000 1011 replaces the functionality of the (8004) Application Identifier, so the AI is not encoded in the EPC.

The filter is used only for fast filtering and preselection of basic logistics types. No values are assigned yet, but EPCglobal anticipates that useful

Table 2-9

GIAI-96-bit
Allocation, Header,
and Maximum
Decimal Values

	Header	Filter Value	Partition	Company Code Index	Individual Asset Reference
GIAI-96	8 bits	3 bits	3 bits	20–40 bits*	62–42 bits*
	0011 0100 binary fixed value	No values assigned	See Table 2-10	999,999– 999,999,999,999 max decimal range	9,999,999–0 max decimal range

Company width + individual asset reference width is 82 bits.

filter values will be determined once there has been time to consider the possible use cases.

Table 2-9 describes the layout for GIAI-96 within the EPCglobal structure.

The GIAI-96 format includes the partition value. This number designates how the single-number *company code + individual asset reference* is to be split.

Table 2-10 describes how to interpret the partition values in GIAI-96.

Table 2-10

GIAI-96
Partition Values

Partition Value	Company Code		Individual Asset Reference	
	Bits	Digits	Bits	Digits
0	40	12	42	12
1	37	11	45	13
2	34	10	48	14
3	30	9	52	15
4	27	8	55	16
5	24	7	58	17
6	20	6	62	18

Example 2-4

If the partition value is 6, the company code is the leftmost 20 bits (6 digits), and the item reference concatenated with the indicator digit constitutes the rightmost 62 bits.

2.2.7 Serial Shipping Container Code (SSCC)

The Serial Shipping Container Code (SSCC) is used to uniquely identify logistical units such as a pallet, carton, case, or even truckload of products. SSCC is an 18-digit numeric data structure and contains the EAN.UCC company code.

The SSCC is used in electronic messages such as the Advanced Shipping Notice (ASN). Information about the products that are contained in a particular carton, pallet, or truck is linked to the SSCC and then read or scanned by the receiving party to record movement of trade items through the supply chain.

An example of SSCC in action might be as follows:

1. Buyer sends purchase order to supplier.

2. SSCC is assigned; supplier begins to pick the order.

3. As the order is picked, the picker scans the GTIN for each product as it goes into the box. The SSCC is cross-referenced to the contents.

4. Supplier creates and transmits ASN to the receiving party. ASN contains all information concerning the shipment and each carton or pallet contained.

5. Upon receipt of the ASN, the recipient can use the information to schedule carriers to deliver products and labor and equipment to unload it.

6. Shipment is delivered.

7. Receiver scans SSCC located on the pallet.

8. Receiver's application retrieves the list of expected contents.

9. Receiver can process the carton or pallet without having to open it.

10. Carton or pallet can now be moved to a location for putaway or cross-docking.

The EPC encoding scheme for SSCC permits the direct embedding of the EAN.UCC standard SSCC codes on EPC tags. The check digit is not encoded. The SSCC-96 encoding is described in Table 2-11.

The filter is not part of the SSCC pure identity but is additional data that is used for fast filtering and preselection of basic logistics types. As of this writing, only 010 has been assigned.

	Header	Filter Value	Partition	Company Code Index	Serial Reference	Unallocated
SSCC-96	8 bits	3 bits	3 bits	20–40 bits*	38–18 bits*	24 bits
	0011 0001 binary fixed value	Only 010, logistical shipping unit	See Table 2-12	999,999– 999,999,999,999 possible values	9,999,999– 99,9999 max decimal range	Not used

Company code + serial reference is 57 bits.

Table 2-11   SSCC-96-bit Allocation, Header, and Maximum Decimal Values

Table 2-12

SSCC-96
Partition Values

Partition Value	Company Code		Serial Reference and Indicator Digit	
	Bits	Digits	Bits	Digits
0	40	12	18	5
1	37	11	21	6
2	34	10	24	7
3	30	9	28	8
4	27	8	31	0
5	24	7	34	10
6	20	6	38	11

Example 2-5

If the partition value is 6, the company code is the leftmost 20 bits (6 digits), and the item reference concatenated with the indicator digit constitutes the rightmost 24 bits.

The header is a fixed value, 0011 0001, designating SSCC-96 encoding. The partition designates how the single-number *company code + serial reference* is to be split. Table 2-12 describes how to interpret the partition values.

2.2.8 Global Service Relation Number (GSRN)

The Global Service Relation Number (GSRN) is a unique numbering scheme defined by EAN.UCC that enables a public or private service provider to track any entity's service requirements and needs over a continuing relationship. GSRN's structure enables the service providers themselves to assign GSRNs, and different service relationship types can be identified. GSRNs have a globally unique Application Identifier (8018). The specification requires that each GSRN be unique for each service recipient, but the exact method of assigning them is left to the issuing organization.

EPCglobal does not specify a mechanism for representing GSRNs in EPC.

2.2.9 Unique Identifier (UID)

The Unique Identifier (UID) is a United States Department of Defense program that enables easy access to information about DoD possessions. A full description of the UID data structure is beyond the scope of this book, but EPCglobal has made special efforts to accommodate DoD applications within the scope of the EPCglobal Network.

2.2.10 General Identifier (GID-96)

EPCglobal has added a new encoding scheme, the General Identifier, called GID-96. Unlike the rest of the encoding schemes in this section, it is not a direct mapping of an EAN.UCC scheme and is brand new for EPCglobal. The GID-96 is a 96-bit identifier composed of four sections:

- GID header
- EPC manager number
- Object class
- Serial number

An example of a GID is shown in Table 2-13.

2.2.10.1 GID Header

The header for the general identity type is fixed at `0011 0101`.

Table 2-13

GID-96-bit
Allocation, Header,
and Maximum
Decimal Values

	Header	General Manager Number	Object Class	Serial Number
GID-96	8 bits	28 bits	24 bits	36 bits
	0011 0101 binary fixed value	268,435,455 max decimal value	16,777,215 max decimal value	68,719,476,735 max decimal value

2.2.10.2 General Manager Number

The general manager number identifies an organizational entity, usually a company, responsible for assigning and maintaining the numbers in subsequent fields.

2.2.10.3 Object Class

The object class is the type (or "class") of object that is being identified. It corresponds to an SKU, or a component parts assembly. The number must be unique for each general manager number domain.

2.2.10.4 Serial Number

The serial number is unique within each object class. The addition of the serial number and the ability to extend applications to keep track of individual items is a major element of the power of new RFID implementations. Now finance, warranty and repair, customer service, and logistics systems have a standardized way to identify and track individual items.

2.3 EPC Gen-2 Identification System

The ID system is EPCglobal's specification for classifying tags and readers and the way they communicate with one another. The communication specification is called the "air interface protocol." The ID system prevalent until December 2004 was EPCglobal's First Generation Classification system, now called Gen-1. The second-generation system, called Gen-2, describes a new set of technologies for how RFID readers and tags communicate. Gen-1 was described in Section 1.3.1.2.

In December 2004, EPCglobal's board of directors approved the final publication of its long-awaited second generation identification system specification. The standard presented a host of new features to the world RFID community, and it promised a rapid path to interoperable new products with new capabilities.

2.3.1 Advantages of Gen-2 ID System

EPCglobal claims for their new Gen-2 standard the following benefits:

- It establishes a single UHF specification, where previously users had to wrestle with several incompatible standards, including EPC Class 0, 0+, and Class 1 and two from the ISO. A single Gen-2 reader can read *all* compliant tags.
- It is designed for global operations. Products built to the Gen-2 standard are expected to work with each other in any area of the world. The standard uses frequency and power in a way that complies with the major regional regulatory environments.
- It provides increased security. Products built to the Gen-2 standard have advanced encryption technology, password protection, and password authorization.
- It provides faster read speeds. The standard provides increased tag read and write speeds for supply chain operations as well as the ability to work in dense reader environments such as distribution centers, warehouses, and so on.

2.3.2 New Features of Gen-2 ID System

The following feature list represents the wish list of major RFID users and manufacturers, and they are incorporated into the standard, so they are included in all compliant products:

- Faster and more flexible read speeds
- Robust tag counting
- Dense reader operating modes to mitigate reader interference
- Reader sessions to manage parallel counting by multiple readers
- Enhanced security and privacy
- Extensibility to higher-function systems

Each of these features is described in the next sections in detail.

2.3.2.1 Faster and More Flexible Operating Speeds

Gen-1 RFID systems read tags at a single communication speed appropriate for "typical" situations, but Gen-2 systems can select from among four different communication speeds. Running at top speed, the system can read over 1600 tags per second. If conditions are noisy, the readers will slow reading down to as low as 100 tags per second in order to achieve reads that are accurate and complete.

Gen-2 also sets aggressive targets for quick tag programming. The specification dictates that tags be written at a minimum of 5 per second and sets a target of 30 per second. At this speed, Gen-2 RFID will work on most high-speed assembly and packaging lines.

2.3.2.2 Improvements in Tag Counting

Tags located at the outer edge of the read zone receive only brief, intermittent bursts of power from the reader. They are, therefore, difficult to read reliably. Gen-2 provides several features that address this challenge: the Q protocol, sleep mode, and a dual state inventory scheme.

The Q protocol, which is also used in wireless networks, uses short, simple query-acknowledgment interchanges between readers and tags. This means that a tag experiencing only brief moments of power can still be read. The reader issues a query and each tag responds with a short, randomly generated number. The reader then issues an acknowledgment that includes a single tag's random number, which prompts that tag to send its ID. This process continues until all tags in the reader's field have been counted. By using short random numbers as the basis for sorting, rather than the longer EPC numbers, all tags can be identified uniquely and counted. Reading the tags is similarly enhanced because the Q protocol breaks up transmissions into very small segments and broadcasts each separately.

Gen-1 tags can be put to "sleep" after they have been read. This helps focus a reader's efforts on the tags in the read zone most difficult to read. To begin a new count, the reader issues a "wake-up" or activation command to ensure that all tags will be awake and ready to be read.

Gen-2 refines this process by introducing a dual-state symmetric inventory scheme. The scheme enables a new count to start without tags needing to hear and take the time to react to a wake-up command. Under this approach, a tag changes state every time it is read. As the reader counts each tag in the A state, those tags move automatically to the B state.

As each B-state tag is counted, its state changes back to A. The process is continued until all tags in the read zone have been read.

2.3.2.3 Dense Reader Operating Modes to Mitigate Reader Interference

Readers operating in close proximity to one another can interfere with each other in several ways. Reader B might inadvertently change the state of a tag in Reader A's read zone. Reader B, transmitting on the same frequency as the tags in Read Zone A, could drown out the relatively weak tag signals. Reader B might stimulate a tag to transmit, colliding with a tag transmitting to Reader A on the same frequency, causing interference.

Gen-1 systems use conventional approaches to enable dense readers to function effectively. In the United States, relatively wide bandwidth and FCC-mandated frequency hopping provide a good solution. In other places, the available bandwidth is narrower. To prevent collisions, the readers must listen before they talk, which works, but is much slower.

Gen-2 improves dense reader operation in several ways. First, it uses available bandwidth more efficiently. The Q protocol sorts tags more economically, as the random numbers are much smaller than the EPCs.

Second, Gen-2 enables a specific mode of operation for dense reader conditions. The reader switches itself to dense reader mode when the number of readers is approximately equal to the number of channels. In dense reader mode, reader transmissions are conducted in channels separate from tag transmissions. This keeps readers from overriding tags' transmissions. It isolates tag responses into side channels where they can be better heard. These are called *Miller subcarrier* rates, and they allow the reader to specify side channels of varying widths, according to the overall environmental noise conditions. The narrower the side channels, the easier it is for the reader to read the tag. Gen-2 also provides for an FM signaling mode that enables faster reads than the Miller subcarriers. This mode is vulnerable to noise, so the reader will only use it when conditions permit.

2.3.2.4 Reader Sessions

Gen-2 introduces the notion of reader sessions to manage parallel counting by multiple readers. Sessions allow different readers to work with the same tag at the same time, each conducting its own "session" with the tag. Readers are assigned any one of four logical sessions for reads. This way, up to four separate readers can be legitimately identifying the same tag without interference and without having to wait for any one reader to complete its sort. Note that this avoids a variation of the dense reader problem, where Reader B might change the state of a tag that was being read by Reader A in the middle of A's sort. The monitoring and allocation of sessions can be managed by readers or from a central control point.

2.3.2.5 Enhanced Security and Privacy

The topic of privacy and the impact RFID systems can have on our personal privacy is very much a factor in the acceptance of the technology by the general public. This is discussed in Chapter 14.

Gen-2 uses 32-bit passwords. These passwords can be used to activate a kill command, which permanently shuts down a tag, as required by some consumer protection legislation. Passwords can also be used to access and relock a tag's memory.

The Q protocol never communicates an entire tag ID over the air at one time. Signals are so scrambled that an eavesdropper listening in would not be able to determine the EPC number.

Gen-2 also introduces optional handle-based commands for securely commanding the tags. When the tag is inventoried, it can randomly generate a 16-bit number, or *handle*, which is used by the reader to enhance the security of read, write, and erase commands.

2.3.2.6 Extensibility to Higher-Function Systems

All of the preceding enhancements are part of the Gen-2 Class 1 specification released in December 2004. But Class 2 systems, whose specification is not yet released, will add larger memory tags, higher levels of read-write functionality, and potentially higher security. Class 3 systems will include a battery on the tag for longer read ranges and the ability of the tag to communicate sensor-based data such as temperature, pressure, shock, vibration, radiation, and others. Tags of all these classes will be readable by EPC readers, making it possible to create more and more powerful functionality in supply chain and other applications.

2.4 EPC Middleware

EPC Middleware serves four functions in the EPCglobal Network. These functions all help bridge the gap between devices and data processing systems, between low-level pools of data and business information, and between remote devices and software applications. The functions are

- Resolving EPC numbers into useful information by accessing EPCIS
- Filtering raw data into useful information by processing it through ALEs
- Managing a disparate group of readers through a common interface
- Transforming data into usable formats and communicating it to the correct applications

2.4.1 Resolving EPC Numbers

EPC Middleware takes the EPC for any item of interest and queries the ONS to locate accessible information about it. ONS is described in Section 2.5.1.

2.4.2 Filtering Raw Data

RFID readers produce a flood of very low-level information. Up to three or more times per second, the reader scans for tags near its antennas. Each time, it reports a list of every tag it sees. This is a raw data stream. It is not business information. Business applications need to know which tags passed through a portal, which were put on a shelf, and which were packed in what carton and sent to which customer. They need to know which containers are located where, which beer kegs were shipped to Customer B and left there, and so forth. This business information can be derived from the raw data stream and supplemented by contextual information such as the reader location, date, and time and perhaps the badge number of the nearby employee, the designator for the container the item is stored in, the order number that initiated the transaction, the name of the company shipping the goods, what other items were in the shipment or on the order, or the flight number the item came in on. A considerable amount of processing is required to get all this information together as it is needed. In general, we call this function *filtering and collecting*. This need was recognized in the early days of the Auto-ID Center, and they gave the name "Savant" to the function that would meet it. Today, we call this function EPC Middleware.

As of this writing, EPCglobal is expected to ratify its first software standard for EPC Middleware. The standard is called Application Level

Events, or ALE. ALE enables the middleware to filter and sort the raw data so any application sees only relevant, significant events. The middleware is the mechanism for making sure that applications are not overwhelmed by a flood of irrelevant data. The ALE standard draws heavily on the work of Steve Rehling, Director of Information Technology for Procter & Gamble. He identified a small number of specific operations (listed next) that would draw business-level information from the raw data stream. More than 60 companies participated in the EPCglobal working group that created the ALE specification. In November 2004, six companies created implementations based on the ALE specification and brought them together for two days of interoperability testing. The operations include removal of duplicate reads, aggregation over intervals of time, grouping, counting, differencing, and filtering (that is, removing certain EPCs based on company ID, product ID, filter value, or other data on the tag).

Through ALE, the application can specify:

- Which locations it wants to read from, where a location maps to one or more readers or antennas
- What interval of time to accumulate the data (either absolute time, or time triggered by external events such as a motion detector)
- How to filter the data
- How to group the results
- Whether to report the complete set of currently visible tags or just additions or deletions

An application can subscribe to an event and receive a customized report or message whenever the event occurs. An event can be the passage of a tag with defined characteristics by a particular reader or set of readers. Any number of applications can subscribe to the same or different events. In this way, different applications get data from the same readers without impacting one anther.

2.4.3 Managing Readers

There is no specification for this functionality, but middleware products generally provide a management framework that enables centralized control of the reader network. This usually includes tracking the reader's read rates, monitoring its health, and changing settings.

If the reader is software-upgradeable, the middleware may be able to drive the upgrades as well.

2.4.4 Reformatting Data

The choices and capabilities distinguish various middleware products from one another, and the ability to reformat data is another capability that is not standardized. However, different application programs you run will need data in different formats, and they may be quite specific in their requirements about record structure, spacing, delimiters, punctuation, leading zeros and blanks, and how metadata is used. EPC Middleware can provide capabilities to deliver the data in the form your applications will need it.

2.4.5 Middleware Vendors

The following vendors offer middleware products: ConnecTerra (www.connecterra.com), GlobeRanger (www.globeranger.com), IBM (www.ibm.com), Manhattan Associates (www.manhattanassociates.com), Microsoft (www.microsoft.com), OATSystems (www.oatsystems.com), Oracle (www.oracle.com), RF Code (www.rfcode.com), SAP (www.sap.com), Savi Technology (www.savi.com), Sun Microsystems (www.sun.com), TIBCO Software (www.tibco.com), and webMethods (www.webmethods.com). New suppliers are continually entering this space, providing new features and functionality.

2.5 EPCglobal Network Information Services Functionality

We have said throughout this chapter that the EPC is a license plate, a number that is meaningless in itself and serves only as a pointer to a database record located in a secure server "somewhere." The EPC serves only as a database lookup key that enables the server to locate and serve the information about the item. The database information itself is provided by the EPC Manager, the organization responsible for the item, and possibly by others as the item makes its way through the supply chain. The database is distributed across any number of physical servers.

EPCIS is divided functionally into three parts:

- Object Naming Service (ONS)
- EPC Information Services (EPCIS)
- EPC Discovery Services (EPCDS)

2.5.1 Object Name System (ONS)

ONS is the directory of information sources available to describe EPCs in the supply chain. ONS contains an entry for every registered EPC, and it has pointers to all of the recognized sources of information about that EPC. The root ONS is managed by Verisign (www.verisign.com), the company that manages the root Domain Name System for the entire Internet. ONS may be implemented with a local cache for any company location. This local cache is used to reduce the need to query the global ONS for each object seen by a reader, since frequently sought or recently sought values can be stored locally. The local cache acts as the first resource for ONS queries and may also manage lookup of private internal EPCs for asset tracking.

ONS points to two types of information: static information and dynamic information. Static information describes the item and gives its permanent characteristics. These might include its name, weight, source, maintenance instructions, and any other information the manufacturer chooses to publish about the item. Dynamic information may be spread physically across several databases. It may describe, for instance, the date and time of its arrival at a given wharf or warehouse. The dynamic information is posted by various organizations as they receive and distribute the item.

ONS is implemented hierarchically, so an ONS record may point to other ONS servers that contain pointers to the information. In this, its structure is similar to that of the Internet itself.

2.5.2 EPC Information Service (EPCIS)

EPC Information Service is the set of data repositories that stores information about every item in the supply chain. The structure of EPCIS is distributed; different parts are controlled, owned, and operated by different companies. Each company is pursuing its own business model, and they are writing and reading data to and from EPCIS in a standardized way. EPCIS makes it easy to secure, find, and share information as needed.

The data is stored in a language called Physical Markup Language (PML). The server on which it resides is sometimes called the PML server. PML is a language for describing physical objects and is a dialect of XML, the standard way to make data readable by different types of computer systems. PML enables organization and storage of any relevant information about a product. For example, it may give location information or the physical properties of an item, or it may describe the number of units in the lot or crate or pallet or dates of manufacture or expiration.

The authoritative reference for PML is the PML Core Specification document available at http://www.epcglobalinc.org/standards_technology/secure/v1.0/PML_Core_Specification_v1.0.pdf.

The middleware, having acquired the EPC from the tag, queries the ONS to find the location of the right PML server. It then downloads the descriptive information from the EPCIS servers. Like various Internet servers, EPCIS servers are owned by private companies that make them available to authorized users on the basis of their own business rules.

2.5.3 EPC Discovery Services (EPCDS)

Discovery Services is a chain-of-custody registration service. The EPC Discovery Service (EPCDS) provides a directory of all EPCIS servers that contain information about a particular item.

2.5.4 How EPCglobal Network Works

The EPCglobal Network collects information by following these steps for every product for every step in the supply chain:

1. The tag's EPC is registered with ONS when the tag is commissioned. The tag remains with the product as it moves through the supply chain.

2. The information associated with this particular item, case, or pallet is added to the manufacturer's EPCIS.

3. A pointer to this information is stored in the EPCDS.

4. When the item, case, or pallet is shipped, the shipment information is registered with the EPCIS.

5. When the product is received at the next point in the supply chain, such as a distributor site, it is automatically read and its arrival is registered in the distributor's EPCIS.

6. Pointers to the information about this item are registered with the EPCDS.

Figure 2-4 illustrates the entire EPCIS process.

This process, repeated for every product in the world, for every step in the supply chain, creates a treasure trove of data that business managers can use to authenticate products, determine accurate inbound and outbound inventory loads, optimize logistics, prevent counterfeiting, prevent loss and theft, correct and control misshipments, and a host of other activities.

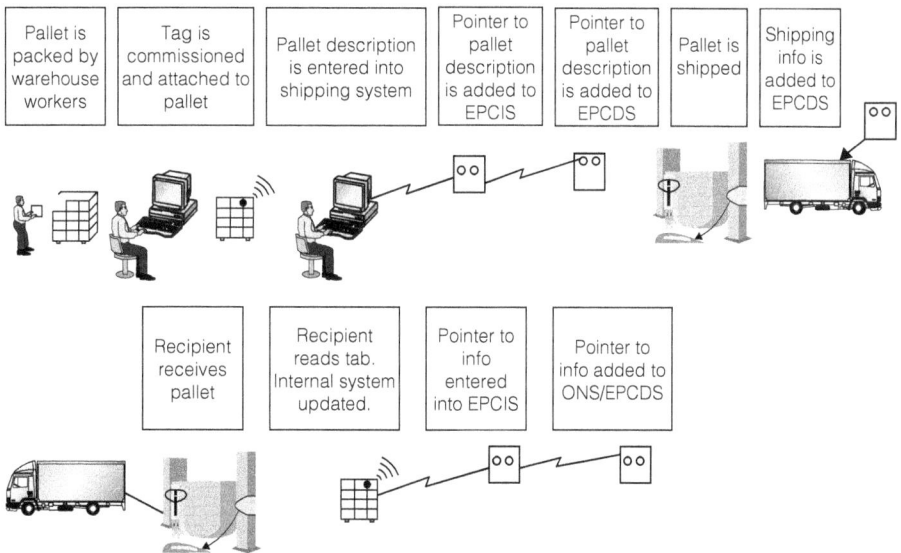

Figure 2-4
EPCglobal process

2.6 Chapter Summary

The EPCglobal organization has grown the original four concepts of the Auto-ID Center to create a force for disruptive change in global trade. The vision of inexpensive tags on billions of items, with descriptive and pedigree information, could hardly be more revolutionary. Each of the ideas—the Electronic Product Code, the standardized identification system, middleware, and information services—began as a simple idea and has been developed into standard products and processes throughout the realm of commerce. Structures from current bar code systems can be rendered in the new formats. The new Gen-2 specification is a standard that will enable users to read tags from all over the world in all locations with standardized readers. The new Application Level Events (ALE) interface represents an unprecedented standardization of data operations to make the RFID data useful to businesses in a host of applications. The information services, the server network, makes the data available to qualified users any place on the Internet.

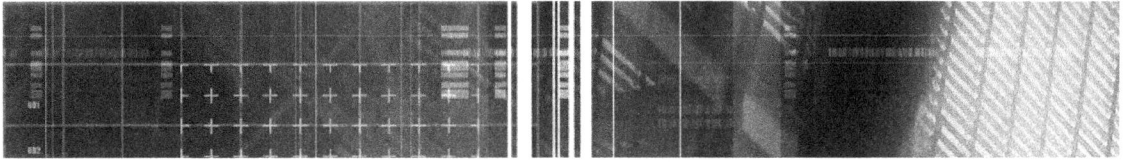

Components of an RFID System

An RFID system is a set of components that work together to capture, integrate, and utilize data and information. This chapter describes each of these components in terms of its function, the various forms it comes in, and the issues the implementer must resolve in order to build a functioning system. The components are as follows:

- Sensors
- Tags
- Antennas
- Connectors
- Cables
- Readers
- Networks
- Controllers
- Data
- Software
- Information Services

Figure 3-1 presents a simple overview of how the components fit together to make a complete system.

Figure 3-1
RFID system overview

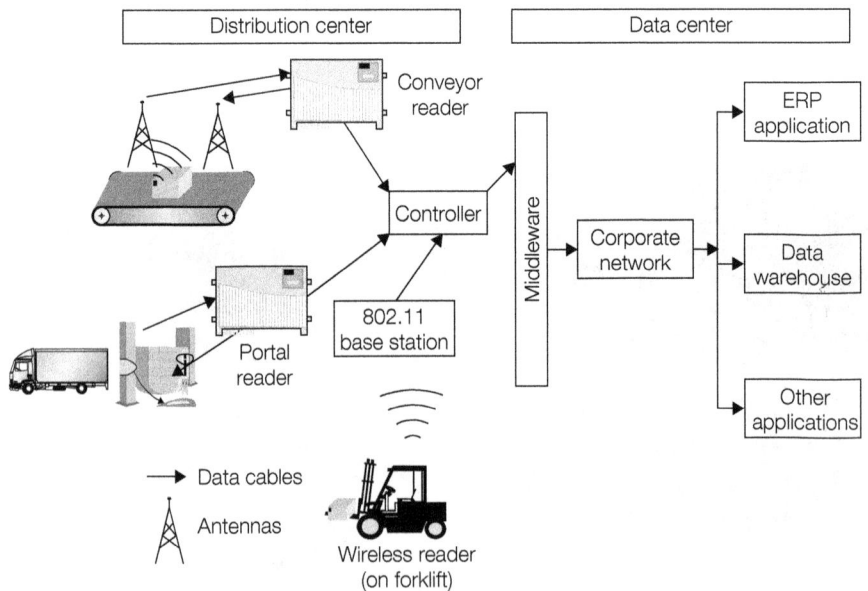

The diagram depicts a simplified distribution center (DC) with a stationary reader on the portal door, another stationary reader on the conveyor system, and a mobile reader on the forklift. The two stationary readers each have two antennas. The forklift reader communicates to a wireless access point. All three readers, along with the pressure sensor in front of the portal door, communicate with a controller, which transmits the data to the middleware. The middleware culls, filters, and structures the data and transmits it over the corporate network in a way that enables the applications to make use of it. The diagram shows the middleware residing in the data center, but its functions are probably spread across several components. Good system design suggests that data be processed as close to the source as possible, and some filtering functions may be performed by the readers themselves in order to minimize network traffic.

3.1 Data

The most important component of the RFID system is the data. Several times per second, readers broadcast a signal, and tags respond. This process creates a list of all the tags in the read zone. That list is called the Raw Tag Database (RTD), and it is usually stored in the reader itself. Optionally, the RTD could be copied and stored elsewhere, and the decision whether or not to do this should be driven by your business considerations. The RTD is raw data. A given tag, passing through the read zone, will probably appear several times in the RTD, so duplicate reads are a virtual certainty. Most modern systems filter out these duplicate reads, but situations still occur when a reader may read a tag multiple times; for example, a driver may load a pallet and drive too close to the reader, generating a host of false positives, or there may be several readers located near one another, and a given tag may respond to more than one of them. This may result from simple proximity, or it could result from a chance reflection as someone walks by with coins in their pockets or a metal clipboard. These examples create multiple data reads, which must be managed or filtered out by the middleware in order to generate a useful base for business information.

Not all of the information in the Raw Tag Database (RTD) is used in every application. For example, the RTD might contain a field telling which antenna read the tag. If you are running an application to check items into a warehouse, knowing which antenna read the tag may not be important. If you are running a system to pinpoint the exact location of a particular container in the yard, it might be critical. Several applications may have access to the same pool of RFID data, but each will have different requirements for data from the RFID system. The order management system will need to know which shipments have arrived and whether the components match the order form. The warehouse management system will need to know which items have arrived and decide where they are to be put on the shelf. Various reverse processes such as returns or customer complaints about short shipments may require you to access data from the past so you can document a response.

Creating value from this torrent of raw data will emerge as the next major challenge of the RFID industry. As this book is being written, implementers are grappling with issues of tag quality, tag read rates, reader compatibility, and placement of antennas, all covered at length in this book. Once these problems are solved, the challenge will be to capture, store, and manage the rich trove of data these systems will produce. Data acquires value by being integrated with other data in useful ways. Designers often assume that the other data is accurate, reliable, and timely. Unfortunately, today the quality of most data in most companies is suspect. Chapter 13 covers this important topic. Data is valuable if it is accurate, can be accessed when needed, and can support beneficial actions. This will require new software capabilities as well as new business processes.

3.2 Tags

An RFID tag is a small device that can be attached to an item, case, container, or pallet so it can be identified and tracked. It is also called a *transponder*. The tag is composed of a microchip and an antenna. These elements are attached to a material called a *substrate* to create an *inlay*. The inlay can be encased in materials that protect it, such as paper, ceramic, plastic, epoxy, or glass, to create useful packages. These packages, the tags, can be designed to function within the harshest environments of cold, wet, heat, exposure to dangerous chemicals, rugged environments, and vibration. Passive tags receive a message from the reader and broadcast a reply. They have no source of energy of their own. Active tags may broadcast continuously, and beacons may broadcast intermittently.

The microchip may be half the size of a grain of rice, but the antenna must be large enough to pick up the radio signal from the reader and to transmit back, so the tag's size is generally determined by the size of its antenna. The microchip contains power conversion circuits, data storage,

Figure 3-2
Chip components

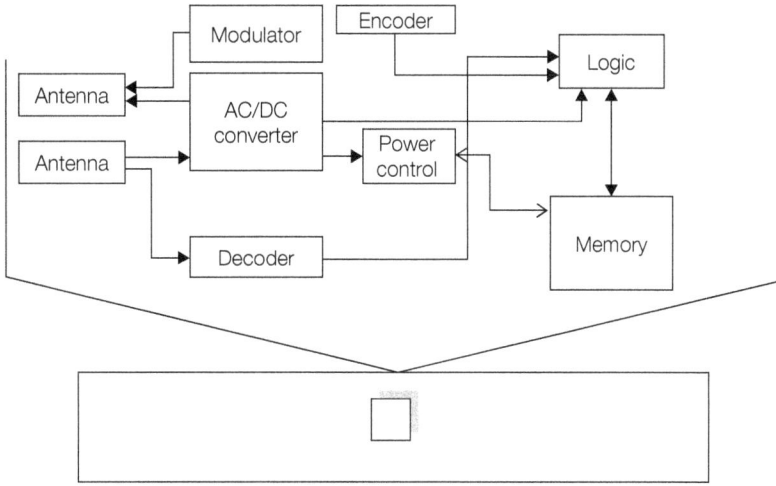

Figure 3-2
Chip components

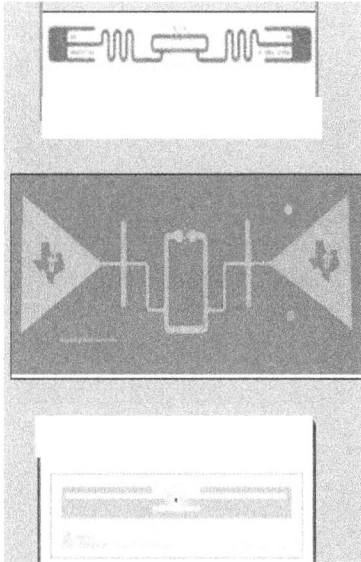

Figure 3-3
Miscellaneous
popular tags

and control logic. Figure 3-2 shows the logical components of tag's computer chip.

Tags come in a variety of shapes, sizes, and characteristics, as shown in Figure 3-3. Different combinations of size, substrate and protective material, cost, and performance characteristics exist to meet the needs of different business applications and environments. Section 3.2.5 provides a list of criteria to help you select a tag for your application.

Who Talks First

There are two mechanisms for initiating the conversation between the reader and the tag. One is called *reader-talks-first* (RTF). The other, logically enough, is called *tag-talks-first* (TTF). EPCglobal's specification calls for RTF communication. With RTF, the reader sends energy to the tags and asks all tags with serial number 0 to respond. If more than one responds, the reader then asks for all tags with a serial number that starts with 01 to respond, and then 010. The reader continues in this fashion until it gets a singular response. Then it interrogates that tag. The tag changes state and does not respond again as the reader continues its query of all tagged items until it has identified all the tags in the read zone. This is called *tree walking*. It is particularly useful when you are looking for a single tag.

TTF systems work differently. When a tag using a TTF protocol enters the reader's power zone it begins broadcasting a short random number, and continues until it receives acknowledgment from the reader. If several tags broadcast at the same time, each waits a random interval before rebroadcasting. That process enables the reader to identify all the tags in its zone. It then queries each one and captures its identifier. Read-only TTF tags can be extremely simple without compromising performance.

RTF protocols require a much wider reader bandwidth than a basic read-only TTF protocol. ISO 18000-6 specifies 200 kHz of bandwidth per channel. In comparison, even a high-performance TTF protocol can operate in a single 12.5 kHz of bandwidth per channel. This is important in Europe and other countries where the bandwidth allocation is narrow.

Some experts say that these technical advantages are sufficiently compelling that there is an urgent need for EPCglobal to change to TTF.

Tags may be custom designed for a particular application, or standard off-the-shelf tags may be purchased.

Tags come in a variety of formats. The following is a partial list of different available tag formats:

- Plastic cards resembling a credit card but slightly thicker
- Flexible labels with adhesive backs

- Coins and tokens
- Paper tags, including some countries' currency
- Key fobs
- Hard tags with epoxy cases
- Wrist bands and wrist "watches"
- Embedded tags (injection molded into plastic products like cases)
- Smart labels—tags sandwiched between two sheets of paper or polyester on which human-readable data and bar code data are printed.

3.2.1 Active and Passive Tags

One way of classifying tags is based on the means by which they transmit messages. Active tags have a transmitter for sending signals to the reader. Their power normally comes from an onboard battery, but it may come from other sources. Passive tags derive their power from the signal they receive from the reader. They derive their power from the reader's radio wave, and they broadcast by encoding their data into the "backscatter" transmitted back to the reader.

Active tags have greater range, data capacity, and processing power. They can work with extremely weak reader signals. They can also incorporate sensors that record and time-stamp such telemetry data as temperature, location via GPS, shock events, tampering events, moisture levels, and radiation. Active tags typically cost more than $20 each, reaching to over $100 each in some cases. Active tags also require maintenance of their power source.

Passive tags, since they draw their power from the reader, require a strong signal from the reader in order to function. In turn, they produce a weak signal back to the reader. Active tags, on the other hand, can work with weak signal levels from the reader, and they can produce strong signals back, driven by their own power source. Additionally, the active tag is continuously powered, always functioning. It can record data even when no reader is present. The passive tag functions only when in the presence of a reader. These characteristics drive very different applications of RFID technology. The profile of inexpensive tags and low-power transmission to a network of portals fits supply chain applications. In these applications, the tags will number in the billions and be read mostly as they pass through well-defined portals or activity centers. The passive tag model does not readily address area monitoring, where tags will number in the hundreds or thousands and need to be read wherever they happen to be. The profile of costlier tags, low-power transmission to just a few readers, and high-power return signals is well-suited for area monitoring but is inappropriate for supply-chain applications.

3.2.1.1 Active Tags

Active tags can be viewed as miniature computers, capable of operating sensors and performing calculations and logic operations, encryption and decryption, and sophisticated two-way wireless communication over substantial distances. Active tag systems can monitor a large area and thousands of tags with only a few readers. The active tags are costly as compared with passive tags, so their use is limited to high-value items and processes. Active tags can be purchased with large read-write data storage and sophisticated data search capabilities. If an active tag has a battery, it has to be maintained and, ultimately, replaced. Active tags may last as long as ten years, but eventually the battery runs out.

Active tags are available on most frequencies where RFID is used, but for many applications, the preferable frequency is in the amateur radio band, right at 433 MHz. At this frequency, the wavelength is about one meter, which enables it to propagate around many obstacles that would halt the shorter UHF radio waves. Active tag read ranges at 433 MHz can reach as high as 300 feet. 433 MHz waves pass through liquids more readily than the shorter-length 915 MHz UHF waves.

3.2.1.2 Passive Tags

Passive tags transmit only when they are in the field of a reader. Otherwise, they are silent. They are much less costly than active tags. Experts generally believe that UHF active tags will reach a cost of 5 cents per tag when purchased in large quantities well before the end of 2008. That will make them economical for some item-level tracking applications. Today, RFID supply-chain systems use passive tags on cases and pallets, and even at these levels, the cost of tags can be daunting. Wal-Mart estimates it receives over 8 billion pallets per year. Recall that in July 2005, tags cost about 30 cents each in quantity. Tagging every pallet would add about $2.4 billion to Wal-Mart's suppliers' costs. By comparison, a bar code label, including the cost of application, costs less than 1 cent.

Passive tags are best used where the movement of tags is highly consistent and controlled, and little security, sensing capability, and data storage are required.

3.2.1.3. Semipassive Tags

A third category of tags has emerged, called semipassive tags. They are sometimes called semiactive tags or battery-assisted tags. Semipassive tags have a battery, but they use it only to run the chip's circuits; they transmit using the same method as passive tags, that is, modulating the backscatter. Active and semipassive tags are both useful for tracking high-value goods that need to be scanned over long ranges. Because the cost of semipassive tags may exceed $1 each, they are still too expensive to place on low-cost items. There are three benefits to the semipassive tags. First, the tags can be read at higher speeds than passive tags. Since they do not need to take the time to energize their circuitry based on receiving transmissions from the reader, they need to be in the read zone for less time than passive tags. The second benefit is an ability to continually monitor and record external conditions. They may use the battery power for various types of sensors. They continue to store the data until they are awakened by a reader. They then respond to a command from the reader to download their stored data. The third benefit is the possibility that semipassive tags will transmit in the presence of opaque materials that would inhibit a passive tag. The semipassive tags' circuitry is more sensitive, since it is not relying on the reader's transmission for its power.

3.2.1.4 Chipless Tags

Both active and passive tags discussed so far work with an integrated circuit chip inside them to store their data and to perform the processing logic. By contrast, a few innovative companies have developed what are called "chipless tags," which use different technologies to store and transmit their data. They have no integrated circuit. Instead, they encode unique patterns on the surface of various materials. These patterns encode the data that is reflected back to readers. Chipless tags are read-only; the data is permanent.

However, chipless tags are significant in that they appear to offer a much lower price point for the tags. The drawback is that no international standards for chipless tags have been established, so none of the prominent mandate bodies—Department of Defense, Wal-Mart, Target, Metro Stores, Best Buy, Albertsons—will accept them as of December 2005. Nonetheless, it has been estimated that chipless tags will comprise up to 30 percent of RFID tags in just a few years.

One chipless tag manufacturer is a company called RFSAW, which produces an entire RFID system (readers, tags, antennas, and middleware) for chipless tags based on its use of surface acoustic wave (SAW) technology. SAW technology solves many of the common problems that arise in RFID pilots and deployments. Read ranges extend up to 100 feet. SAW is not sensitive to metal or liquid and are smaller than IC-based tags. SAW tags operate at 2.45 GHz, which is not accepted by Wal-Mart, other retailers, or the Department of Defense. However, the frequency is internationally recognized and generally available for RFID.

3.2.2 Tags Classified by Programming Method

Tags acquire their data in a variety of ways. The initial entry of data into the tag is called *commissioning* the tag. Pre-encoded tags are commissioned by the manufacturer when the chip is made. WORM (write-once-read-many) tags are encoded by the customer one time—and only one time—during the workflow. Read-write tags have their data read and written to them several times throughout their life. Complicating matters, some tags partition their memory, making different portions available to different queries with different passwords. Some partitions may be over-written, and others may be read-only.

3.2.2.1 Pre-encoded Tags

The tag manufacturer can encode a unique number into the circuit when it is manufactured. Customers may order the chips with customized numbers, but they cannot change them themselves. Pre-encoded tags are mostly used in closed loop applications and in small pilot projects. At some point in the workflow, a link is recorded in a database between the number in the tag and the item it is attached to. Pre-encoded tags support the very simplest of workflows.

3.2.2.2 Write-Once-Read-Many (WORM)

Unlike pre-encoded tags, write-once-read-many (WORM) tags are commissioned by the customer, usually when they are first put into service. This first write definitively establishes the data contents of the tag and then it cannot be changed. The process is logically equivalent to printing a label. Commissioning is typically done when the identity of the object to be tagged first becomes known. WORM tags enable you to put your own number or other information into the tag at any single point in the production and distribution process, and that permanently establishes it. These tags are also called *field-programmable* tags. Most smart labels incorporate WORM tags.

3.2.2.3 Erasable Tags (EEPROM)

Electronically erasable programmable read-only memory (EEPROM) tags can be written many times by a qualified piece of equipment. This enables the tag to be rewritten with new information at different points in the work process or to be reused. EEPROM tags will hold their data values for up to about ten years without accessing a power supply. The oxide layer of the EEPROM deteriorates after that, so tags approaching the ten-year longevity mark should be replaced or discarded.

3.2.2.4 Read-write Tags

Data can be added to read-write tags by a simple reader at any point in their lifetime. This type of tag brings important capabilities to your application; it enables the tag to acquire information as it moves through a business process, store it, and provide it back to readers later on. A work-in-process application is a good example where read-write tags are useful. The tags can collect and record data such as date, time, who worked on it, what machines were used, and environmental conditions as the unit proceeds through the production process. When the process is complete, the tag contains a complete record of every step taken, and the record can be downloaded at the end. Another example of read-write tag usage might be monitoring of fresh food as it moves from the farm to the consumer. Again, the tag can monitor external conditions and record them with a time-stamp. We have already seen one food company have their contractor replace marginal refrigeration units on some of their trucks to reduce spoilage due to temperature variation. The data on the tag enabled them to locate exactly where the marginal refrigeration units were in their supply chain.

Active read-write tags may be manufactured with sensors. They meet users' needs to continuously monitor the contents and the environment for conditions that might indicate tampering, spoilage, theft, or other forms of deterioration. Passive read-write tags are also available, and they are described in the Gen-2 specification.

Security might be a problem in your application if the tags are capable of being written at any time. Malicious vandals could implant erroneous or spurious data. The Gen-2 specification includes a description of a password scheme to protect tags from unauthorized writing. You will need to carefully spell out the security requirements for the data on the tag in your application, consider the threats and select tags, and design dataflows and workflows with the appropriate level of protection.

3.2.2.5 Database on a Tag

A company called IDComm has developed proprietary algorithms and a form of extreme compression to create tags that can contain the entire data record pertaining to a particular item. They call this a "database on a tag." The records are encoded when the tag is created, and they are updated in the field by their SmartWare software. The data is encrypted; some tags can be permanently locked. IDComm sells the system so the tags can replace paper documentation and, unlike paper, they are waterproof. They are not used to meet the retailers' compliance mandates.

3.2.3 Tags Classified by Physical Form Factor

Besides being classified by how they acquire data, RFID tags can also be classified by their physical form factor. The tags are available today in four different forms. They can be purchased as inlays, adhesive inlays, labels, or as converted products.

Inlays include just the RFID microchip and the antenna mounted on a substrate. Inlays are typically sent to converters to be transformed into products. They are not used as tags themselves.

The adhesive inlay format simply adds a pressure-sensitive adhesive to the back of the inlay. These are used when no special protection for the tag is needed. Smart labels are adhesive inlays attached to thermal transfer labels for use in printers. Converters laminate or encapsulate the inlays, making a specialized package. The package can be made of paper, plastic, rubber, glass, or some other material and can be custom designed, molded, or laminated.

In addition to these variations, the industry is seeing interest in printed electronics, which are tags manufactured by printing antennas, circuits, and even batteries right on the packages. This has the possibility of reducing the cost of RFID tags by 90 percent or more.

3.2.4 Selecting a Tag for Your Application

Most experts agree that RFID tag selection is very important, can be difficult and time consuming, and should not be short-changed. In your planning, you should allocate the time to do it carefully; it is essential to the successful operation of your project. Unexpectedly, even tags operating in the same frequency band differ in important ways. Odin's Benchmark study has shown that different manufacturer's tags—even those with identical operating specifications—differ widely in their performance in different circumstances. So you will have to test your tags at your locations and on your packaging and with your equipment before finalizing your decisions.

NOTE   *Odin Technologies has published a report on tests of eight passive UHF Class 1, Class 0, and Class 0+ tags. The report measures communication link margin, distance performance, speed performance, and orientation sensitivity.*

Do not expect one tag to work for all your packages. Different products, different configurations, different profiles in terms of the reader configurations, speed of the packaging line, and effects of opaque materials, all will likely drive the tag choice for each of your packages. At the same time, a "best practice" is to have a single tag for as many different packages as possible. As you test your various SKUs (see Chapter 12), give preference to tags that you already are using to reduce the complexity of dozens of separate procurement, inventory, and distribution activities for dozens of different tags.

The set of tag decision criteria falls into two categories. One set drives the decision about your system in general. These are questions of which frequency you will be working on, whether you need active or passive tags, what distances you will need to accommodate between readers and tags for reading and writing, and so forth. The second category is exactly which tags you will need for your system. Section 3.2.5 describes those criteria.

3.2.5 Tag Selection Criteria

This section lists and describes the criteria you can use to determine what tags will best suit your application. Specific product references are included in this chapter, but these are not recommendations. They are intended only to provide general information about what is available on the market.

3.2.5.1 Size and Form Factor

Consider first the size and nature of the space available for the tag. Tags are small, but they can overwhelm small articles. Maxwell Corp's Heliport system operates at 13.56 MHz and can attach the tag to very small items such as test tubes, specimen bottles, vials, and prescription drug containers. Smart labels can often duplicate the size and form factor of the bar code labels they replace. For clothing, special tags are available to fit behind the labels sewn into shirts, sweaters, and skirts.

3.2.5.2 Requirement for Human-readable Information

Will the application require or benefit from human-readable and bar code data in addition to the data stored on the tag? Keep in mind that read rates may not always be 100 percent, and you will need to plan and support a workflow to deal with packages with quiet tags or locations in the supply chain that are not RFID enabled. If you will need human-readable or bar code data, smart labels can be used. Smart labels are described in Section 3.2.6.

3.2.5.3 Durability

Determine whether the tags need to function in a harsh environment or be subjected to unusual stresses. Special tags are available for low-temperature, high-temperature, moist, or other harsh environments. Special low-frequency tags are available that can be inserted into a cow's second stomach. Other tags are available that will function submerged.

3.2.5.4 Reusability

Will the tag need to be reused? Active tags may be too expensive in most cases to be thrown away and can be reused. Passive tags are usually considered inexpensive "throw-away" items and they remain with their item—or its packaging—as it moves through the supply chain to the customer and, ultimately, to the landfill.

3.2.5.5 Orientation Sensitivity

What requirements exist for the tag's orientation to the reader field? You will learn in Section 3.3.1.2 that antennas function best when the orientation of tag and reader antennas are identical. Odin's Tag Performance Benchmark found that various tags exhibited different levels of sensitivity to orientation. This means you will need to test the various tags you are evaluating at different orientations to see how their performance varies in your application.

3.2.5.6 Communication Distance

Your system design must ensure that read ranges and write ranges are sufficient to your needs. Even within a given frequency, different manufacturers' tags have different levels of performance across a given distance. Bear in mind that false positives, such as reading tags farther from the reader than you intended, are as dangerous as false negatives, that is, missing tags that should have been read.

3.2.5.7 Opaque Materials

Chapter 1 described how opaque materials such as metal, liquid, sodium, and graphite will impact the tags' performance. Tags from different manufacturers, even operating in the same frequency range, differ in their sensitivity to opaque materials. You will need to test each package that you are going to tag to make sure you can get the performance you will need.

3.2.5.8 Supported Communication Standards and Protocols

The application description will determine which standards you will need to follow. But there are subtleties in different manufacturers' implementations of various standards that may impact the tag's performance in your application. One answer is to select tags certified to work with your selected readers. But certification should not be a substitute for real testing at the site where the tags will be used, on the exact packages to be processed. Bear in mind that the Gen-2 specification is a menu of capabilities, not a definitive prescription. Different tags will provide different sets of capabilities which may or may not work with all readers.

3.2.5.9 Storage Requirements

Will the tag be a "license plate tag" or will it have to store more data? In either event, how much storage will be needed in the tag? Chapter 2 described the workarounds that enable Gen-2 tags to work with 64-bit tags, but these older, smaller tags are being phased out of production. For supply chain applications where you are going to store license plate numbers only, 96-bit tags are the smallest practicable. For other applications, you will need to carefully document the on-tag storage requirements and accommodate them in your tag selection.

3.2.5.10 Tag Performance

Tag performance is sometimes measured in terms of how far from the reader the tag can be read, but several other performance factors need to be considered. One measure is the number of tags that can be read at the same time. Measuring this performance is complicated when you need to read the tags while they are in motion, up to 600 feet per minute. You will need to define exactly what are the measures of performance your application depends on and then select tags that can meet them. Consider the number of tags that can be read per second, the speed of data transfer, and the distance at which tags can be read. If your application involves writing data to the tag, consider the same three measures for writing of tags.

Another performance factor is the failure rate. Until recently, tag manufacturers routinely informed customers to expect about a 20 percent failure rate on batches of tags. In November 2005, R and V Group of Chattanooga, Tennessee, and Japanese electronics manufacturer Omron announced that they would guarantee 100 percent performance of its ruggedized RFID labels. The cost was 19 cents per label in quantities of 50,000 or more. This announcement has reintroduced the discussion of tag quality into an RFID marketplace that had been dominated solely by the issue of lower costs.

3.2.5.11 Reader Compatibility

Which readers are going to read the tag? Select tags that are certified with the readers that will be reading your tags, and then test them attached to your product package at your read site as part of your selection methodology.

3.2.5.12 Security Requirement

Evaluate whether the information on the tag is sensitive or not. Hostile forces can probably do little damage by being able to read a license plate number, unless they have access to the database that stores the actual information. But any medical data or personal identification data needs to be carefully protected and is even governed by state and national laws. Privacy advocates are concerned about the ability to track persons based on the license plate number of tags in their clothing or other purchased articles. The ability of tags to require passwords for reading or other functions means you will need to specify your security requirements carefully and determine what tag functions and contents should be protected. You will also need to design workflows and dataflows to manage the passwords, keep them safe, and make sure they are available when needed.

3.2.5.13 Commission Logic: Pre-encoded Tags, WORM Tags, and Rewriteable Tags

Pre-encoded tags are useful in small pilots and in closed loop applications where you will not be sharing data outside the application itself. But even in these applications, if there is any possibility of scaling the application to work with others in the future, you can save yourself the trouble later on by making the early effort to embrace international standards and by commissioning the tags yourself as they are being put into use. Keep future growth in mind as you decide between pre-encoded tags, WORM tags, or rewriteable tags.

3.2.5.14 Capabilities and Functions

Tags vary in capabilities, from a simple 1-bit security tag to sophisticated computers that can be programmed to evaluate their environment and provide alerts and messages. Tags can record and time-stamp and location-stamp a variety of events. All of these capabilities come at a cost. Document carefully what functions you will need on your tags. If your application envisions delivering tags to end-customers, you will need a way to decommission the tag at the point of transfer.

3.2.5.15 Cost

What tag cost makes the application feasible? Some RFID experts say that the 5-cent tag makes pallet-level tagging feasible and the 1-cent tag makes item-level tagging feasible. This type of generalization is interesting, but you can and should calculate the tag cost level that makes your application feasible in your exact circumstances. It is important to note that tag costs are coming down rapidly, so you can start your implementation with the expectation that reductions will occur. But you should know the tag cost your application needs and how various tag cost assumptions affect your project's ROI calculations.

Table 3-1 introduces several prominent tag suppliers.

Company	Comments
Alien Technology (www.alientechnology.com)	Alien manufactures EPC-compliant tags for pallet-, case-, and item-level tagging. The company claims a patented low-cost manufacturing process and offers an extensive variety of tag designs. Alien also offers pressure-sensitive RFID labels, as well as inlays that can be converted into labels by their partners.
Applied Wireless (www.appliedwireless.com)	Applied Wireless active tags operate at 916.5 MHz and continuously transmit ID information. The range for these products extends from up to 1200 feet, and the battery lasts up to two years.
Avery Dennison (www.rfid.averydennison.com)	Avery Dennison claims the broadest portfolio of UHF EPC tags in the industry. One, the AD-220 tag, is a UHF Class 1 tag based on the Gen-2 EPC protocol. It features 96-bit read-write memory and an operating frequency of 902–928 MHz. It is produced with the Impinj Monza chip and is small enough to fit into a 4" × ½" label. The company also offers the AD-410 designed for use in carton, tray, and pallet applications. The AD-210 is designed for use with various case contents. It is not Gen-2 compliant.
Escort Memory Systems (EMS) (www.ems-rfid.com)	Escort Memory Systems offers over 100 different models of tags. Tags are available for various standards including ISO 14443, I-Code, ISO 15693, and EPC-global. EMS's patented tags can be read through obstructions such as water, wood, plastic, and more. Their high-temperature tags survive temperatures up to 415° F.
Impinj (www.imping.com)	Impinj manufactures micro chips that other companies purchase and use to manufacture tags. The company was the first major supplier to announce the availability of actual Gen-2-compliant RFID chips. Monza chips implement all mandatory features and several optional commands of the Gen-2 protocol. Orientation insensitivity is available with Impinj's dual antenna configuration.
Inkode (www.inkode.com)	Inkode has developed a system for chipless tags. Tags are available at the rate of $5 per hundred, and Inkode claims to be able to produce over 500 million tags per month.
Inksure (www.inksure.com)	In February 2005, Inksure patented a system for chipless tags.

Table 3-1   Tag Suppliers

Company	Comments
KSW Microtec (www.ksw-microtec.de)	KSW Microtec of Germany offers tags that can track supply chain integrity for food, pharmaceuticals, and blood bags. The tag has 1024 bytes of memory and will not lose data if the battery runs out of power. It is also programmable with a number of commands. The device has a battery monitor that can tell whether there is enough battery left to make a trip or delivery. The tag will cost around $5 on orders of a million or more.
Magiix (www.magiix.net)	Magiix sells Bolus, a temperature-sensing passive tag that lodges in a cow's intestinal tract. The Bolus reports the cow's temperature to a specially designed reader-recorder to provide early detection of disease, estrus, and pregnancy.
Precisia (www.precisia.com)	Precisia, a subsidiary of Flint Ink Company, is in the business of printing very small antennas, which can be attached to low-cost chips. Called the FlexWing, the antenna measures ¾" × 1-½". Precisia provides three separate formats for the FlexWing: as a printed antenna that can later be attached to an RFID chip to create an inlay, as part of an inlay manufactured by one of Precisia's partners, or as part of an RFID label converted by a Precisia partner. FlexWing claims a read range up to 20 feet with this antenna. Precisia provides Gen-2-compatible electronics, as well.
Printronix (www.printronix.com)	Printronix produces certified RFID smart labels that combine the technology of thermal-transfer, pressure-sensitive labels with RFID tags.
Philips (www.philips.com)	Philips announced the world's first Gen-2-compliant chip. It was scheduled for mass production in the third quarter of 2005. Philips also produces the HiTag chip, which is low frequency, long range, and secure. Philips also produces MIFARE chips and cards for contactless smart card systems.
Rafsec (www.rafsec.com)	Manufactures high-quality antennas, attaches ICs to them, and delivers them mainly in die-cut or continuous roll format. Rafsec's tag consists of a chip and an antenna on a thin polyester film, a permanent adhesive on the back, and an over-laminated layer protecting the inlet on the top. The pre-die-cut tags are delivered in reels with a siliconized backing paper. Several inlet dimensions are available. Rafsec offers Gen-1 and Gen-2 tags.
R and V Group (www.randvgroup.com)	R and V Group is a certified label converter (manufacturer) that uses Omron tags. They also offer custom smart labels.
RFSaw (www.rfsaw.com)	RFSaw has developed a system for chipless tags based on surface acoustic wave (SAW) technology.
RSI ID Technologies (www.rsiidtech.com)	RSI ID Technologies offers smart labels and active RFID solutions including asset visibility solutions.
Texas Instruments (www.ti-rfid.com)	Leader in the field. Also manufactures the chip that goes inside many other vendors' tags.

Table 3-1   *Tag Suppliers (continued)*

3.2.6 Smart Labels

Smart labels are a popular version of the RFID tag, with low cost and other attractive features. They consist of shipping labels with RFID tags embedded in them. Smart labels are printed with text and with bar codes, so humans and existing bar code readers can read them. Thus suppliers can attach smart labels to their products, comply with retailer mandates, and continue to use their existing bar code readers and work processes.

Figure 3-4 illustrates a common smart label format.

Generally, smart labels are the easiest, lowest costing, and least disruptive way to initiate RFID for mandate compliance. Smart labels are used to "slap and ship" and for other implementations. Slap and ship is useful in several scenarios.

Smart label systems have error recovery and validation built in. This is because nearly all encoder/printers read the label after they write to it. They will overstrike any labels where the read is unsuccessful. In addition, the labels are human-readable, which means that it is straightforward to recover from failed tag reads farther down the supply chain, and they can be read in places where RFID readers are not available.

Smart labels are available in bulk roll quantities in a variety of sizes and types. Formats often allow smart labels to substitute for existing bar code labels. Smart labels are available that are designed to meet particular standards (military specifications, freezer-grade, pharmaceutical chain-of-custody labels, and so on).

Figure 3-4
Smart label

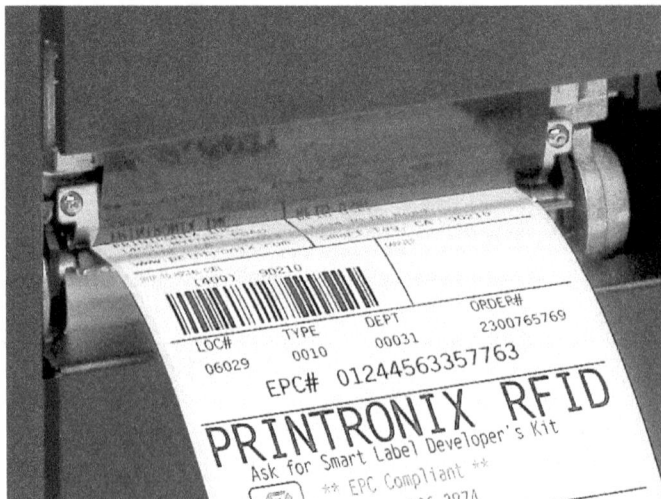

3.2.7 Surface Mount Technology (SMT)

Surface mount technology (SMT) is a manufacturing technique that uses unique soldering materials and mounting techniques to implant the components on pliable materials. One company that produces these tags is called Automated Assembly. They say their SMT tags are able to survive environments traditional tags cannot. SMT particularly solves problems caused by flexing of cards in customer's wallets and other harsh environments.

3.3 Antennas

Antennas are common in everyday life, but not many people know the technical definition of what they do. Antennas convert energy between flowing electricity and broadcast radio waves. Both readers and tags have antennas. Readers generally produce electricity modulated to contain data and instructions and feed it to an external antenna. The antenna converts the electricity to radio waves and broadcasts them. The tags' antennas receive the radio waves and convert them back to electricity used to power the embedded integrated circuit chip and to decode the data and instructions. The tags feed the electricity to their antenna to power the circuits and broadcast their coded response, which the reader reads.

Antennas for each of the four RFID frequency bands are very different from one another. For low-frequency (125 KHz) and high-frequency (13.56 MHz) systems, the wavelength is very long. They use magnetic induction to communicate, so antennas are typically coils of copper wire. High-frequency antennas must be made very precisely, with little tolerance for error. Ultra-high-frequency reader antennas are more tolerant of minor deviations, and they emit radio waves that can travel relatively longer distances. Their wavelength is about 33 centimeters.

3.3.1 Antenna Characteristics

Antennas' fundamental characteristics are

- Impedance
- Polarization
- Gain and effective radiated power
- Bandwidth
- Appearance

3.3.1.1 Impedance

Impedance is the resistance of an electrical component to alternating current. It is measured in ohms. Impedance is important in RFID implementations because the impedance of the antenna must match the impedance of the device circuits in order for it to function. If the impedance of the components is the same, usually 50 ohms, power will pass from the reader to the antenna easily. If there is an impedance mismatch, then some of the power will be returned to the reader. In extreme cases, this could damage the reader, but modern units have protection against this particular disaster.

Antennas' impedance may change based on the frequency to which they are tuned and whatever materials are in the close environment. Opaque materials in the environment close to an antenna will affect its impedance.

3.3.1.2 Polarization

Radio waves travel in a particular direction through space. The polarization of the field is perpendicular to the direction the wave travels. Antennas can be designed so the polarization is constant as the wave moves through space. These antennas are called linear-polarized antennas. Linear polarized antennas are either horizontally or vertically polarized, as shown in Figure 3-5.

Antennas can also be designed so polarization rotates in a plane perpendicular to the direction of the wave propagation. These antennas are called circular-polarized antennas, as shown in Figure 3-6.

Polarization is very important to the RFID implementer. The best power transfer between two antennas takes place when their polarizations are aligned. Thus the best performance occurs when the orientation of the tag can be controlled and you use a linear-polarized antenna. A circular-polarized antenna derives only half the power from the radiated signal, but that performance is constant regardless of the orientation of the tag. When your work process allows you to guarantee the orientation of the tag, use a linear-polarized antenna and make sure the reader antenna and tag antenna are parallel. This will ensure optimum read rates. However, the more common work process will require you to work with tags oriented in a variety of directions, such as a conveyor belt moving numerous packages of varying size, shape, and placement on the conveyor. In these cases, a circular-polarized antenna is the best choice.

Figure 3-5
Wave pattern from
linear-polarized
antenna

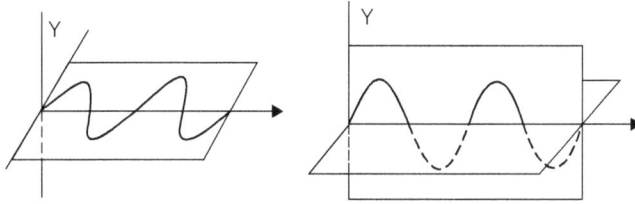

Figure 3-6
Wave pattern from
circular-polarized
antenna

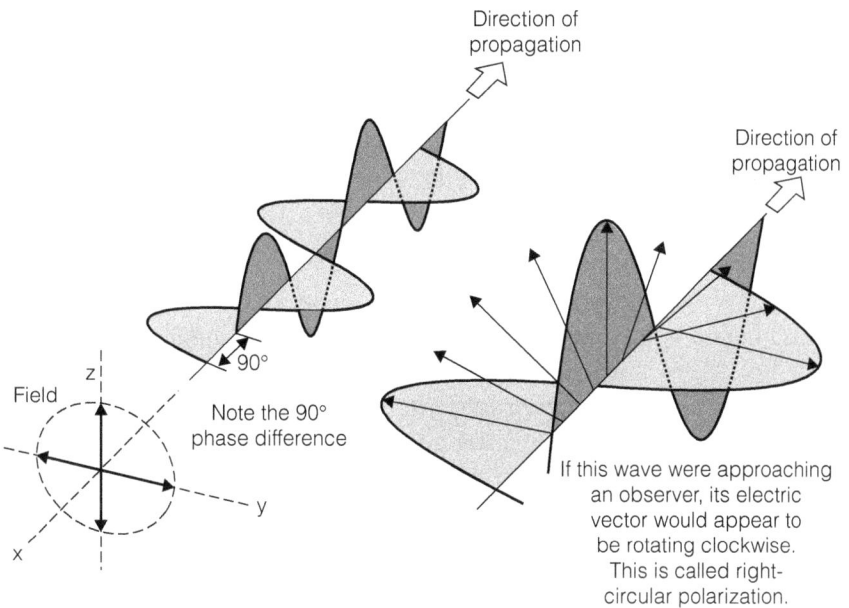

Direction of
propagation

Direction of
propagation

Field

z

90°

Note the 90°
phase difference

y

x

If this wave were approaching
an observer, its electric
vector would appear to
be rotating clockwise.
This is called right-
circular polarization.

3.3.1.3 Gain and Effective Radiated Power

Antenna performance is measured in what is called *gain*. The illusion
persists that an antenna magically creates power when it exhibits gain.
This is not true. *Gain* does not compare output with input. Antenna gain
compares output in a given direction against a standard. In many cases
the standard is the basic, one-half wavelength dipole antenna that radiates equally in all directions in a characteristic donut shape, as shown in
Figure 3-7. The effective radiated power (ERP) is calculated as the power
applied to an antenna multiplied by its gain in a given direction.

Gain is measured in decibels (dB). The decibel is a ratio of two power
measurements. It is a logarithmic measurement; every three dBs represents a doubling of the power output. While we have defined gain as the

Figure 3-7
Dipole linear
polarized
antenna pattern

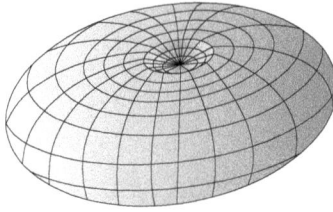

Figure 3-8
Directional
antenna pattern

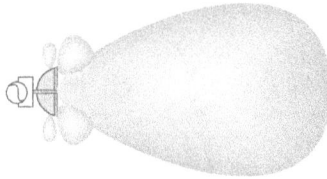

comparison of output in a given direction with a half-wave dipole antenna, many vendors and regulators use a different standard for comparison. They express gain compared with a theoretical construct, the so-called isotropic antenna instead of the half-wavelength dipole. The measure is called Equivalent Isotropically Radiated Power (EIRP). EIRP is reported in a figure called dBi.

An isotropic antenna is an antenna of zero dimensions (that's why it is theoretical!), which radiates equally in all direction, like a light bulb (see Figure 3-8). Engineers in the EIRP camp report the gain in dBi. dBi and dB are not directly comparable, and you should not select an antenna based on a dBi figure larger than one expressed in dB. The United States Federal Communications Commission (FCC) expresses its regulatory limitation on power output in EIRP. The FCC allows operation up to 4 watts EIRP. This is 1 watt plus 6dB of gain.

The antenna achieves gain because it focuses the power in a given direction, so a higher power radiated in one direction requires that a lower power be radiated in another direction. The antenna's data sheet usually discloses the direction of maximum radiation.

The reported dB gain is available only at the optimal direction. Before you make your selection, make sure you will be able to mount the antenna to position the read zone and read tags at that optimal location relative to the antenna. The higher the gain, all other things being equal, the narrower the read zone.

3.3.1.4 Bandwidth

Some people are surprised that bandwidth is listed as a characteristic of the antenna. But the physical characteristics of an antenna—its size and shape—determine the frequency(s) at which it will perform the best. The design of antennas is a complex topic, worthy of a separate book. But antenna size and shape can limit the performance of your system. Different antenna sizes and configurations behave differently at various bandwidths.

3.3.1.5 Appearance

Sometimes the appearance of an antenna is important. This is particularly relevant in access control applications, where the antenna must fit in with the décor of the surroundings or be unobtrusive.

3.3.2 Reader Antennas

Reader antennas are employed in both portal and conveyor applications. The following sections describe both.

3.3.3 Portal Antennas

Portal antennas read tags passing through a door or gateway. The tags may be mounted on items, cases, pallets, or totes. The tags may be as low as 3 feet off the ground and as high as 10 feet or more. The doorways are usually 12 to 14 feet across. Since antennas are generally placed on both sides of a portal, each antenna must establish a horizontal read zone of 6 to 7 feet. Your objective is to select and place antennas that drive the highest number of successful tag reads, the lowest number of false negatives (that is, missing tags that should have been read), and the lowest number of false positives (that is, reading tags that should not have been read).

Your application may require that you establish the direction the tag was traveling when it was read. This is not possible with current RFID technology, so some kind of sensor such as pressure mats or photoelectric cells, integrated with your middleware will be required.

Typically, you create a portal read zone with two to four inward-facing antennas. The antennas' read zones should overlap to ensure good coverage.

Figure 3-9
Patch antenna

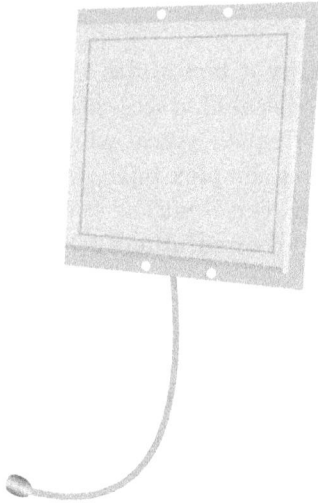

The antennas should be canted, as well, so they are not facing one another. Canting also reduces the interference one reader will cause units nearby. Symbol readers and some others use separate antennas for transmit and receive, so there may be as many as eight separate antennas covering a portal. The transmit and receive antennas may be packaged in the same housing. All of the antennas are not operating at the same time. The pairs of antennas are multiplexed, so at any instant only one pair of antennas—one transmit and one receive—is operating. It is important to configure the multiplexer so that each of the corresponding pairs of transmit and receive antennas is covering the same read zone at the same time. It is usually best to position paired transmit and receive antennas on the same side of the portal. Otherwise, the relatively stronger reader signal could drown out the weaker tag signal.

A patch antenna is widely employed in portal applications. It is about 30 centimeters to a side, and it is convenient to mount and to adjust. Typically today's patch antennas have gains of 6–8 dBi, and most of the power is directed within a beam width of 90–100 degrees. Figure 3-9 illustrates a typical patch antenna.

3.3.4 Conveyor Antennas

Conveyor systems, like the one shown in Figure 3-10, are installed to automate the movement of items, packages, and materials from one place to another.

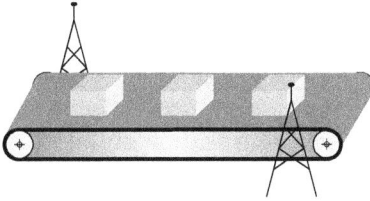

Figure 3-10
Conveyor with
RFID antennas

3.3.4.1 Accumulation Conveyors

Accumulation conveyors, such as the one shown in Figure 3-11, help create unit loads. They can be used to collect a group of items into a container, pallet, or shipment. RFID technology can help automate this function, linking an item to its business information such as customer, region, order number, or priority to enable the system to optimize the shipping strategy. The conveyors themselves may have sensors that measure weight and dimensions to feed the shipping strategy system. Some accumulation conveyors can rotate or flip items for more efficient identification or packing.

3.3.4.2 Transportation Conveyors

Transportation conveyors, shown in Figure 3-12, route items through the facility. RFID-enabled automatic identification will support automatic routing, eliminating errors and reducing the human workload.

Figure 3-11
Accumulation
conveyor

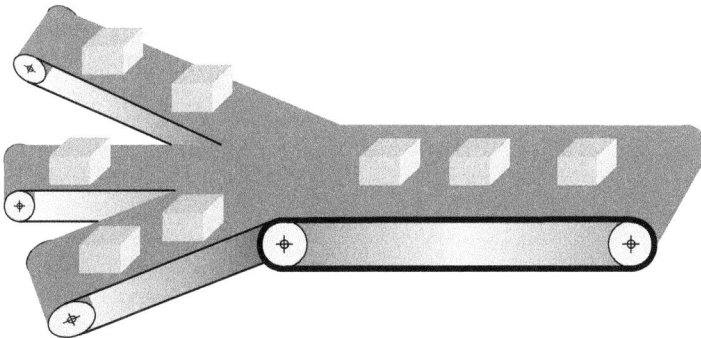

Figure 3-12
Transportation
conveyor

3.3.4.3 Sortation Conveyors

Sortation conveyors, as shown in Figure 3-13, disperse items, packages, or pallets out to different bins or other destinations for storage or use. These systems depend on perfect identification of the items. RFID systems provide reliable identification and enable them to work automatically.

3.3.4.4 RFID in Conveyor Systems

RFID can perform a number of functions in a conveyor system. It can support the conveyor function itself, identifying articles and activating gates to route the article to its correct destination. It can monitor the conveyor systems and report the amount, location, and identity of inventory located on the conveyor. It can also report the location of individual cartons or cases of product in real time. If goods are perishable or subject to degradation by temperature, moisture, or other environmental effects, sensors located in the tags can record and report these problems as well.

There are challenges for an RFID reader in a conveyor application. The system must be able to read tags regardless of where they are located on the package and regardless of orientation. If the application requires writing to the tag while it is traveling, there is always the danger of encoding the wrong tag. In designing your system, consider the various sizes and heights of items passing the reader and avoid any requirement that smaller packages be located in the center of the conveyor belt to get a successful read rate.

On a moving conveyor belt, the readers must be able to read the tags reliably regardless of speed of conveyance, orientation, angle, or location of packages on the conveyor. As discussed previously, circular-polarized antennas can read tags regardless of orientation. Dual-dipole

Figure 3-13
Sortation conveyor

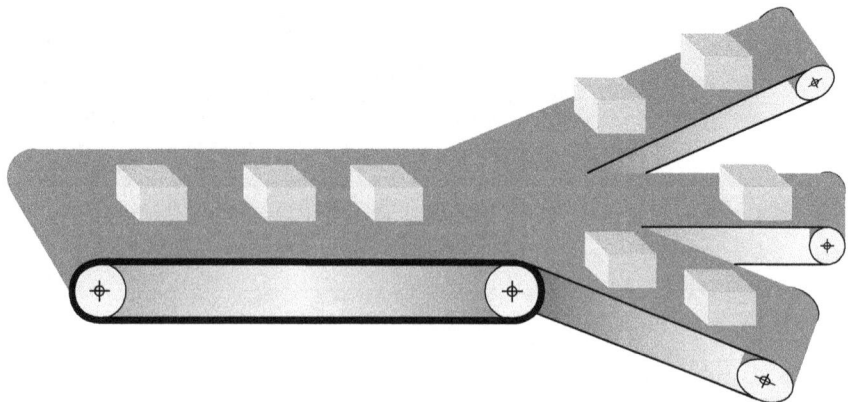

tags can be read at any orientation, but they are large and expensive. One issue that may impact the reader's ability to read tags reliably is the composition of the conveyor's rollers. Metal rollers in your conveyor can create problems for the readers, and you should replace them with plastic rollers if possible.

3.3.4.5 Planning Antenna Placement in Conveyor Systems

Antennas monitoring a conveyor system, as shown in Figure 3-14, must be placed so as to optimize read performance of the entire range of possible tag locations, conveyor speed, and variations in tag performance, contents of packages, and presence of radio interference. Chapter 12 describes step-by-step protocols for accomplishing this. Directional antennas limit the read zone to a single item on the conveyor belt. The main factors for successful installations are read speeds, conveyor belt speed, antenna read zone, and number of tags in a read zone at one time. Good antenna selection and placement will give you maximum read speeds.

Ultimately, you will need to select antennas with characteristics to match the requirements of your location. Then you will need to position the antenna to generate complete coverage of the spaces a tag can occupy. The next step is to engineer the highest possible number of times the reader can query a tag as it passes. For Gen-2, the least amount of time to send a SCROLLID or SCROLLALLID command and receive the tag's response is 62.5 microseconds, so the theoretical limit is about 1600 tag reads per second in North America. Lower data rates are expected in Europe and the Far East, due to the narrower bandwidth available.

Figure 3-14
Conveyor system
read zones

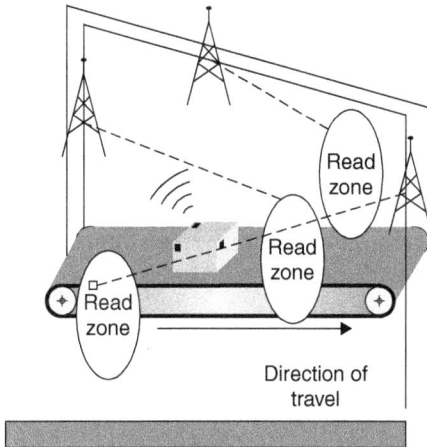

Older Class 0 tags have a theoretical read rate of about 1000 tags per second; Class 1 tags' theoretical read rate is about 500. If a conveyor is traveling 3 meters per second, the ideal Class 0 reader would attempt 3 reads per centimeter. (The ideal Class 1 reader would try 1.5 reads per centimeter.) If there are two antennas (one on each side), each antenna will be in service only 50 percent of the time, cutting the read rate for any particular tag in half; if there is a third antenna above the conveyor, the read rate drops to a third. As a result, real-world read rates are much slower than published theoretical read-rates.

These relatively detailed explanations support the suggestion that antenna selection and placement must be carefully considered in order to read tags reliably. This is accomplished by selecting, locating, and orienting the antennas to ensure complete coverage of the possible tag locations. It is also important that the read zones be as long as possible along the axis of conveyance to grant the highest number of queries possible for any given tag. When the antenna is slanted relative to the conveyor, the read length increases. The read zone length should be configured to allow at least 12 attempts for each tag, given the conveyor speed and reader type.

3.3.5 Tag Antennas

A tag consists of an integrated circuit chip and an antenna. There are numerous configurations of antennas for tags, and the various configurations exhibit different performance and cost characteristics. There is an infinite variety of geometries that can function as antennas, and they all offer enhancement in one application and deficits in another. Designing antennas is an art form as much as a science. Some of the UHF and microwave antenna configurations found in tags are

- Dipole antenna
- Dual dipole antenna
- Folded dipole antenna

A dipole antenna is half the length of the radio wave. A dipole antenna tuned to 915 MHz will be 16.4 cm in length. The dipole antenna is relatively simple and inexpensive.

A dual dipole antenna has two dipoles in it. The dual dipole is less sensitive to tag orientation, but it is more costly to manufacture.

A folded dipole looks like a dipole antenna with a wire folded back on itself. Television antennas are folded dipole antennas. Folded dipole antennas work well across a broader range of frequencies than do simple or dual dipoles.

LF and HF tag antennas must be a coil (usually copper) to capture the magnetic wave broadcast by the reader.

Today, tag antennas are encased in the tag along with the microchip. There is, however, considerable research and development being conducted based on a process called *printed electronics*. As these techniques mature, it will become possible to print antennas directly onto the substrate or indeed on the product packages themselves. This may drive the costs of RFID tags low enough to become competitive with bar codes, although this is highly speculative. The conductive inks will be more costly, and the printing will need to be more precise.

3.4 Connectors

Connectors are used to connect cables to the readers and antennas.

The United States' FCC requires that manufacturers provide inconvenient connectors to their antennas. This is intended to discourage customers from illegally boosting power output by connecting unauthorized antennas. The inconvenience might arise from the configuration of the connector or from threading the locknuts opposite from the normal direction.

RF connectors are more specialized than you might think. The geometry and size of the connector are an important part of the impedance calculation, and any changes will introduce undesirable variance. Recall that matching impedance between the antenna and the unit is very important, and the balance is delicate.

RF and microwave connectors are precision-made parts and are easily damaged or deformed. You should handle them with care and keep them clean. They can be cleaned with a solvent such as alcohol, but take care that the solvent not touch the dielectric spacers or resistive materials; both can be damaged irreparably by solvent.

The connection between the cable and the reader will require a ferrite choke. This device prevents conducted interference from limiting your range or performance.

Table 3-2

Cable Losses
for Common
50-ohm Cables

Cable Type	UHF Loss dB/meter
RG58/U	0.66
RG59/U	0.36
RG8/U	0.30
RG174/U	1.00
RG188/U	0.98
RG213/U	0.31

3.5 Cables

The cable connects the reader to the antenna, and it bears some discussion. Good practice dictates that cable loss be limited to 2 dB. Table 3-2 details UHF loss for common cables. Cheaper, thinner cables will tend to exceed this criterion. Cables must match impedance with the antennas, so they should measure 50 ohms. Cable lengths in the United States are governed by the FCC (Part 15), so you will be required to use the cable lengths (neither longer nor shorter) specified for your reader.

3.6 Readers

A reader uses its antennas to stimulate tags, read their data, and transmit it via a network to a host computer. Readers can also commission a tag and write data to its memory. Thus the term *reader* should properly be "reader/writer." In the simplest case, the reader transmits a simple query, and any tags within its field transmit their contents. In today's more sophisticated cases, the reader sends authentication information and commands coded in the radio waves. The receiver in the tag detects the signal, authenticates the message, and uses its own antenna to send back its response.

This section begins with a discussion of reader performance. It then goes on to describe the various configurations of readers available and where they will be used in your installation.

3.6.1 Reader Performance

The obvious consideration in selecting readers is their performance. Reader performance is very important, but it is difficult to describe in isolation. Actual reader performance is highly dependent on the environment, the antenna, and the performance of other system components. The environment consists of levels of interference in the environment, number of tags in the read zone, impacts of nearby readers and other RF sources in the area, and how well the system has been engineered to deal with opaque materials. The other system components that will impact reader performance are the cables, tags, and antennas. Chapter 12 provides a methodology for determining and engineering RFID system performance.

Reader performance is usually measured expressed in terms of the reader's effective read range. In point of fact, this is just one element of reader performance. Reader performance consists of the following elements. The testing protocols in Chapter 12 describe how to test for each parameter.

- **Identification range**   The distance at which 100 percent of a tag population can be identified
- **Identification rate**   The number of tags that can be identified and counted per second
- **Read range**   The distance at which 100 percent of a tag population can be read
- **Read rate**   The number of tags that can be read per second
- **Write range**   The distance at which unique identifiers can be written to 100 percent of a tag population
- **Write rate**   The number of tags that can be written to per second

Readers come in three configurations: fixed (stationary), handheld, or mounted.

3.6.2 Stationary Readers

Stationary readers, shown in Figure 3-15, are usually mounted on a wall, doorway, or other kind of portal, or they may rest on a table or desktop. They are attached to one or more antennas that cover the portal through

Figure 3-15
Two stationary
readers

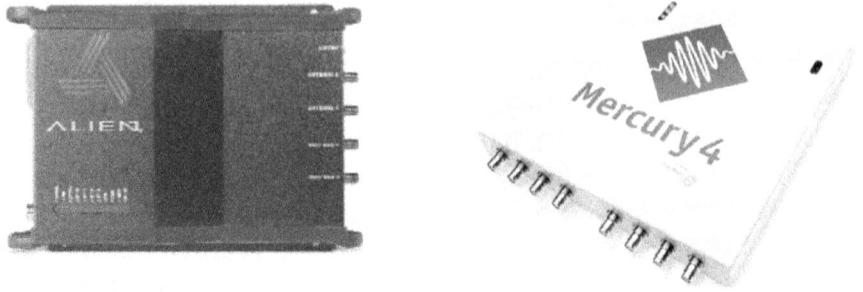

which pass the goods to be tracked. The reader reads each tag passing through the portal and stores it in the real-time database, which can be queried by the computer for various applications.

3.6.3 Handheld Readers

Handheld readers can be used to locate items in a distribution center or to read tags that do not go through the portal. They are battery powered and communicate over a wireless connection to an access point. Some handheld readers are complete, self-contained computers (see Figure 3-16). Others are devices that can be attached to a handheld computer made by such companies as Palm or Psionic, or a Windows CE computer.

Figure 3-16
Handheld reader

3.6.4 Mounted Readers

Mounted readers may be attached to a forklift or truck, as shown in Figure 3-17, to record the movement of goods from one place to another. Mounted readers may interface with tags embedded in the ground or mounted on walls, so the computer knows their location. They also read the tags attached to the article, pallet, or cartons they are carrying.

Readers operate in two modes: autonomous mode and interactive mode. In autonomous mode, the reader reads the tags that enter its read zone and immediately transmits the data record to the host. In interactive mode, the reader stores the tag records until the host calls for it to be uploaded. The record for each tag may include

- The tag identifier
- The date/time the tag was first read
- The number of times the tag was read
- Which antenna read the tag
- The name of the reader

Figure 3-17
Reader mounted on forklift

In interactive mode, the reader receives commands from the middleware, application, or a user and executes them. The commands may tell it to enumerate and send the list of tags currently in its read zone or to change the reader's configuration settings.

The host computer(s) can run one or more applications that periodically download the tag list. In this way, the host system can continuously update its internal model of the operation. Note that the host can also continuously monitor the health of the reader network and provide early warning of deteriorating performance by readers or antennas.

3.6.5 Reader Selection Criteria

The following list summarizes considerations for selection of a reader:

- Antenna attachment options
- Tags to be read
- Reader performance
- Environment
- Connectivity and power requirements
- Space available for unit
- The ability to upgrade
- Control functions
- Drivers
- Total cost of ownership

3.6.5.1 Antenna Attachment and Control Options

Stationary readers will accept external antennas, as differing conditions and differing applications may require different antennas. Consider whether you will need multiple antennas attached to the same reader, in which case they will need to be multiplexed. Some readers have multiplexers built in; others require a separate unit. Compare the number of antennas the reader can handle against the number you will need at your reader site.

Handheld readers (and tags) typically have their antennas built in.

Readers differ in the levels of control they afford over the antenna attachment. For example, they may allow you to set different power levels for different antennas. Or, they may allow you to control the polling.

Thus, for example, if the read zone for Antenna 2 is particularly challenging, you may be able to set the reader to poll Antenna 2 two or three times in each polling cycle, polling Antennas 1 and 3 only once.

3.6.5.2 Tags to Be Read

Obviously, the reader must be able to read all the tags that you expect it to see. In some applications, it is likely that the same reader will be processing different tag protocols. The best solution to this problem is to have what is called an agile reader to read multiple different protocol. Note that *multiple* protocols does not mean *all* protocols. In a closed loop application where you control all the tags and readers, you may not need an agile reader.

3.6.5.3 Environment

Readers and encoder/printers may need to be located in harsh environments and/or outdoor locations. Escort Memory Systems is one vendor that supplies readers for demanding environments such as scorching paint ovens or post office parcel tracking applications that must read and write dozens of labels at the same time. A number of companies provide handheld readers and submersible antennas for the disc drive industry.

3.6.5.4 Connectivity and Power Requirements

There are two ways to connect a reader to an application. Each has its advantages and drawbacks. Serial readers connect to an application through a serial link. Network readers behave like any device on the network.

Serial readers use a serial link to connect to an application. The link can be either an RS-232 or an RS-485 serial connection, as shown in Figure 3-18. The RS-232 protocol is older. The RS-485 protocol allows cable lengths up to 1200 meters, and it can be used as a bus, enabling a single cable to connect multiple devices. The serial link is very reliable, but lengths are limited, and there is no structure for managing the large numbers of readers that enterprises expect. There are other drawbacks as well, including the lower transmission speed of the serial ports and difficulty of upgrading the readers when necessary.

To support the vision of networks of readers working together to generate an enterprise class data resource, the readers will need to connect to their applications and be coordinated with one another through a network. Ethernet has emerged as the most commonly used connection method.

Figure 3-18
RS-232 and RS-485
connectors

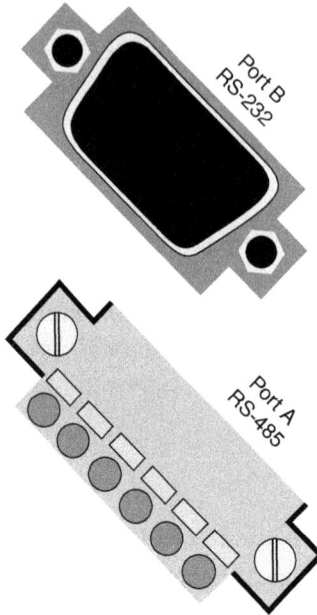

Port B
RS-232

Port A
RS-485

Readers can connect via Ethernet cables or through a wireless Ethernet connection. A PCMCIA slot is available on some readers so you can plug in a wireless Ethernet card. Common Ethernet interfaces include coaxial cable, wireless, and twisted pair. Ethernet over twisted pair offers its own power source, called Power Over Ethernet (POE), which may keep you from having to install AC power in new places for your fixed readers.

Reader connectivity is the subject of intense interest right now. See Section 3.8 for a discussion of RFID controllers, which are connectivity managers for reader networks.

3.6.5.5 Space Available for Unit

There may be limits on the space available for the reader, which may need to fit in small working spaces. Keep in mind forklifts, men carrying heavy objects, and other dangers and the need to keep your equipment out of harm's way.

One solution is a commercial rack specially made for RFID readers and antennas. Built of heavy-gauge metal, it has facilities for mounting antennas, storing readers, and channeling cables.

3.6.5.6 Ability to Be Upgraded

Some readers will have their control logic hard-wired into the unit. This means you will have to replace the unit when new standards or new functionality are required. Other units have their control functions stored in an EEPROM. This means the unit can be upgraded by reprogramming or replacing the chip. This is a significant advantage over hard-wired units, but it means that someone has to physically and manually work on the unit whenever it needs to be upgraded. Some readers advertise themselves as "software upgradeable." This means that they can be upgraded over the network, without anyone touching them. Two companies, ThingMagic and Alien Technology, describe their readers as small computers just running software. This means that the units can be upgraded over the network, without anyone having to touch them; the upgrade can be completed via the network from anywhere.

3.6.5.7 Control Functions

Many implementers simply plug in the reader and use the factory settings. They wonder, then, why their read rates are low. The professional installer can work with the reader control functions to tune a reader's performance. The fact that readers differ in the details of the control functions they make available makes this an interesting area to explore as you are evaluating your reader choices.

Some readers offer named "profiles," which constitute a group of settings to represent ideal performance in a particular environment (for example, "Warehouse Profile"). One of Symbol's readers, for example, allows you to reset the antenna type, scan period, retry limit, air interface, and antenna gain with a single selection. Consider the control functionality that the reader makes available and whether it can be set through a browser interface, through the middleware you plan to use, and any other access mechanisms.

Readers may also offer circuits for synchronization with various devices on your assembly line.

Middleware products offer reader control functionality. Vendors such as GlobeRanger (iMotion), OATSstems (Senseware), ConnecTerra (RFTagAware), and Sun Microsystems (EPCInformation Server) all offer a consistent interface for controlling the disparate set of readers you are likely to acquire. Be sure that whatever readers you are considering are supported by your middleware control software.

3.6.5.8 Drivers

To ensure that your readers can work with your RFID middleware, you will want to investigate the availability and compatibility and scope of the drivers. Try to get an idea how quickly new requirements are accommodated and, conversely, how often you will have to update your drivers. Determine any charges for these updates.

3.6.5.9 Total Cost of Ownership (TCO)

The numbers of readers that you will ultimately require could be very large. Even a few tens of dollars are significant when you are purchasing thousands of machines. In other venues, customers have found the maintenance and support contracts burdensome. Investigate thoroughly these cost elements:

- Unit cost
- Installation cost
- Contract maintenance cost
- Repair costs
- Cost to train your personnel on support, configuration, and maintenance
- Interface/integration cost, if any
- Requirements for custom interfaces that will have to be upgraded when new software versions are released
- Special costs, if any

3.6.6 Nearby Readers

Nearby readers may interfere with one another. A reader can transmit about 100,000 times more powerfully than a passive tag can, so unless you take steps to prevent it, nearby readers will drown out tag transmissions. This is a problem that may not show up in your pilot, since pilots are often conducted with just a single reader. However, nearby reader interference is a real problem and has bedeviled several installations. These are the methods you can employ to deal with it:

- Control the direction of your transmissions
- Manage the power output
- Manage when readers are active

- Place physical shielding between portals
- Utilize Gen-2 dense reader protocols

3.6.6.1 Control the Direction of Your Transmissions

Chapter 12 describes how to use a spectrum analyzer to create contour maps for your readers. Contour maps show the shape and extent of the antenna's transmission and read zones. They also show the direction and strength of the reader's signal. Just pointing antennas away from each other is the easiest step to reducing interference between nearby readers. Use the contour maps to position and orient your antennas.

3.6.6.2 Manage Power Output

Many installers make the mistake of solving tag reading problems by simply raising the power threshold of their readers. In practice, however, increasing reader power can create as many problems as it solves. As the power output of the reader increases, the likelihood of interfering with nearby readers—or other RF equipment—increases. It also increases the likelihood of false positives, such as reading tags going through a nearby reader. The best practice is to operate your readers at the lowest power setting that will give satisfactory tag reading performance.

3.6.6.3 Manage Reader Activation Periods

Front every portal with a detector that notifies the reader of the arrival of a pallet and activates it. Set the system to have the reader active for as short a time as practical.

3.6.6.4 Place Physical Shielding Between Readers

If you have a problem with readers interfering with one another, you may wish to place physical shielding between them. The shielding can be made of metal fences or screens. The openings in the screen must be no larger than about 3 inches for UHF systems. The screens must extend beyond the line of sight in order to eliminate diffraction of the transmissions around the edges of the screen.

3.6.6.5 Dense Reader Protocols

The Gen-2 specification has elaborate mechanisms to reduce interference from nearby readers. Different mechanisms are specified for different jurisdictions, depending on the bandwidth available. In North America, with its relatively generous bandwidth allocation, frequency hopping and a large number of channels are used. In Europe, with its narrower bandwidth allocation, listen-before-talk is used. This severely degrades system performance for some applications.

Gen-2 specifies a *Q algorithm* to sort the tags in a read zone. The Q algorithm requires tags to send a short identification number to the reader so it can identify all the tags in the read zone and then sequentially interrogate each one for its longer EPC number. The reader changes the "state" of all the tags to off and then sequentially turns each one on, reads its EPC, and then turns it off so it will not interfere with subsequent reads.

Gen-2 also introduces a new radio-signaling mode that can isolate tag responses into a side channel, where they can be heard more clearly. These are called *Miller subcarriers*, and they permit the reader to create channels of varying widths, depending on the noise conditions in the environment. An alternative to Miller subcarriers is also provided, called *FM0* signaling mode. FM0 is faster but more susceptible to noise.

A third Gen-2 function to optimize performance in noisy environments is the reader's ability to operate in three different modes. The reader can detect the environment and change modes as conditions warrant. The three modes are: single reader, multireader and dense-reader. Multireader mode is utilized when the number of tags in the read zone is equal to or less than the number of channels available. Dense-reader mode is utilized when the number of tags is greater than the number of channels available. The reader will operate in the fastest mode that conditions permit.

A fourth Gen-2 function verifies that tags have been written correctly. This function is not specific to dense-reader mode, but it improves effectiveness of the entire system.

A fifth Gen-2 function is called Sessions. Sessions are important when the possibility exists that a single tag will be interrogated by more than one reader at the same time. In many cases this is not a problem, as the application only needs to know that an item has been read by one of several readers. But the Sessions feature provides a solution when the application needs to know which specific reader has read the tag. It also addresses the possibility that Reader B will change the tag's state while it is being read by Reader A, thus impairing the effectiveness of the Q algorithm. Each Gen-2 tag can maintain a session, that is, a separate set of communications, with up to four different readers. Changes any reader makes in the tag's state are effective only for that session, and they do not interfere with other readers' sessions.

Thus, up to four readers can develop coherent communications with any single tag.

The dense-reader features are optional in the Gen-2 specification. If your application has need of their functionality, you should not assume they are available in any reader merely because it is advertised as Gen-2 compliant. Also, different readers will provide varying performance levels in different environments. As always, you should test these functions before finalizing your selection of readers.

3.7 Encoder/Printers for Smart Labels

Smart labels are described in Section 3.2.6. Encoder/printers are specialized devices that write data to the tag, print the bar code, and print the information in human-readable forms on the smart label, all in the single operation. Encoder/printers combine the functions of reader, for commissioning the tag, and printer, for printing the text and the bar code on the label. The reader tests the tag after it is written, to confirm that the tag is functioning.

The simplest workflow is as follows:

1. Operator identifies item to be tagged.
2. Host computer provides the data to the encoder/printer.
3. Encoder/printer encodes the tag and prints the label.
4. Encoder/printer (automatically) verifies the tag was written correctly.
5. Human applies tag to the pallet or case.
6. Host computer records the tagging event and details about the item tagged.

This workflow describes a "slap-and-ship" solution (see Chapter 11). Equipment is also available that can automatically attach the labels to boxes on a conveyor belt. For example, the Markem Cimjet, shown in Figure 3-19, is designed to be mounted on a conveyor belt and connected to a network, power, and air. It is then ready to automatically print, encode, and apply smart labels to boxes.

Encoder/printers come in a wide variety of form factors, reflecting the variance in label size, speed requirements, connectivity, and ability to fit various environments. Some examples are discussed in the next sections.

Figure 3-19
Automatic applicator
(Picture courtesy of
Markem Corporation)

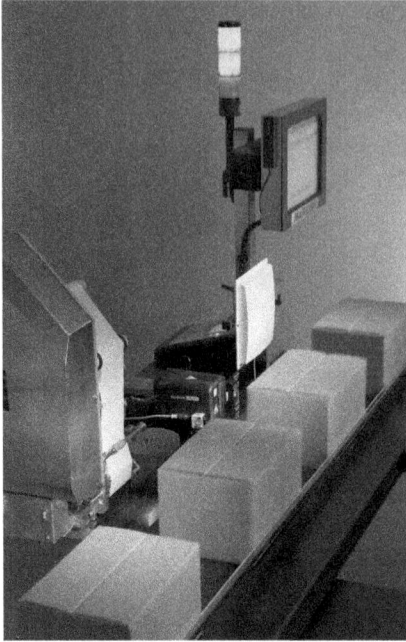

Figure 3-20
Printronix
encoder/printer

Intermec and several other companies offer printers that can attach directly to the network, needing no host computer for the connection. Their printers may be upgraded to add RFID encoding capabilities. Zebra has a line of RFID-enabled printers for various applications. Printronix also sells RFID printers, as shown in Figure 3-20.

The work process and business software must generate the data, and the computer infrastructure must get it to the encoder/printer to write it to the tag. The encoder/printers on the market vary greatly in cost and capability. Section 3.7.1 summarizes considerations for selection of an encoder/printer.

3.7.1 Encoder/Printer Selection Criteria

Selecting the right encoder/printer is not difficult. It is just a matter of carefully documenting your requirements and considering the alternatives. Several alternative strategies exist, which you should consider. First, if many of your proposed sites are going to upgrade existing bar code read stations, you should consider upgrading the equipment already there instead of replacing it. Contact your supplier and see if there is an RFID upgrade module available. Second, do not be constrained by the standard offerings of major vendors. Numerous alternatives exist. Customizing your readers may enable you to add significant functionality to your application at a reasonable cost. As always, consider the total cost of ownership (TCO includes routine maintenance, software upgrades, warranty terms, and so on) before making any final decision. In choosing your encoder/printer, use the reader checklist (see Section 3.6.5). The considerations in the next sections are additional factors to consider when specifically selecting an encoder/printer.

3.7.1.1 Label Width

The width of your desired labels is a function of the space available on the package and the amount of text you will need to print. Any standardization you can make across various package configurations will pay dividends in throughput and prevent your operators from having to change stock too often. Various manufacturers offer encoder/printers for various sizes. Here is a representative sample:

- Zebra offers 2", 3", 4", 5.5", 6", and 8.5" widths.
- Intermec offers sizes from 0.5" (13 mm) to 4.7" (120 mm).
- IBM offers the Infoprint 6700, which prints 4", 6", and 8" labels.

3.7.1.2 Label and Adhesive Material

If you have special requirements for survivability or harsh conditions, they will influence your choice of encoder/printer. Most encoder/printers use direct thermal printing to print on specially treated paper. Make sure the papers meet your survivability requirements. Survivability issues include moisture, heat, cold, dirt, and rough handling.

3.7.1.3 Number of Labels Printed per Day

Make sure the printer's duty cycle will meet your needs.

3.7.1.4 Print Resolution

Printers from different manufacturers make various print resolutions available. The following list shows some examples of what is on the market:

- Zebra offers 200, 203, 300, and 600 dpi resolution printers.
- Intermec offers 203 dpi, 406 dpi, and 300 dpi.
- IBM offers interchangeable print heads on the same machine, supporting 203 and 300 dpi.

3.7.1.5 Print Speed

Intermec offers printers ranging in speed as low as 1.2 inches per second up to 9 inches per second. IBM claims up to 10 inches per second.

3.7.1.6 Failure Strategy

It is inevitable that some tags will fail the encoding step. Encoder/printers should detect quiet tags and mark them visibly for nonuse. Work processes must be devised to deal with failed tags, including tracking the percentage of bad tags in a given batch.

3.7.1.7 Total Cost of Ownership

The TCO elements for encoder/printers are the same as for RFID readers (see Section 3.6.5.9), with the addition of costs for supplies.

3.8 Controllers

RFID controllers are a relatively new class of hardware. They interface between a group of readers and the computer network and provide infra-structure connectivity to the various sensors and devices of a complete RFID solution. For example, Arcom's RFID Edge Controller product provides connectivity with IBM's WebSphere RFID Premises Server middleware. Infra-structure devices such as readers, label printers, pallet controllers, and conveyor controllers and discrete devices such as proximity switches, light towers, and alarms can all be interfaced to one or more controllers, providing a common interface to the host computer system. Some companies—for example, Alien—build controller functions into their readers. However, the power of a stand-alone controller is that it can aggregate, filter, transform, and manage a network of readers and other devices, performing tasks outside the scope of a single reader.

Startup Reva Systems has published an architecture called Tag Acquisition Network (TAN) to address the issue of managing networks of readers. Their product is a device called the Tag Acquisition Processor (TAP). It is a controller to provide integration with enterprise applications without needing middleware. The TAP would link into an enterprise's wired or wireless local area network. Individual readers are administered through an enterprise network manager system or assigned IP addresses, just as printers are managed through an enterprise system today. A reader-to-network standard interface would allow users to add, remove, and diagnose problems with readers through a centralized server software or a piece of equipment, just as a router is used to manage large banks of computers. It would provide a way for data to be sent to applications without requiring that the applications be bound to middleware.

The Internet Engineering Task Force (IETF) points out that existing and proposed ISO and EPCglobal protocols deal with communications between tags and readers, and they manage the readers with middleware. But a protocol is required, they say, for connecting the networks of readers to the enterprise network. IETF has begun work on what they call the Simple Lightweight RFID Reader Protocol (called, inevitably, SLRRP—pronounced "Slurp").

3.9 Software

There are two categories of RFID software: RFID middleware and application software. This section describes both.

3.9.1 RFID Middleware

As you contemplate the task of converting the raw tag database that readers create into useful, meaningful information, the importance of the middleware starts to become clear. Unfortunately, few industry analysts have been able to describe exactly what middleware is. Every vendor understands it somewhat differently, and no vendor is willing to concede any scrap of functionality to their competitors, just in case they might want to develop it themselves in the future. So vendors claim to provide products that are "complete," and it's up to you to figure out what you need and who can provide it.

If you are just implementing a tag-and-ship operation (see Chapter 11), the data management and data integration needs are limited. This is especially true if you are working with a limited number of SKUs and a single point of integration into just a warehouse management system. An investment in middleware may not be required at this time. The challenge escalates quickly as you move to a larger number of products, tagging stations, facilities, users, and integration points.

Middleware must work within the constraints of your corporate standards and security policies. We will not discuss these constraints further here. But the middleware you select must interface with your warehouse management system(s), your ERP system, and, potentially, your supply chain partners' systems. We will consider these interfaces in a generic way in this section. Each company will have its own unique set of circumstances and requirements. Middleware must provide the following functionality:

- Device management for both readers and other devices
- Data collection and integration
- Data acquisition and structuring
- Data filtering and routing
- Line coordination and control
- EPC allocation, both global and local
- Visibility and reporting, both business information and device status and performance
- Track and trace applications for recall and shipping
- Graphics creation

We will now look at each of these functions in greater detail.

3.9.1.1 Device Management

Most middleware products provide a framework for reader management. The middleware provides a single point of control for the network of readers purchased from a variety of vendors. With varying degrees of flexibility, they provide access to the reader's settings and allow you to change at least some of them. Middleware may also enable you to perform software upgrades on the readers without having to touch them, if your readers are software-upgradeable (see Section 3.6.5.7), and they may provide monitoring functions so you can see how the readers are performing. There are a number of challenges for your middleware in the area of device management. First, it would be comforting to believe that you will be able to acquire a homogeneous set of readers to manage, but this is unrealistic. As time goes on, it is inevitable that you will be managing readers from different vendors, different models from the same vendor, and different versions of the firmware and control software of the readers. It is inevitable, also, that even identical readers will be set up differently at different locations. These variations and settings will have to be managed and maintained when readers are replaced, moved, or upgraded. Middleware helps you manage these variations.

As RFID data moves back into the enterprise, a whole new set of devices will need to be managed. The middleware will need to manage bar code scanners, printers, scales, sensors, validation stations, employee badge readers, Programmable Logic Controllers (PLCs), and others as they become part of the IT infrastructure. These devices are not built to the specifications of an international standards body, so drivers for all these devices will have to be provided, maintained, and updated when needed.

3.9.1.2 Data Collection and Integration

The middleware needs to collect data from multiple disparate systems and integrate it back into other information systems. EPCglobal standards provide some help in this regard, and the growing acceptance of XML and web services means this will be easier to do.

3.9.1.3 Data Acquisition and Structuring

The middleware should enable you to work with the data in productive structures and map it into the needs of your applications. EPCglobal has released its first software specification, called Application Level Events (ALE). The ALE specification creates a useful structure called *events*. For example, an event might be "a tag from a Company in the Supply Chain passed a Reader in a Warehouse in Alabama." ALE-compliant middleware enables you to subscribe applications to the events. The middleware notifies the application and sends a configurable table of data when the designated event takes place. The ALE specification is discussed in Section 2.4.2. Consider whether ALE meets your needs for interfacing your various applications with the RFID data, and if it does, select middleware that can support it.

3.9.1.4 Data Filtering and Routing

The volumes of data produced by numbers of RFID readers will have to be filtered and reduced. Middleware has a role in this, shared with smarter readers and controllers. The middleware is distributed; good practice as well as economics dictate that the filtering be performed as close to the source of the data as possible. This suggests that the middleware will be distributed across several servers. Then you will need your middleware to enable you to route the data to the right person or application.

3.9.1.5 Production Line Coordination and Control

Applying variable information in real-time on the production line will require autonomous, integrated line coordination and process control. Ensuring that uniquely programmed tags are placed on individual cases and pallets, to say nothing of individual items, will require control of conveyors, PLCs, sensors, reject mechanisms, queuing, and other components. This will require each line to run autonomously while being networked within a greater system to ensure data integrity and scalability in the production process.

3.9.1.6 EPC Allocation of EPC Numbers

The assignment and management of Electronic Product Codes (EPCs) must be managed. During a pilot, with a single station and a very few product lines, the requirement for EPC allocation systems is not apparent. But when you are in full production, you will need systems to provision the various encoding stations with the numbers they will need, appropriate to the jobs they are doing. Some stations may work with the same product line all the time, but others will undoubtedly work with different product lines at different times. Also, some product lines will be produced—and tagged—at multiple locations so management of your "inventory" of EPCs will need to be managed by the middleware.

The EPCglobal organization controls the EPC allocations to the manager (that is, company) level. The enterprise (that is, you) control(s) them to the pallet, case, and item levels. Your middleware must enable you to make sure no duplicates are created and each item is tagged correctly. It must also record the event when the tag is commissioned and transmit it to EPCIS. When you consider the changing set of definitions, a structure must be in place to manage them, to make sure new object identifiers are entered and accessible and applied correctly. This may be particularly challenging when you have multiple sources of supply for the same SKU.

3.9.1.7 Visibility and Reporting

RFID data will be used to make automatic processes smarter and to enable management to respond more quickly to changing events. This assumes that reports are available that present the information in an actionable way. The reports may come from applications that have sub-scribed to RFID events, but they may also come from the middleware itself. One area that is solely the responsibility of your middleware is monitoring the performance of your systems. The middleware should warn you of devices in peril before they fail and certainly notify you of any actual failures. Policies may require rerouting or halting of lines or other appropriate response.

3.9.1.8 Track and Trace Applications

In the end, RFID, when fully deployed should enable you to track and trace where any item, or any group of items, is in the supply chain. You will be able to tell where it has been, how it got to its current location, how long it took to get through each step along the way, and potentially, what conditions it encountered. This information will enable a host of new applications and new capabilities within existing applications. The middleware provides a standardized way for these new applications to get access to the RFID data stream.

3.9.1.9 Graphics Creation

It is simple for a single station to print the labels with the correct graph-ics, legends, messages, and variable information. As multiple packages begin to pass through an encoding station, different messages and differ-ent graphics will be required. Even if only one SKU is processed through any one station, the messages, legends, and graphics may change over time, season, or circumstances. The middleware should provide a way for the reader to quickly identify and access the correct graphics and mes-sages as well as the variable information for a given job.

3.9.1.10 RFID Middleware Summary

It is clear from the preceding sections that RFID middleware functional-ity and capability will drive a significant portion of your project's success. Your ability to easily access the RFID data stream, manage your readers, allocate and track your serial numbers, and use the data in actual produc-tion applications all will be made possible or difficult depending on your middleware's capabilities. Take the time to make your selection carefully and test each critical function in your own environment before assuming that it will work for you.

3.9.2 Application Software

The RFID implementation will create significant new software requirements. The focus thus far in the marketplace has been on the hardware side and the challenges of getting radio frequency communications working reliably. But the focus will inevitably turn to software because that is where the next wave of value creation will take place.

Sophisticated companies already use standard ERP applications from vendors such as SAP, JD Edwards, Oracle, PeopleSoft, or Microsoft to help manage their supply chain activities. Manugistics is a leader in this field, and both Wal-Mart and the United States Defense Logistics Agency (DLA) use their ERP system. Sophisticated companies also use Manufacturing Execution Systems (MES), Warehouse Management Systems (WMS), Supply Chain Management Systems (SCMs), and Finite Forward Scheduling (FFS) packages. In addition, some are starting to work with what are called Advanced Planning Systems.

These existing software operations and initiatives are providing the opportunity for the next wave of value creation. As the RFID installations move past tag-and-ship, they will begin to create data that the applications can use. The role of RFID middleware has already been discussed, but the applications themselves have a role as well.

The various ERP applications differ in their ability to work with RFID data at this time. As item-level tagging comes into wider usage, the ERP applications and their underlying databases will have to be upgraded to accommodate it. Several other vendors offer software, specifically enabled for RFID, for specific applications.

For retailers, distributors, logistics companies, and manufacturers, a decision will have to be made in each case whether to upgrade or to replace the legacy systems. The decision to upgrade means following your vendor's roadmap. This may be the most economical and pain-free path. The decision to replace the system means a considerable expense and effort, but it may open you to other benefits.

Once the integration with legacy systems is completed, a new wave of value creation will begin, as entirely new applications emerge to take advantage of the data.

RFID implementations outside the supply chain will be done using custom or commercial software for the specific application. Tracking hand tools, opening doors, tracking and routing baggage in an airport, taking cash at a baseball game, or managing prisoner information will all be supported by dedicated application programs either purchased off the shelf or custom written for the end user.

3.10 Chapter Summary

Chapter 3 presented a detailed description of the various components of an RFID *system*. You will need to know this material very well, and you will probably find yourself returning to this chapter often, until it is familiar to you. Knowledge of the data, tags, antennas, connectors, cables, readers, and controllers, as well as the software and information services, is the essential underpinning of a successful RFID project.

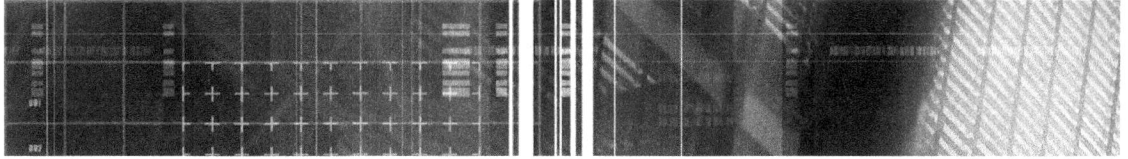

Bar Codes and RFID Tags

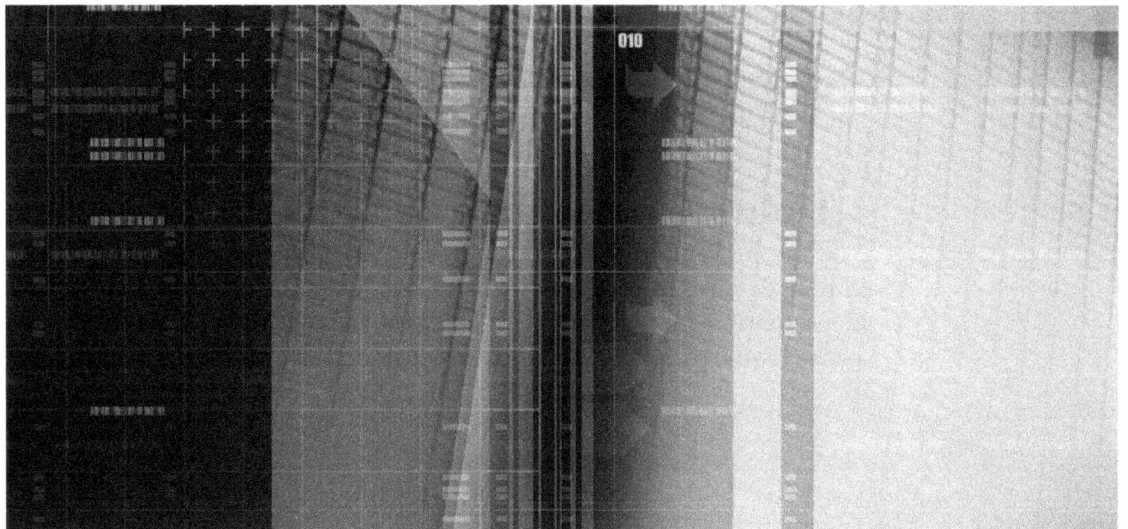

Functionally, RFID tags are sometimes viewed as a very powerful upgrade to the bar code. It is natural, therefore, to consider whether the tags will replace the bar codes throughout the world and what their application will be in many industries. The idea of RFID as a replacement for bar codes has been circulated and published in recent years, as the potential of RFID technology is embraced by different companies in various industries. *SecureID News*, an industry trade publication, recently proclaimed: "New RFID tag standard poised to replace bar code on every consumer product." *The Star* headlined, "RFID may replace bar code soon." As a result of recent articles such as these in various publications, many people think that RFID will replace bar codes. This is a very unlikely scenario. In the future, bar code technology will co-exist with RFID technology. Each technology has its own strengths, weaknesses, and associated costs. Billions of bar codes are produced and used every year. Billions of dollars have been invested in bar code systems and in systems that depend on them for input. Bar codes are so inexpensive—especially when they are printed directly onto packages and labels—that the cost to migrate to RFID tags at this point is prohibitive. Bar code systems will not be replaced any time soon.

Bar codes and RFID have very different advantages and disadvantages. There are sound reasons to replace bar codes with RFID tags in specific circumstances. There are also sound reasons to use bar codes in others. Like so much in life, the decision regarding which technology to use is not so simple and always depends on other relevant criteria.

This chapter will describe the advantages and disadvantages of bar codes and RFID. It will equip you to make an objective determination of which to use in any given circumstance. It begins with a definition of bar codes and a look at the various encoding techniques the systems employ. It will also provide you with data to use in converting from bar codes to EPC tags if necessary. (For additional information that will be useful in this conversion, see Chapter 2.)

4.1 Bar Codes Introduced

A bar code is an automatic identification technology in which numbers, and sometimes characters, are encoded in a pattern of vertical bars, spaces, squares, and dots so they can be read by a bar code scanner.

The scanner translates the pattern into data and transmits it to a host computer for processing. Bar codes are used to identify products, documents, shipments, tools, equipment, and so on.

4.1.1 Reading Bar Codes

Bar codes are read by specialized devices called scanners. There are five different scanner technologies currently available: wand scanners, laser scanners, CCD scanners, camera scanners, and MEMS scanners. Each employs fundamentally different processes.

4.1.1.1 Wand Scanners

Wand scanners are the simplest and least expensive type of bar code scanner. They are also the most durable because they have no moving parts. They must come into direct contact with the bar code to read it. They are small and ideally suited for use with laptop computers or very low volume applications.

4.1.1.2 Laser Scanners

Laser scanners are the most common type of bar code scanner. They can read a bar code from about 6 to 24 inches away. Long-range laser scanners can read up to 8 feet from the label, and there are some very expensive models that can read labels up to 30 feet away. Laser scanners are available as handheld "guns," and also as countertop or fixed-position configurations. Laser scanners can read labels very quickly and can compensate for curvature of the label or bad label quality.

4.1.1.3 CCD Scanners

Charged-couple Device (CCD) readers employ an array of several hundred small light sensors at the front of the reader. When the image of the bar code is projected onto the sensors, they generate a pattern of electricity, which can be read and decoded. CCD scanners have a fast scan speed, but their range is very limited, often less than 3 inches from the label. They are well suited for point-of-sale applications, but their short read range makes them unsuitable for warehouse or industrial applications.

4.1.1.4 Camera Scanners

Camera readers use the new technology developed for digital cameras. Camera readers are smaller, faster, and cheaper than the others, and they can read 2D images. One negative aspect of camera scanners is the fact that they are more sensitive to label quality than laser scanners.

4.1.1.5 MEMS Scanners

A new class of scanner has been released in the marketplace based on MEMS (micro-electro-mechanical systems) technology. This technology includes sensors, gyroscopes, actuators, motors, and pumps on a silicon substrate that can be smaller than a grain of sand. MEMS products include inkjet printers, projection systems, vehicle airbags, medical imaging, and battlefield displays. MEMS is not nanotechnology because it is not engineered at the molecular level.

MEMS scanners do not use a motor to move the mirror, which eliminates an important source of failure. Instead, the mirror is etched into silicon and uses a method of frictionless oscillation. MEMS mirrors are smaller and faster than the ones used in laser printers. MEMS can scan a single label about 500 times per second, which increases the likelihood of reading damaged, low-contrast, or torn labels.

MEMS scanners can read PDF417 and other stacked 2D symbologies, but they cannot yet read Data Matrix or other matrix-style codes.

4.1.2 Commissioning Bar Codes

The process of encoding data onto a bar code is fundamentally different from commissioning or adding data to RFID tags. Since bar codes are printed once, they contain only static information. This is a major difference between the bar code and RFID technologies. RFID tags are available that can accumulate and store data throughout their production lifetime, and the data can be changed from time to time. In contrast, bar codes are printed once and the contents remain static.

The bar code is rarely encoded at the item level, so it is feasible to print them, for example, on packages along with the other product-specific data, information, and graphics. A million boxes intended to hold 48 Huggies diapers each will all have the same UPC code, so the boxes themselves can be printed with the same bar code directly onto the side of the box. In cases where there needs to be a unique bar code, thermal bar code printers are available from manufacturers such as Zebra, Intermec, Symbol, and others.

4.2 Bar Code Symbologies

The exact mapping of bar code patterns to characters is called a *symbology*. Several hundred symbologies have been developed over the years, and about 50 are in use today. Different symbologies have differences in their efficiency, their density, their ability to represent different degrees of information, and the error checking they offer.

Bar codes are typically used to encode a few digits, a "license plate" for the item they are attached to. But there are symbologies that can encode letters as well as numbers, and there are symbologies that can encode hundreds or even thousands of characters. This section will describe the most widely known bar code symbologies.

4.2.1 Uniform Product Code (UPC)

The UPC is the most widely known and recognized bar code symbology in the United States. It encodes only numerical digits. It does not encode letters or punctuation. This code is managed by the Uniform Code Council (UCC) in the United States and was created to identify consumer goods in trade. As was discussed in Chapter 2, the UCC has merged with EAN to become GS1, the parent of EPCglobal.

The UPC comes in two formats. The most widely known is UPC-A. Less recognized is UPC-E, which is a short-form representation of the same data.

The UPC-A format, shown in Figure 4-1, is a 12-digit, numeric symbology. UPC-A symbols consist of 11 data digits and 1 check digit.

UPC-E consists of six data digits and one check digit. As shown in Figure 4-2, it represents the same information as the UPC-A but with a set of convoluted rules to compress the data. In most cases, UPC-A can

Figure 4-1
UPC-A example

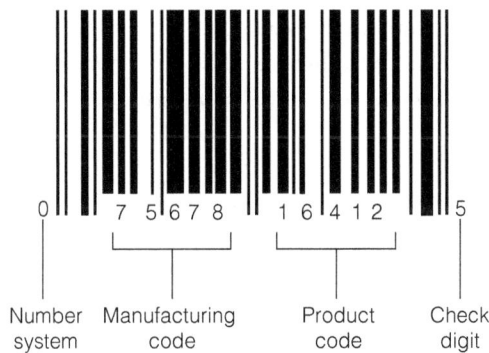

Number system   Manufacturing code   Product code   Check digit

Figure 4-2
UPC-E example

Original
UPC-A →
bar code

0   4 2 1 0 0   0 0 5 2 6   4

Equivalent
UPC-E →
bar code

0   4 2 5 2 6 1   4

Figure 4-3
UPC-A bar code with
2-digit supplemental
message

4 5

0   2 2 3 3 4   5 4 5 4 5   0

be converted to UPC-E by suppressing extraneous zeros. Since UPC-A allows room for 99,999 products, and very few manufacturers have that many products for sale at any point in time, the system works quite well in practice. UPC-E is used on small packages, where there is not enough room for the larger UPC-A to fit.

UPC numbers are assigned to specific manufacturers by the Uniform Code Council (UCC). Manufacturers are free to assign their own product numbers. To apply for your own UPC number or for more information, you can contact the GS1-US at 8163 Old Yankee Road, Suite J, Dayton, OH 45458, (937) 435-3870.

Both UPC-A and UPC-E allow a supplemental 2- or 5- digit number to be appended to the main bar code symbol (see Figure 4-3). This supplemental message was designed for use on publications and periodicals. If you enter a supplemental message, it must consist of either two or five numeric digits. The supplemental is simply a small additional bar code that is added onto the right side of a standard UPC symbol.

4.2.3 European Article Number (EAN)

The EAN is the European extension of the UPC. EAN-13 is derived from the UPC-A, with an additional digit. This digit, along with the twelfth

Figure 4-4
Example EAN-13
bar code

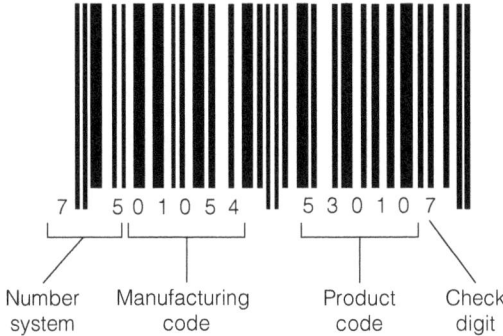

7 5 0 1 0 5 4   5 3 0 1 0 7

Number   Manufacturing   Product   Check
system   code   code   digit

digit, usually represents the country. Adding country codes to the design improves the UCC's ability to deal with international trade items. In Figure 4-4, note the 2-digit number system code at the beginning (75). This is usually interpreted as a country code.

The EAN-13 symbology is also used by the publishing industry to represent ISBN numbers for books. ISBN is a bar code in EAN-13 format; its first three digits are 978.

4.2.4 Code 39

Code 39 is the most widely used bar code outside of retail. Code 39 Normal encodes all digits and uppercase letters and the space, the characters $, /, %, -, and the period. In addition, the asterisk (*) is used as a start and stop character. Code 39 Full ASCII encodes all 128 characters of the ASCII character set. This includes upper- and lowercase letters. Unlike the UPC codes, Code 39 can be as long as necessary. It is a variable-length code, used widely in manufacturing. In Code 39 (also called "Code 3 of 9"), each character is made up of 9 bars, 3 of which are wider than the others, as you can see in Figure 4-5.

Figure 4-5
Code 39 example

TEST8052

Figure 4-6
Code 128 example

4.2.5 Code 128

Code 128 is a variable-length, high-density, alphanumeric symbology. It has 106 different bar and space patterns, and each pattern can have one of three different meanings, depending on the character set employed. You can see an example in Figure 4-6.

Special characters at the beginning tell the reader which of the character set is initially being used. To change character set in the middle, there are three shift codes available. One character set encodes all uppercase and ASCII control characters. The second set encodes all upper- and lowercase characters. The third set encodes numeric digit pairs 00 through 99. Code 128 also employs a check digit for data security. In addition to ASCII characters, Code 128 also allows encoding of four function codes (FNC1, FCN2, FCN3, and FNC4). The meanings of function codes FNC1 and FNC4 were originally left open for application-specific purposes. Recently, an agreement was made by the Automatic Identification Manufacturers Association (AIM) and the European Article Numbering Association (EAN) to reserve FNC1 for use in EAN applications. FNC4 remains available for use in closed-system applications. FNC2 is used to instruct a bar code reader to concatenate the message in a bar code symbol with the message in the next symbol. FNC3 is used to instruct a bar code reader to perform a reset. When FNC3 is encoded anywhere in a symbol, any data in the symbol is discarded.

4.2.6 PDF417

The preceding symbologies are called one-dimensional codes. PDF417 is a high-density, two-dimensional bar code symbology. As shown in Figure 4-7, two-dimensional codes carry far more information in the same physical space, and the symbol can sustain greater areas of damage to the area where it appears and still be read. PDF417 consists of a stacked set of smaller bar codes. It can encode the entire 255-character ASCII set. PDF stands for "Portable Data File" because it can encode

Figure 4-7
United Nations
Charter in PDF417

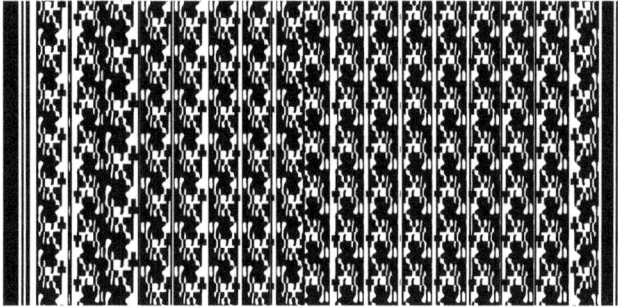

as many as 2725 data characters in a single bar code, and any number of bar codes may be placed on the item and logically linked together. There is no theoretical limit to its capacity.

The complete specification for PDF417 provides many encoding options including error detection and correction options, data compaction, and variable size, and aspect ratio symbols. The symbology was published originally by Symbol Technologies, Inc. to fulfill the need for higher density bar codes but is now in the public domain. PDF417 is governed now by the Association for Information Management (AIM) and is the two-dimensional bar code most often used. It is used at trade shows to encode the contact information on badges. It is also used to encode calibration data and operating instructions on equipment and notifications on hazardous waste containers.

The low-level structure of a PDF417 symbol consists of an array of code words (small bar and space patterns) that are grouped together and stacked on top of each other to produce the complete printed symbol. An individual code word consists of a bar and space pattern 17 modules wide. The user may specify the module width, the module height, and the overall *aspect ratio* (overall height to width ratio) for the complete symbol. A complete PDF417 symbol consists of at least 3 rows of up to 30 code words and may contain up to 90 code word rows per symbol with up to 928 code words per symbol.

The code words in a PDF417 symbol are generated using one of three data compaction modes currently defined in the symbology specifications. This allows more than one character to be encoded into a single data code word. Because different data compaction algorithms may be used, it is possible for different printed symbols to be created from the same input data. The symbology also allows for varying degrees of data security or error correction and detection. Nine different error correction levels are available with each higher level adding additional overhead to the printed symbol.

4.2.7 Data Matrix

Data Matrix is a code developed by a company called Acuity CiMatrix and has been placed in the public domain. It is a 2-D matrix code designed to store a large amount of information in a very small space. A Data Matrix symbol can store between 1 and 500 characters. The symbol is also scalable between a 1-millimeter square to a 14-inch square. That means that a Data Matrix symbol has a maximum theoretical density of 500 million characters to the inch. In Figure 4-8, "RFID Implementation Guide" is rendered in Data Matrix format.

Data Matrix has several other interesting features. Since the information is encoded by absolute dot position rather than relative dot position, it is not as susceptible to printing defects as other bar codes. The coding scheme allows the symbol to be read correctly even if part of it is missing. Each symbol has two adjacent sides printed as solid bars, while the remaining adjacent sides are printed as a series of equally spaced square dots. These patterns enable the scanner to determine both orientation and printing density of the symbol.

The most popular applications for Data Matrix is the marking of small items such as integrated circuits and printed circuit boards. These applications make use of its ability to encode about 50 characters of data in a symbol 2 or 3 mm square, as well as the fact that the code can be read with only a 20 percent contrast ratio.

The code is read by CCD video camera or CCD scanner. Symbols between one-eighth-inch square to seven inches square can be read at distances ranging from contact to 36 inches away. Typical reading rates are five symbols per second.

Figure 4-8
"RFID Implementation Guide" in Data Matrix format

Using Electronic Product Codes in Bar Code Formats

One company in Britain has deployed a complete EPC track-and-trace solution for the administration of hemophilia products to patients in Dublin, Ireland. The application makes use of EPCs, which are closely associated with RFID technology, but it actually uses the Data Matrix bar code. The application generates a unique EPC for every bottle of Clotting Factor Concentrate (CFC), the medicine used to treat hemophilia.

The company developed the solution after hundreds of hemophiliacs were accidentally infected with hepatitis and HIV after consuming contaminated products. This was termed "one of the most catastrophic medical complications of the twentieth century," by Dr. Barry White, Director of the Irish National Center for Hereditary Coagulation Disorders. Practitioners observed that there were considerable difficulties in identifying who had received the infected CFCs and in recalling all the contaminated products.

This use of the technology project demonstrates how the EPC number can be embedded in a Data Matrix code rather than a RFID tag. Nonetheless, the company, Domino Printing Sciences of Great Britain, plans to extend the application by adding RFID tags. This will enable higher levels of automation and freedom from bar code's line-of-sight restrictions.

4.3 Bar Code Advantages

Bar codes offer significant advantages when compared to the manual processes they usually replace. They speed up data collection; using laser readers, you can scan many bar codes in a very short time. Also, bar codes are far more accurate than manual data entry. The overall accuracy of bar code reading is one error per three million reads. Bar codes are a relatively simple and cost effective way to encode small amounts of static data reliably so they can be read and sent to a computer for identification, tracking, reporting, or analysis purposes.

When compared to RFID, bar codes have significant advantages as well:

- Cost
- Ubiquity and acceptance

- Human readable
- International acceptance
- Privacy
- Material type
- Mature technology

4.3.1 Cost of the Two Technologies

The current cost of RFID tags struggles to approach the five-cent/tag mark. The cost of bar codes hovers well under one cent. Thus the tag manufacturers struggle to reach a cost five times higher than that of the bar code. A handheld bar code scanner is less than $400, while the cost of a handheld RFID reader is over $800. Similarly, a stationary scanner is around $700, while a stationary UHF reader costs over $900. Also, the UHF reader generally requires at least two additional antennas, which may cost up to $500 each. Even if the equipment were identically priced, the ongoing cost of the tags themselves would significantly affect the return on investment (ROI) model supporting the technology at this time.

4.3.2 Ubiquity and Acceptance

Today, there are billions of bar codes attached to products every year. These systems are cost effective and efficient for the companies that use them. Replacing the existing bar code infrastructure would entail changing those systems and a significant investment in new capital equipment. The systems cost involves not only the information systems that process the bar code data, but also the infrastructure to keep the scanners functioning, the connectivity issues, and the supplies. Bar codes have been deployed throughout the supply chain in most industries and are a well-known, well-understood, ubiquitous technology.

4.3.3 Human Readable

Bar codes are usually human readable. This is not to say that humans can read bar codes, but the labels are typically printed with their numbers and letters at the bottom, and humans can read them if required. This means that when the scanner fails to read the bar code properly, or the process or the specific station of the supply chain does not have a bar code reader, the nearby human can read the data manually, and the process is only minimally interrupted.

Note that smart labels are available that provide the same element of human readability for RFID tags as exists for bar codes. See Chapter 3 for a complete description of smart labels.

4.3.4 Bar Code International Acceptance

There are no international restrictions on the use of bar code labels. The "transmissions" of data via bar code raise no regulatory issues any place in the world. RFID, as we have seen, uses frequencies that are heavily regulated in most of the industrial world.

4.3.5 Privacy

Bar codes raise no privacy issues. They have been in use for over 30 years, and no privacy groups object. In part, this is because the bar codes are not implemented with item-level serialization; they only identify the type of item which is labeled. Also, their use is visible, open, and obvious. There are no fears of surreptitious reading.

4.3.6 Material Type

Bar codes can be affixed to metal containers with liquids or any type of packaging without having their performance affected. Their performance depends solely on the ability of the label to adhere to the material and the ability of the scanner to "see" the label. Bar codes are not limited to adhesives; they can be etched into metal and attached with screws or rivets, or they can be printed right on the packaging. If the attachment of the bar code holds and the reader can get line-of-sight access to the label, the bar code will function.

4.3.7 Mature Technology

Bar codes are widely in use throughout domestic and international supply chains. The technology for working with them is very well developed, robust, and supported. Its reliability is documented and beyond question. The systems that work with them may not provide all the flexibility and manageability one could wish for, but they do get the products out the door, the shipments billed, and the inventory maintained.

4.4 RFID Advantages over Bar Codes

RFID enthusiasts claim numerous advantages for their technology as compared to the older bar code systems. In many cases, these claims are accurate, but in others the claimed superiority exists only under certain circumstances. The following subsections state advantages that proponents of RFID claim their technology has over bar codes. This discussion looks at each advantage to see under what circumstances it is true.

4.4.1 Dynamic Data

RFID read-write tags' data contents can be changed and added to after they have been commissioned. Bar codes' data cannot be changed once they are printed. Also, read-write tags have the capabilities to record process or environmental data such as temperature, moisture, radiation, or time of tampering.

4.4.2 Line of Sight

Direct line-of-sight access is not required to read a tag, but bar code scanners must have good line-of-sight access to the labels. This can have significant impact on the workflow in many circumstances, as it means that the RFID tags can often be read without human assistance.

4.4.3 Read Range

Proponents claim that RFID tags can be read over much greater distances than bar codes. Bar codes usually must be within a few feet of the scanner in order to be readable, although there are long range scanners available.

UHF tags can be read routinely at distances up to about 20 feet, and active tags can be read at distances as great as 300 feet.

4.4.4 Storage Capacity

RFID tags can store more data than bar codes can represent. As described previously, the most popular bar codes store only about 11 characters of useful data, although there are bar codes that can encode hundreds or thousands of characters of data. Nonetheless, with their electronic memories, there are tags that can hold substantially more data in smaller spaces.

4.4.5 Multiple Reads

Multiple RFID tags can be read at one time as they travel through a portal. Only one bar code can be read at a time through that same portal. Often, the bar code scanner is the bottleneck in a system, the element that limits the speed of the entire process.

4.4.6 Survivability

RFID tags can be designed to survive hostile environments that would destroy a bar code label, or at least make it illegible. RFID tags will function when painted over, buried in dirt, or covered with mud and snow. These are situations in which bar codes become unreadable.

4.4.7 Programmability

RFID tags can be programmed to perform calculations, record sensor readings, and perform logical operations. An active tag can function as a small computer. A bar code is only a static storage mechanism.

4.4.8 Accuracy

RFID tags are said to be more accurate than bar codes and less likely to produce erroneous reads. This advantage, widely claimed, is not supported in actual usage.

There are special circumstances, such as baggage handling, where the error rates are much higher for bar codes. In ideal cases, the accuracy rate for UPC readers is about one error in 400,000–800,000 reads. Accuracy for PDF417 readers ranges between one error in 10.5 million and one in 612.4 million. This is very accurate. In particular, in applications such as baggage handling, where the environment works against line-of-sight reading of labels, RFID will handily out perform the bar codes. In other applications, the bar codes could be more accurate. But overall, RFID systems, properly installed and configured, can match bar codes' accuracy but not surpass it.

4.4.9 Serialization

RFID tags are able to support item-level identification; bar codes support only package-level (SKU) identification. This statement is true, but it refers to the manner in which bar code applications are usually constructed rather than to limitations of the technology itself. It is possible to construct bar code systems that support item-level serialization.

4.4.10 Read Rates

The issue of read rates is much more complex than simply comparing equipment or theoretical read rates. RFID systems support automated workflows, so they will need to read tags much faster than bar codes will. The workflows around bar codes are usually built around human speeds, while RFID systems assume—and require—a much higher throughput. Bar code systems typically achieve their rated speeds, while RFID systems installers are struggling to meet theirs. Once the systems are optimized, however, RFID systems offer a much higher rate of throughput.

4.5 Chapter Summary

As you read this chapter, it should be clear that RFID tags will not replace bar codes any time soon. The RFID cost differential is too high compared to bar coding at this time for general deployment. The existing bar coding systems are too successful, reliable, and widely deployed, and the privacy issues are nonexistent. This does not mean that RFID is doomed to failure. The technology is still in its infancy. Costs are falling. Awareness is growing of the high capabilities that currently exist in RFID technology. These advantages are being effectively communicated and valued by businesses every day. The software supporting RFID is becoming less costly and more capable with greater functionality.

RFID does not have to replace all bar code tags to be successful. RFID, working with other technologies such as databases and the Internet, clearly has wide applicability and enormous power to transform business processes. Earlier adopters such as Wal-Mart, other retailers, and the U.S. Department of Defense (DoD) will not relent. They are determined to utilize RFID technology to wring out of their supply chain systems the costs associated with shrinkage, human errors, out-of-stock situations, asset tracking, and so on. They are determined to gain visibility over their work in process, finished inventory, materials in transit, equipment utilization, and so on. In all these endeavors, RFID has shown itself to be an irreplaceable part of the solution.

The view of the RFID technology as simply an upgrade to bar codes is misleading. If companies merely replace bar codes with RFID tags, there is very little payback. RFID is not an application in and of itself. The payoff, as we have said many times, lies in generating new data and putting it to use through the integration of RFID into an organization's ERP and asset tracking systems, and ultimately, its purchasing, logistics, and stocking decisions and sharing it with trading partners.

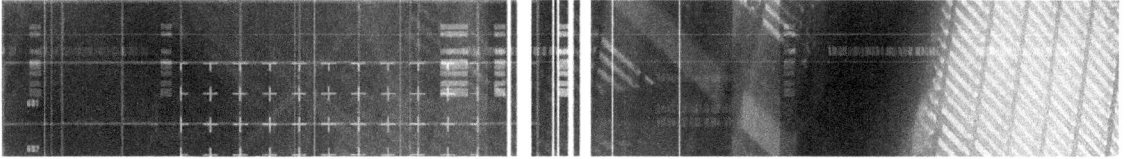

PART II

Applications

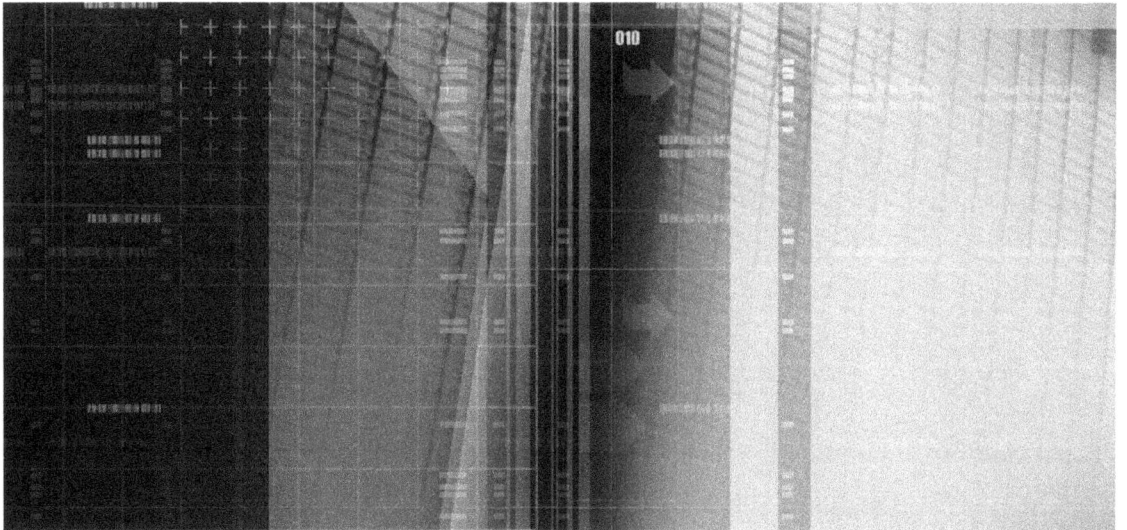

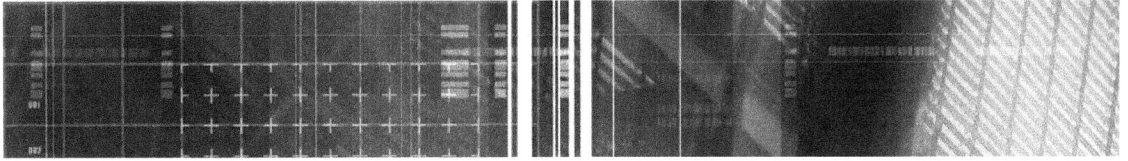

CHAPTER 5

RFID
Applications

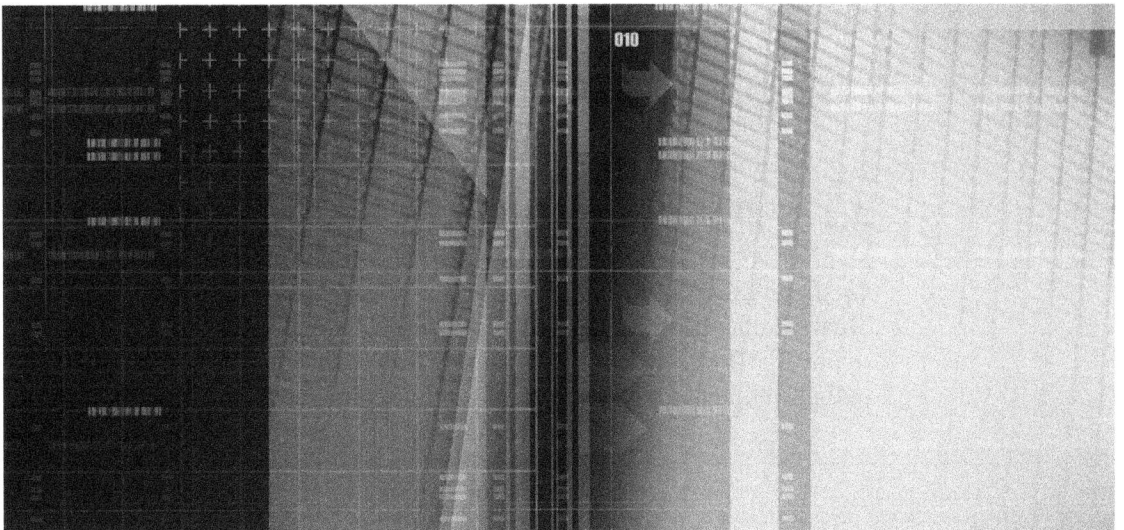

Part of the excitement around RFID is the wide variety of its applicability. The applications of the technology are limited only by our imagination, and it seems that new applications are being defined each day. This chapter describes first the "signature" applications that placed RFID in the public eye and made it a celebrity. (We discuss nine of these "signature applications.") Second, it categorizes the range of RFID applications and discusses each in detail.

In many of the cases, participants report difficulty in documenting a return on investment (ROI) from their RFID initiatives. This topic is covered in Section 5.3.

5.1 Signature Applications

Several signature applications have taken root in the public consciousness. They define for the public a set of benefits they might see and costs they might incur from RFID technology. These signature applications are

- The retail initiative for supply chain RFID, begun by Wal-Mart
- The McCarran Airport initiative for airline baggage tracking
- The Boeing initiative for manufacturing optimization
- Container positioning at the Port of Singapore
- Tagging of OxyContin at Purdue Pharmaceutical
- The Speedpass program at Mobil Oil Company
- The express payment program at American Express ExpressPay
- Passports with implanted tags issued by the United States Department of State
- Students tagged at Bratton School in Lawrence, California

5.1.1 The Retail Initiative

On June 11, 2003, Linda Dillman, the Chief Information Officer of Wal-Mart, announced that the retailer's top 100 suppliers would be required to put RFID tags carrying Electronic Product Codes (EPCs) on cases and pallets (but not individual items) by January 1, 2005. Wal-Mart stated that a major goal of the initiative was to reduce out-of-stock conditions so consumers would have a higher probability of finding the products that they want on the store shelves. Analysts have said that by implementing RFID, Wal-Mart could reduce its inventory costs by 5 percent and its logistics costs 7.5 percent. In fact, it has been estimated that Wal-Mart could save as much as $8.35 billion per year in supply chain efficiencies, which is more than the total annual revenue of half the companies on the Fortune 500 index.

This announcement hit the fledgling industry with the force of a tidal wave and RFID catapulted to public attention. Wal-Mart's mandate required:

- 100 percent of all cases and pallets sent by those suppliers were to be tagged. The target was later reset to 65 percent, as experience revealed problems with tag-read rates.
- The next 200 suppliers would be required to tag their shipments in January 2006, and all suppliers would be required to tag their shipments by the end of 2006.
- EPCglobal First Generation Class 0, 0+, and Class 1 tags were to be used initially, with a transition to Gen-2 as those tags became generally available.
- Advance Shipping Notices would be required, and they would be communicated to Wal-Mart using RetailLink or EDI. RetailLink is Wal-Mart's proprietary electronic marketplace.
- Tags would be encoded with a GTIN and a serial number, a serialized GTIN.
- Single-item cases such as a large-screen television set or a bicycle must be tagged.

Most of the giant retailer's largest suppliers announced they would comply. An additional 37 companies announced that, although they were not strictly required to comply by the first target date, they would like to be part of the project. Wal-Mart admitted all of them to the project. It appears that there were three separate motivations for the volunteers. Some wanted to take the opportunity to gain experience in what was clearly a disruptive technology. A second motivating factor was the desire to gain competitive advantage and to enjoy the greater efficiency and cost savings the technology promised. A third was undoubtedly the desire to curry favor and retain or grow the business with an important customer.

During the year after the announcement, several other vendors followed suit. Home Depot, Best Buy, Albertson's, and Target announced similar mandates. In Europe, Metro Stores and Tesco announced their own programs.

In January 2005 Wal-Mart had to announce that only around 30 of their top suppliers were shipping RFID-tagged cases, but by the end of January, 94 suppliers had shipped tagged product to Wal-Mart, and by the end of March, the company confirmed a 100 percent compliance rate; 100 percent of their top 100 suppliers were shipping pallets and cases with tags.

As the timeline marched on, various reports and reactions surfaced. The suppliers were experiencing difficulty in achieving acceptable read rates. If a significant portion of the tags fail to broadcast, the benefits of the technology are lost. Many of the problems centered on cases with opaque materials such as metals and liquids. Other problems arose from ineffective adhesives. Gradually, these problems have been reduced to acceptable levels.

Companies differed in how they responded to the mandate. Companies such as Procter & Gamble, Gillette, and Kimberly-Clark are embracing the technology to track their products within their own companies. They are beginning to implement applications to reduce their own costs and their own cycle times. Other companies are taking a less strategic approach. Citing immature standards, low tag-read rates, high tag costs and uncertain return on investment (ROI), some companies are adopting the strategy known derisively as "slap-and-ship" (see Chapter 11). Some columnists with a sense of humor have noted the sensitivity of some readers to tag placement and orientation and renamed the strategy "position-tags-carefully-and-ship," a much more expensive proposition.

As public debate continued, it was suggested that the ROI for RFID investment would flow to the retailer, rather than to the supplier, who bore most of the costs. There were several reasons given for this view. First, both suppliers and the retailers would have to purchase readers and upgrade their IT infrastructure. But those are one-time costs. The cost of purchasing and attaching tags, however, is ongoing, and those costs are borne largely by the suppliers. Wal-Mart did not allow the suppliers to pass those costs on. Second, according to consulting firm A.T. Kearney, RFID technology enables manufacturers to improve tracking and inventory visibility, enhance labor efficiency, and improve fulfillment. However, most manufacturers are, according to Kearney, "well beyond the basics when it comes to supply chain efficiency," so they will not realize a significant benefit.

Retailers, on the other hand, will realize significant savings and process improvements.

In February 2005 Wal-Mart described the process changes enabled by their new RFID system. Simon Langford, the company's manager of global RFID strategy, said the company installed readers at the receiving docks at the back of the building, near the trash compactors, and between the back room and the retail floor. Readers record the arrival of cases with tags by reading the tag affixed to each case. They then read the tags again as the goods are brought out to the sales floor. The company uses its existing point-of-sale system, which is not using RFID, to subtract the number of units of a particular item that are sold to customers from the number of cases brought to the sales floor. Software monitors which items will soon be depleted from the shelves and generates a list of items that need to be picked from the back room to replenish the store shelves. "By reading the tags on the cases that are brought out from the back room, we're able to see what items have actually been replenished," says Langford. Before this system was implemented, the company had to rely on people to walk the floors and record manually what had been picked from the back room. Suppliers can get updates on the location of their goods within 30 minutes of the goods' movement.

In October 2005 the University of Arkansas RFID Research Center announced the results of a 29-week study of Wal-Mart's RFID. The study design focused on 24 stores. Twelve were RFID-enabled and twelve were not. The non-RFID stores were designated "control" stores. The research was structured to isolate the impact of RFID and to exclude any other process improvements. The researchers found a 16 percent reduction in out-of-stocks at the test stores as compared with the control stores. In addition, it showed that out-of-stock items with EPCs were replenished three times faster than comparable items using bar code technology. Just as important, Wal-Mart saw a significant reduction in manual orders, resulting in a reduction in excess inventory.

Linda Dilman said of the study, "This (RFID) is no longer a take-it-on-faith initiative. This study provides conclusive evidence that EPCs increase how often we put products in the hands of customers who want to buy them, making it a win for shoppers, suppliers, and retailers." Skeptics point out that reductions in out-of-stock do not necessarily lead to increases in sales. Other experts point to studies that show that 48 percent of out-of-stock reductions result in increased sales. No one has yet calculated whether the 48 percent out-of-stock reduction creates an ROI for the cost of the RFID system, although the University of Arkansas promises such a calculation in the future. Beyond improvements in in-stock, Wal-Mart also sees benefits in overall inventory reduction, which is a significant driver in cost reductions.

5.1.2 McCarran Airport

McCarran International Airport is the seventh busiest in the United States, and traffic is growing more than 10 percent annually. McCarran serves the growing Las Vegas metropolitan area.

Officials at the airport have identified baggage processing as a significant concern. The Transportation Security Administration (TSA) has mandated screening for all airline bags. The officials conducted extensive research to find alternatives to conventional baggage handling and tracking processes and discovered the numerous advantages of RFID smart-label technology. Airports had been considering using RFID technology to track baggage for more than ten years. Tests were done at several American airports including San Francisco, Atlanta, and Jacksonville, FL. Several European airports currently use RFID tags embedded in reusable trays that carry the luggage.

The key point for McCarran was the increased accuracy reported by the RFID tests. On their own bar code system, McCarran was experiencing up to 30 percent misreads. This was due to bar codes being incorrectly oriented and sometimes falling off the bags. Incorrectly read bags had to be manually rerouted. The projected savings in manpower alone would more than offset the higher cost of the tags. McCarran selected tags from Matrics (now owned by Symbol Corporation). They selected FKI Logistex of Danville, KY to integrate the Matrics RFID hardware with their automated baggage-handling equipment. Swanson Rink of Denver designed the system. ARINC of Annapolis, MD integrated the software.

Phase I of the installation involved only three airlines: Alaska, AirTran, and Champion. Employees of these companies started placing RFID tags on bags as they were checked in. The tags were Class 0 tags, encoded by the manufacturer. Readers on the conveyor system read the tags and then routed each piece of luggage to the appropriate airline, or, if necessary, to another security screening station. Readers mounted on the conveyors routed each bag through an online explosive-detection system.

Phase II saw the upgrade of the tags to smart labels. The labels were now Class 1 tags and were encoded upon check-in with a key that uses a central database to associate each bag uniquely with the passenger's data. The system routes the bag first to the inline explosive detector and then to the passenger's airplane. By the second quarter of 2005, the system was expanded to all 30 airlines that use McCarran. The airport now reports that baggage accuracy has increased to 99.5 percent.

It is interesting to note that no U.S. airline has yet adopted RFID for tagging luggage. Delta ran two pilots for several years and concluded they were satisfied with the results, but no deployment resulted. In part, the fact that no U.S. airline has put in a system is a result of the U. S. airlines' unfortunate financial circumstances, where three of the top five airlines are currently in bankruptcy. It's also a result of concerns about the financial return on the investment. Bearing Point has conducted a study and concluded that benefits only accrued to an airline losing more than five bags per thousand passengers and spending $75 in cost per mishandled bag. (Delta said they were losing four bags per thousand, but that their cost was about $100 per bag.) The International Air Transport Association (IATA), a consortium of about 265 airlines, has been building the foundation for an industry-wide baggage handling system. Their specification calls for use of EPCglobal UHF standard tags.

5.1.3 Boeing

Boeing has announced plans to introduce RFID tags on maintenance-significant parts of the 787 Dreamliner, their newest airplane design. Boeing expects that RFID technology will improve configuration control and help airlines reduce costs by managing parts maintenance and repair histories. The manufacturer plans for the tags to contain unique identification for each part, as well as maintenance and inspection data in accordance with industry standards. The parts to be tagged will be line-replaceable units and life-limited parts as well as on-board emergency equipment.

Shortly after this announcement, the United States' Federal Aviation Administration (FAA) reminded Boeing personnel that tags were approved only for ground operations, not for airplanes in flight. The FAA then ran tests for six months and approved the use of tags for in-flight operations. The tests showed that passive RFID devices do not adversely affect the simultaneous operation of any aircraft systems or interfere with safety in flight.

Boeing has specified the use of 64-kilobyte passive tags that would have read-write memory, allowing mechanics to access the part's serial number and date of manufacture. The tags would also include links to maintenance and fault-detection manuals and contain notes from mechanics about the part's service history. The tags would have to be resistant to changes in pressure, temperature, and humidity. Boeing also specified that the tags function when applied to metal parts. The company has selected HF tags in the 13.56 MHz frequency range, since this frequency is standard in all countries of the world. Noting that the read range for HF tags is shorter than that for UHF tags, the company will work with airlines to study UHF tags as well.

Boeing and its chief commercial rival, Airbus, have both mandated that their suppliers ship equipment with RFID tags. They have collaborated, and both agreed on an update to Spec 2000, IATA's industry-wide standard for machine-readable bar code labeling of parts. The update defines how the tags will be formatted and used.

One problem is that the tags Boeing and Airbus have specified do not exist at this time. Boeing's statement on this topic is quoted: "You tell the chip and inlay manufacturers what you want and expect them to figure out a way to respond.... You tell your suppliers what you expect in terms of their compliance and you forge long-term partnerships with your valued suppliers."

Boeing spokespeople pointed to the key advantage RFID tags have over bar codes in this application: unlike bar codes, the tags on equipment behind bulkheads can be read without having to remove panels.

5.1.4 Port of Singapore

The Port of Singapore Authority (PSA) tracks thousands of cargo containers daily. It also manages arrivals and departures of up to 50 ships per day. Manual processes were error-prone and sometimes resulted in containers being lost for hours or even days. The result was delayed departures and incomplete shipments.

PSA installed thousands of RFID tags in the asphalt to create a multi-dimensional grid. A centralized system manages placement and location of containers as they are offloaded onto the port shipyard, based on three-dimensional coordinates derived from the unique ID codes in the tags. Readers are located on the forklifts and trucks, and additional tags are located on the containers themselves. The location of every container is tracked in three dimensions as it is put away. The result is that the port office always knows where any container is at any point in time, and loss rate on containers has gone to nearly zero.

5.1.5 Tagging Pharmaceuticals: OxyContin

OxyContin is a very controversial prescription drug. Legitimate users hail it as a life-saving pain-management drug. At the same time, it is highly abused, often stolen and counterfeited, because of its addictive nature. OxyContin is Purdue Pharmaceutical's most successful drug. Sales are over $1 billion a year and climbing 20 percent annually. To help protect against counterfeiting and theft, Purdue last year began a pilot program to put RFID tags on labels of 100-tablet bottles of the drug.

In doing this, Purdue is following the guidelines of the United States Food and Drug Administration (FDA) (see Chapter 7). The system Purdue has implemented creates an electronic pedigree for each bottle as it makes its way through the supply chain from manufacturer to various wholesalers and to the pharmacy. An RFID reader can identify the 48 bottles in each carton shipped to distribution centers; the reader reads the tag and records the specific bottle's serial number. The pedigree gives retailers (initially, Wal-Mart and HD Smith) visibility on the authenticity of the drugs they are receiving. Without this system, it is relatively easy for secondary wholesalers today to falsify the origin of a prescription drug in the supply chain. RFID will make it virtually impossible for counterfeit drugs to enter legitimate distribution channels.

Purdue faced two key challenges in setting up this system. First was the small amount of space available on the bottle. The smaller the antenna, the shorter the read range. The second challenge was to reliably read tags on bottles containing liquid drugs. Purdue worked with Symbol Corporation and with other pharmaceutical companies for over two years to develop a tag that would fit with the label and be read reliably by the reader. Getting the pilot up and running took another year. The solution is a 1-inch square Class 0 tag (915 MHz) from Symbol on each bottle of the drugs shipped to pharmacists. The company does not tag bottles sold to consumers. Purdue is working closely with law-enforcement agencies and has donated 100 handheld RFID readers to federal law-enforcement cargo-theft units and state and local task forces targeting prescription drug thefts. Once the readers read the EPCs on the bottles, the officers e-mail the identification number or phone them to Purdue, which researches the pedigrees and identifies the origin of the drugs in question.

Secondary wholesalers today can now easily falsify the origin of a prescription drug. RFID will make it more difficult for counterfeit drugs to enter the legitimate supply chain, but only if the technology is widely deployed. One analyst has said that to significantly improve the detection and elimination of counterfeit drugs, the industry will have to use a reader every time a drug changes hands.

Spokesmen for the company say their application "is not slap and ship. RFID data is now fully integrated into our manufacturing systems." Tags are placed on the bottles as part of the final manufacturing/packaging process, and 48 bottles are then placed into a shipping carton. Only boxes that produce 48 accurate reads leave Purdue's shipping docks.

Purdue has integrated the management of the unique EPC number on each RFID label into its SAP ERP system. But as of this writing, there were no mechanisms defined to share this data with Wal-Mart or HD Smith, the two retail partners in the project.

Purdue estimated that they spent about $2 million on the pilot.

5.1.6 Speedpass

In 1997, Mobil Oil Corporation introduced Speedpass, an RFID system to speed payments at the gas pump. The company expected to have about 30 million customers by 2005, but the actual number was about 10 million.

Speedpass grew out of a 1993 Mobil study that aimed to identify key customers and determine what drives loyalty and incremental spending. The answers, Mobil concluded, were "speed, smiles, and strokes." This was interpreted to mean convenience, friendly service, and recognition for the best customers. The "Speed" group wanted to create the simplest and fastest buying experience in the world. The system was envisioned as a cross between electronic toll collection systems and the Hertz Gold program, which keeps regular customers' accounts online to speed car rentals and returns.

Speedpass operates on the low frequency RFID band, at 134 KHz. It can accomplish a complete authorization in just a few seconds. It is programmed to know which customers prefer a receipt. The RFID system is less costly to merchants than credit card purchases, for which merchants must pay a percentage to the card processor, but the initial setup cost at each store is about $15,000.

5.1.7 American Express ExpressPay

ExpressPay from American Express is a contactless payment initiative that offers a quick way to make everyday purchases. The system uses a chip built into a keychain attachment or embedded into an American Express credit card. The system conforms to the ISO 14443 standard (see Section 1.3.2), communicating the account information in magnetic stripe format, so the transaction processes as a traditional credit or debit card transaction.

ExpressPay is an alternative to cash for making purchases where moments count and convenience is paramount. Such venues as quick-service restaurants, convenience stores, supermarkets, drug stores, gas stations, corporate cafeterias, concerts, and sporting events all are characterized by consumer desire for quick transaction cycles. No signature is required for ExpressPay.

ExpressPay can be funded in two ways. One option is to link it to an American Express charge or credit card for payment. Individual charges are recorded directly on the cardholder's monthly billing statement. A second option is to prepay up to $600 monthly using any major credit, debit, or charge card. The card will reload automatically from the same payment source when the stored value drops below $20.

ExpressPay is faster than other payment methods. On average, ExpressPay transactions are 53 percent faster than paying with payment cards with no signature and 65 percent faster than cash. ExpressPay customer spending increased about 20 percent compared with cash spending. Consumers appreciated the convenience of the payment method. Eighty-seven percent agreed that it was better than cash and said they would use it daily. Merchants benefit from the improved transaction speed.

ExpressPay was first, but now it has been joined in the marketplace by offerings from American Express's traditional rivals, Visa and MasterCard. These two credit card giants have taken different approaches to the marketplace. Visa has chosen not to give its RFID-tagged card a name; instead, a special logo will appear on the card identifying it as contactless. MasterCard, on the other hand, has set up a separate card called PayPass. All parties in the credit/financial services industry agree that readers must be interoperable among the major vendors, so a merchant can purchase a single reader and process cards from all of them.

In addition to the three card issuers, Chase Manhattan Bank has announced that they will offer their own RFID payment service, called Blink. Chase will offer Blink on both Visas and MasterCards that it offers. 7-11 stores and Sheetz restaurants announced they would accept Blink credit cards.

5.1.8 The United States Department of State Passport Initiative

The United States Department of State announced in October 2004 that they would embed RFID chips in United States' passports issued after October 2005. The chips would contain the individual's name, address, and digital photograph. They would also contain biometric identification information. The State Department hoped that the addition of the chips would make passports more secure, more difficult to forge, and harder to use if they are stolen. They noted that the contactless chips could be read by anyone with a proper reader. The Department announced that diplomats and other State Department officials would be issued the new passports as early as January 2005, and other citizens applying for new passports will get the new version starting in April 2005. While none of the information on the chip would be encrypted, the chip would contain a digital signature that would verify that it was created by the government. The department decided not to encrypt the data because of the risks of sharing the method of decryption with 180 other countries.

Privacy advocates argued that putting RFID chips into passports was "a terrible idea." "If 180 countries have access to the technology for reading this thing, whether or not it is encrypted, that is a very leaky system," said attorney Lee Tien of the Electronic Frontier Foundation. Other experts argued that besides identity thieves, commercial travel companies, including hotels would capture the data in order to upsell their customers. The vulnerability of the system was hotly debated. The State Department argued that the chip would be readable only within a few centimeters. But one public demonstration showed that a reader could read a chip from 3 feet away. (That was a UHF chip broadcasting on the 915 MHz frequency, but the demonstration swayed the public and even many experts.) An adversary willing to build his own reader and violate his country's broadcast regulations could read chips from a much greater distance. That opened the door to "skimming," the practice of reading the chip without the passport holder knowing his document was being read.

The International Civil Aviation Organization (ICAO) created the specifications for countries adopting RFID passports and called it Basic Access Control (BAC). The ICAO specification is based on ISO Standard 14443 (see Section 1.3.2), broadcasting at 13.56 MHz. The BAC system encrypts the data on the RFID chip, making it unavailable to any reader without the key. To obtain the key, the passport officer needs to physically scan the machine-readable text printed beneath the photo in a particular font.

The reader then hashes the data to discover the key to authenticate the reader.

In response to the criticism, the State Department delayed implementation until October 2006. They also decided to adopt a modified version of BAC and to put metal fibers in the passport cover to prevent anyone from surreptitiously reading a passport as long as it was closed. These fibers will create what is called a "Faraday Cage," which will prevent surreptitious reading of the tag data. The Department rejected calls to encrypt the data stored on the chip. Instead, the transmission stream between reader and tag is encrypted. The Department also rejected calls for smart-card chips that would require contact between tag and reader.

While generally applauding the State Department for listening to its critics, experts point out that several flaws in the plan remain.

- The system is being implemented without significant testing. The nature of RFID applications is that there are flaws that show up in early phases of the implementation and require at least some amount of fine-tuning.
- While the Faraday Cage will prevent most instances of skimming, it will not prevent eavesdropping or surreptitiously reading the data as it is being read by a legitimate reader.
- While the data on the chip is encrypted as it is broadcast, the collision-avoidance scheme of ISO 14443 requires each chip to first broadcast a Unique Identification Number (UIN) in the clear. Only in this way can the reader arbitrate the competing signals when more than one tag is in the read zone at the same time. This UIN can be eavesdropped readily and used later to uniquely identify the passport.
- The designers apparently considered only the use of the passport in airports, but the ICAO guidelines envision use in e-commerce, as well as hotels and other businesses travelers frequent. Security in these types of establishments is likely to be weak.
- The proposed workflow envisions optically scanning the text beneath the picture in order to create the hash code so the chip's contents can be read. This would seem to negate the benefits of wireless communication conferred by RFID and makes this author wonder: why accept the drawbacks listed above?

The State Department has had Congress require European Union member states to implement biometric authentication technology into their passports by October 26, 2005. With the extension of their own deadline, the question arose whether to extend the EU's deadline, as well.

5.1.9 Brittan School

A small California startup company called InCom developed a system called InClass to automate student attendance tracking. The system involves passive RFID tags attached to student ID badges and UHF readers installed in classroom doorways. The school district was interested in the project to automate taking of attendance in classes and recording attendance every day, as is required by the state Department of Education. The pilot was conducted at Brittan Elementary School in Sutter County. The two founders of InCom are employees of the school district.

The system works with a unique, 15-digit identification number assigned to each student and encoded on the tag. As the student walks through the doorway, the reader interrogates the tag, receives the number, and reports the student's attendance and time of arrival. Students who arrive late ("tardies") are identified. The reader transfers the information to a central server where it is collected, stored, and compared with the list of enrolled students. A list of present, absent, and tardy students is downloaded over an 802.11 wireless network to a PDA for the teacher to verify. Once the teacher has confirmed the list is accurate, the teacher sends an approval so the records can be submitted to school administrators.

The pilot coincided with issuance of photo IDs to students, although the two were not related. The ID cards show the student's picture and full name but do not contain the tag. The holder for the ID card does contain the tag and is a device patented (pending) by InCom.

Brittan received a flood of media attention because an unspecified number of parents (once reported to be six) raised questions over the use of RFID technology in the school. The parents were supported by the Electronic Frontier Foundation (EFF), the ACLU, and the Electronic Privacy Information Center (EPIC). The parents' letter made the following complaints:

- Parents were not notified of either the ID card or the RFID tag implementation.
- "The most glaring concern was the name and grade of the student plainly visible on a badge that was required to be worn outside the clothing at all times."
- "We were concerned about privacy issues, especially tracking children in bathrooms."

As a consequence of the parents' complaints and media uproar, InCom announced that it would end its pilot program. The company continues to market its program to other school districts. In an interview, a company spokesperson underlined the importance of parental notification prior to any implementations. There is no public record of any purchases of the system. It appears that parents were more upset about the badge and the monitoring of the bathrooms than the RFID attendance system itself. But organizations resisting RFID, often for very good reasons, successfully used the controversy to advance their own agendas.

5.2 RFID Application Types

We divide the applications of RFID technology into four areas. The distinction is based on what the systems fundamentally do, and we explore the numerous ways these functions are utilized. The types of applications are

- Tracking and tracing of items
- Electronic payment
- Access control
- Telematics

5.2.1 Tracking and Tracing of Items

Tracking and tracing items is the area with the greatest application of RFID technology. It takes advantage of the technology's ability to uniquely identify an object and report that identification, along with the time, location and sensor-based information such as temperature, moisture, tampering, vibration, and radiation. The ability to link this identified object with records in a database and to trigger automatic actions such as notification or ordering from suppliers vastly increases the power of the application. The breadth of this application lies in the wide variety of types of items that can be tracked. A sampling of the various types of items that can be tracked is shown in the following list. Each type of item has its own value proposition for tracking, its own methods, and its own flavor of technology.

- Airline baggage (see Section 5.1.2)
- Works in the manufacturing process
- Items in inventory

- Equipment
- Shipping containers
- Goods in trade, at the container, pallet, case, and item level (see Section 5.1.1)
- Patients in hospitals
- Prisoners
- Small children and animals (see Section 5.1.9)
- Controlled materials such as drugs, tobacco, and hazardous materials
- Library books
- Parts in airplanes, military vehicles, and equipment

5.2.1.1 Works in the Manufacturing Process (Work In Process—WIP)

What is tracked? This application includes tracking subassemblies and components as they move through various stages of the manufacturing process and combine into final finished products.

What is the process? Tag each subassembly as it is manufactured. As this is done, associate the unique identification number with a line item on the work order that describes the item. It may also associate the item with the tools, equipment, and personnel scheduled to do the work and verify that the correct item is in the correct place at the correct time with all resources necessary to complete the task. Additionally, the data collected can tell the exact time spent on each action, which will support process improvement and cost recognition activities.

What are the benefits? Automatically tracking and tracing manufacturing components reduces errors, shrinkage, and the time spent in counting and recording. The system automates record keeping where in-process documentation is required. It can conduct its activities in environments where other methods cannot function. It may incorporate sensors and automated equipment that can report its settings (voltages, flow rates, and so on) to automate record keeping and ensure compliance with laws, regulations, or business mandates. Having a pedigree for each finished unit assists in managing quality, returns, and recalls.

What are the drawbacks and issues? ROI on this application is difficult to calculate. The ROI is generated by improving business processes, reducing labor for data collection tasks, or other initiatives to improve productivity.

Examples of RFID in the Manufacturing Process

An Italian pharmaceutical company's manufacturing process involves sterilization at high temperature for a period of time. No human, and no bar code, could withstand this environment, but RFID tags can. Regulations require the company to document the sterilization. The previous method of capturing this information, workers standing at the door counting items in and out of the ovens, produced unacceptable error rates. The RFID system has ensured delivery of accurate and complete records, and it has relieved 2–3 employees for more productive work.

A manufacturer of cut salads was losing customers due to short shelf life. RFID tags enabled them to see where in the process temperatures were rising and locate and replace faulty equipment.

International Paper has smart lift trucks and a tag on every roll of paper in its plants. The lift trucks have readers that connect wirelessly to the host computer. The rolls are 70" thick, so it makes sense to avoid mispickings and misroutings, as well as to make sure that a roll meant to be delivered on a first stop isn't buried deep in the trailer. Planners can plan the manufacture and disposition of each roll and then track its movement against the plan.

What is the technology? EPCglobal Class 4 active tags with large data memory broadcasting at 915 MHz can be used.

5.2.1.2 Items in Inventory

What is tracked? Retailers and distributors stock inventory in order to have goods available for sale. Manufacturers stock inventories of parts and raw materials to be used in the manufacturing process. Organizations such as airlines maintain inventories of critical parts they may need in a hurry. All of these organizations need to be able to track receipt, location, use, and reorder of these items.

What is the process?

1. Attach a tag containing a unique identifier to an item to be monitored.

2. Record the item in a database.

3. Record the presence or absence of the item on a periodic basis. If absence occurs, the system may take some automatic action, such as notifying someone, or it may order a replacement from the supplier, based on processing rules.

4. Record the movement of the item as it passes strategically located portals with readers. If the portal has a means of identifying the associated person, either by reading their RFID-enabled identification, or by capturing the card information of the person removing the object, record the responsible person as well.

What are the benefits? Reductions in inventory levels are likely with improved visibility. Losses due to misplacement or theft of items are reduced. Labor associated with record keeping is reduced. Improved visibility of process flow enables improvements. Immediate online access to manufacturer's information and data about the item enables replenishment to take place more quickly. Reorder of critical items can be automated.

What are the drawbacks and issues? Cost and ROI are big issues. Most companies use bar codes to perform this inventory control function, and they have systems already in place. Bar codes' costs are less than 1 cent, while passive RFID tags today cost between 10 and 50 cents each. The cost of replacing these bar code systems, along with the infrastructure that supports them, could be significant.

What is the technology? There are several technology solutions for inventory. One approach leverages the EPCglobal framework and utilizes UHF passive tags. This approach relies on counting items as they pass through portals. When an inventory count is required, the workers can walk the aisles with handheld readers and read the tags of items on shelves. A much-discussed extension of this application is called *smart shelves*, which are shelves with built-in antennas and readers that continually monitor and report the identity of each item present.

A second approach uses active tags. This is justified when the items themselves are sufficiently valuable to justify their expense. This application continually polls the area and reports the identity of each item present.

5.2.1.3 Equipment

What is tracked? Fixed assets; office equipment; manufacturing equipment; transportation equipment; food carts in airplanes; passenger loading-ramp dollies outside airplanes; farm and construction equipment; roadside equipment such as signs, guardrails, fences, traffic signals; hospital equipment such as defibrillators, wheelchairs, heart-rate monitors.

What is the process? Put a tag on the item to be tracked and associate its number with a record in the database. Systems with active tags may monitor the presence of equipment every few seconds and issue an alert when an expected item is no longer in the read zone. These systems are called real-time locator systems (RTLS). Passive-tag systems will broadcast only when within the read zone of a reader to support regular counts of fixed assets. Tags may store maintenance information accessible through the reader to include both maintenance history and documentation to support maintenance personnel. Tags may also include LED readouts so that the item gives a visual cue when it is being searched for. This is particularly useful when a number of tags are likely to be in the read zone at the same time.

What are the benefits? Reduced labor costs are incurred when taking inventory. The system will perform real-time monitoring and generate alerts when an item is removed. Lower maintenance costs are incurred due to ready availability of information. Users value quicker location of lost assets. Tags are readable in circumstances where bar codes become illegible due to wear, dirt, and other conditions. Accessibility is also an issue, as bar codes require line-of-sight access while tags can be read even if you can't see them.

What are the drawbacks and issues? Cost is an issue. Many of these functions are performed satisfactorily with bar code readers today at lower cost per tag. Equipment to be tagged is likely to contain metal, which interferes with UHF and microwave transmissions.

What is the technology? There are several competing technologies for this application. Active tags may play a role. Several vendors offer products that utilize (existing) 802.11 WiFi networks, and therefore require no separate network installation for the readers. Vendors with location-aware WiFi networks include Wherenet, AeroScout, PanGo, and Ekahau.

Other vendors offer systems with dedicated readers and tags in unlicensed frequencies such as microwave (2.45 GHz). These vendors have two ways to determine the location of an asset. One is called TDOA (time difference of arrival). In this method, the tag is read by different readers, and they calculate the location based on the time difference of the signal's arrival. The second method calculates RSSI (received signal strength indication) and performs better in walled environments such as hospitals and distribution centers. TDOA performs better in unobstructed, outdoor environments such as yards or hangars.

Four vendors offer products for equipment tracking: Radianse, RFCode, Savi and WhereNet. Radianse focuses exclusively on healthcare and is a leader in 433-MHz systems for tracking humans. Its tags are disposable and inexpensive. RFCode offers passive tags as well as active ones. The passive tags, of course, are less capable and less costly. Savi offers tags, readers, and software to enable a global installation. Their products operate at LF and 433.92 MHz frequencies. WhereNet holds the leading position, with deployments in automotive manufacturing, transportation, logistics, and military markets.

5.2.1.4 Reusable Containers

What is tracked? Reusable containers are used throughout the supply chain to transport goods from supplier to retailer. These items range from reusable pallets to huge containers, stainless steel fruit containers, or beer kegs.

What is the process? Some tags are embedded in the containers when they are manufactured. Others are attached prior to use. The tags are rewritten each time the containers are filled, recording data regarding the contents, shipper, destination, and so on.

What are the benefits? The active RFID tags can function as sensors to detect the presence of explosives or radioactive transmissions, as well as temperature or moisture changes that might lead to product degradation. They can broadcast notification to readers or merely record the event in their internal memory for later retrieval. The location of the item can be determined, which reduces the time and expense of searching for misplaced items.

One company, Trenstar, makes a business out of purchasing the reusable assets of a company and leasing them back. This gets the assets off the books of the company and relieves them of the costs and risks of managing them. They charge the company a fee for each use. Trenstar embeds the containers with RFID tags in addition to the bar codes, and they closely track the location and use of these assets. Trenstar has provided this service to Carlsberg Brewery for their beer kegs and to Kraft Foods for their stainless steel fruit containers.

What are the drawbacks and issues? The cost of these tags is several dollars per tag or higher if additional functionality is required. The cost may not be justified in all commercial applications.

What is the technology? Active tags operating at 433 MHz is the generally used technology. In April 2004, the FCC announced two changes to signal its support for the security of shipping containers. The maximum signal level was increased for 433–434.5 MHz RFID readers, and the maximum duration limit was increased from 1 to 60 seconds.

These systems are governed by three ISO standards. ISO 10374 describes standards for UHF and microwave read-only tags. These frequencies do not work well for containers, so the standard has never been implemented. ISO 18185 describes passive and active RFID tags used as container seals. ISO 23359 describes read-write tags for freight containers.

5.2.1.5 Goods in Trade (Item Level)

What is tracked? Goods in trade are all goods sold by any company to another or to the consumer. The goods may be tracked at the item level, the case level, or the pallet level; this section deals with item-level tracking.

What is the process? Items are tagged by the shipper before they are staged for shipment (see Chapter 11). The tags are read as they are loaded on the outbound truck or train and the shipper sends an ASN (Advanced Shipping Notice) to the receiver. The tags are read each time the goods change hands or means of conveyance until they arrive at their destination. In the event the shipper wishes to make use of the data for their own operations, the goods are tagged earlier in the supply chain. At the destination, the recipient can read the tag to receive the unit into inventory with less physical handling.

What are the benefits?

- Reduces shrinkage.
- Enables retailer to improve management of inventory.
- Reduces out-of-stocks. As items are sold, the information is transmitted back to the manufacturer so replenishment activity can begin.
- Reduces time spent counting, sorting, putting away goods.
- Gains greater visibility on movement of goods, so retailer can improve his own processes.
- Reduces errors. By some counts, over 30 percent of orders have an error.
- Improves management of promotions.
- Reduces costs. By some measures, logistics cost savings could equal up to 11 percent of America's GDP, so systems that reduce the cost and inefficiencies can be highly beneficial.
- Enables retailers to support high value activities to improve customer experience.

What are the drawbacks and issues? Tagging and tracking at the individual item level is very challenging. This is due to the wide variety of materials, shapes, and sizes in packaging and trade. Also, the quantity of data that will have to flow through the systems could swamp the networks and computers.

Consumer groups voice significant privacy issue concerns; they worry that tags can be traced after the customer leaves the store.

Cost is the most serious factor. Many consumer goods are sold at thin margins and will not bear any additional cost.

What is the technology? Item-level tracking has been conducted mostly at 13.56 MHz using inexpensive Class 1 tags. Tags at the pallet and case level are 915 MHz UHF Gen-2 tags. Portal, vehicle-mounted, and hand-held readers are used. Several manufacturers have released UHF products for item-level tagging. EPCglobal has a working group called Item Level Tagging Joint Requirements Group that has identified seven use cases for item-level tracking and is evaluating all RFID frequencies for their applicability.

5.2.1.6 Goods in Trade (Cases and Pallet Level)

What is tracked? Many companies have opted for case-level and pallet-level tagging. This consists of putting RFID tags on the cases and pallets that contain the goods. This enables them to start working with the technology, achieve some significant benefits, and avoid the higher cost and greater technical demands of item-level tagging.

What is the process? Today, nearly all pallets and cases are bar coded, and there are longstanding systems in place to read them and record the data. The most common process for tagging cases and pallets utilizes smart labels, which duplicate the functionality of the bar code labels but includes an RFID tag, for compliance sake. This method is described in Chapter 11. The smart labels are printed and commissioned using an encoder/printer attached to the network. The software keeps track of each unique identification number and stores it along with the description of the pallet or case in a database. In many cases, it sends out an ASN automatically when the pallet or case is shipped.

What are the benefits? In many cases, the database can store the information describing the contents, so the system can record item-level movements even without item-level tracking. Thus most of the benefits cited in Section 5.2.1.5 can be achieved, albeit with less certainty.

What are the drawbacks? The main drawback of this particular application is that it duplicates the functionality of the bar code systems already in place, adding costs but not adding much value by itself. In some cases, the ability of the tags to store greater quantities of data has reduced the costs of applying and managing bar codes.

What is the technology? This application has extensively used the EPC-global Generation 1 technology, and is migrating to Gen-2 as the chips and equipment become available. For higher-value pallets, and in the military applications, active tags operating at 433 MHz have been used.

5.2.1.7 Patients in Hospitals

What is tracked? Patients in hospitals are given wristbands that contain RFID chips.

What is the process? At Jacobi hospital in New York, the system improves access to patient records. Upon check-in, the patients are given wristbands encoded with their Patient ID Number. The basic information—name, date of admission, date of birth, gender, and medical record number (MRN)—is printed on a label and inserted into the pouch of the plastic band. The MRN is printed as a bar code as well. When nurses need information from the database or need to make an entry, they use a tablet computer with an RFID reader.

At Beth Israel Deaconess Medical Center in Boston, the system tracks the location of patients, surgeons, and medical equipment. The active tag system transmits signals to wall-mounted readers so the location of any of these persons or items can be readily determined.

At Washington Hospital Center in Washington, D.C., the system tracks the location of equipment, patients, and staff. The tags contain only a unique ID number.

Alexandra Hospital in Singapore began a new tracking system in its Accident and Emergency (A&E) department in the wake of the Severe Acute Respiratory Syndrome (SARS) scare. The system tracks all patients, visitors, and staff entering the hospital. Each is issued a card embedded with an RFID chip. The card is read by sensors installed in the ceiling, which record exactly when a person enters and leaves the department. The information is stored in a computer for 21 days and then deleted.

What are the benefits? The Jacobi system replaces a manual access system, where nurses had to have direct line-of-sight access to the labels in order to get the patient identification number. This sometimes generated a significant discomfort to the patient, as the bar code may be hard to access around the blankets and equipment connections. RFID tags can be read without disturbing the patient. The system reduces the time spent on paperwork, as reports are filed automatically. It reduces errors, as it monitors administration of drugs and other care. Jacobi compared RFID with a bar code system, but bar codes can wrinkle and tear, making them unreadable. Also, the patient disturbance was a factor. Emergency treatment is faster because personnel have instantaneous access to patient records. Staff no longer has to return to the nurse's workstation to obtain patient data. Plus, doctors and nurses can use the new platform to order lab tests, enter notes on treatment, or update medication instructions right from the bedside.

The Beth Israel and Washington Hospitals focus on location of equipment; locating personnel and patients are secondary.

The Singapore system enables health care workers to keep tabs on everyone who enters the A&E department. That way, if anyone is later diagnosed with SARS, a record of all other individuals with whom that person has been in contact can be immediately created. Other hospitals in Singapore are expected to adopt similar technology.

What are the drawbacks and issues? Privacy is a concern, as patients worry about their private information being broadcast. Also, a hospital is already very dense with radio frequency broadcasts. There is a concern that new RF systems will either disrupt existing ones, or be impacted by them.

What is the technology? At Jacobi, the system uses 13.56 MHz passive read-write tags. At Washington Hospital Center, the installation is the first ultra-wideband (UWB) installation, with active tags. The UWB system operates in the 6.2 GHz band. It is said to minimize the potential for self-interference caused by reflection of RF signals between readers and tags within a building. That is because UWB transmits very short-duration radio waves that are finished before they can reflect off walls and ceilings. The system does not interfere with LF and UHF equipment already in the hospital. The tags are tracked by unique identification numbers emitted by each tag every second. With four UWB antennas placed at the perimeter of a room, the system can track tags in three dimensions located within an 18" sphere. The tags also transmit information on battery status, whether the tag has been tampered with, and the status of the medical device to which it is attached. The system is designed by Parco Wireless (www.parcomergedmedia.com).

Alexandra Hospital uses active tags broadcasting at 433 MHz. The system was developed by Singapore's Defense Science and Technology Agency.

Beth Israel has deployed their system over 802.11 WiFi system that was already in place at the hospital.

5.2.1.8 Prisoners

Tagging and tracking people has a sinister overtone, but there are special, limited circumstances where it can be highly beneficial. Parents who have lost a child at an amusement park wish they could push a button and find them. Caregivers who have lost an Alzheimer's patient express the same wish. Kidnap victims in Mexico want police to be able to locate them by pushing a button. Prison guards need an accurate way to identify each prisoner, to know where they should be located, and to be alerted when they are someplace else. The Transportation Security Administration (TSA) would like to keep track of all persons temporarily in the United States. In addition, in the face of quick outbreaks of infectious diseases, hospitals are finding themselves called upon to keep track of persons exposed to disease carriers and to report this information after the fact. Nonetheless, except in very restricted circumstances, most free people do not want unseen systems tracking their movements without their knowledge and consent, and many are suspicious of ulterior motives. Tagging prisoners, however, has advantages.

What is tracked? Prisoners are tagged with tamper-evident wristbands. Guards and administrative personnel carry belt-mounted transponders.

What is the process? Upon check-in, prisoners are assigned a unique identification number. The number is encoded on the wrist tag that they wear. The wristband looks like an industrial-size wristwatch. The tag broadcasts an alarm when any tampering occurs. Guards and staff carry a belt-mounted unit. Every two seconds, the tag broadcasts its identification number picked up by readers scattered throughout the prison. The application software compares the location of each prisoner with the list of areas in which he is permitted at that time and issues an alert if any prisoner is in a prohibited area. Guards and administrators can query the system to find the location of any prisoner. The system counts the prisoners every two seconds and issues an alarm if any are missing.

What are the benefits? Guards probably do not know each prisoner by face, and the prisoners themselves are not a reliable source for their own identification. The system provides an unambiguous method for guards and administrators to identify a particular individual and to track and determine his location. The system reduces operating costs by automating many of the monitoring functions such as headcounts (normally taken 5–10 times per day) and internal investigations. It deters violence on the part of inmates who know they are being monitored. Reports from the system support medicine and meal distribution, adherence to predetermined time schedules, restricted area management, monitoring of specific locations, recording of arrivals and departures, and reduction of escape attempts.

What are the drawbacks and issues? Cost is a concern. Civil rights and privacy advocates are uneasy about the tagging and monitoring of prisoners, but so far none have taken a strong stand against it.

What is the technology? The system uses active tags and proprietary algorithms for counting thousands of tags rapidly and for pinpointing the location of an individual tag. The tamper-detection technology is licensed. One complete prisoner management system is marketed by the TSI division of Alanco Technologies in Scottsdale, AZ.

5.2.1.9 Animals

What is tracked? Animals are tracked for three purposes: pets are tagged so they can be returned to their families, populations in the wild are tagged and tracked to support programs of wildlife management, and food stocks are tracked to enable public bodies to deal with concerns about health and safety.

The United States Department of Agriculture published the National Animal Identification System (NAIS) specification in draft form in October 2003, and it was well received by the industry. The purpose was to be able to respond to an outbreak of Mad Cow or other disease within 48 hours by identifying all animals at risk and moving to quarantine or destroy them. It became law the following year, with compliance voluntary until January 2008. NAIS specifies two numbering systems: a Premises Identification Number (PIN) and an individual Animal Identification Number (AIN). *Premises* refers to any location where animals can commingle. Thus premises could be farms, ranches, grazing areas, veterinary clinics, animal exhibition centers, livestock markets and auctions, and slaughter and processing establishments. Each premises has its own PIN. Over 79,000 premises have been registered. Each individual animal of the following species is registered: camelids (llamas and alpacas), cattle and bison, cervids (deer and elk), horses, goats, poultry, sheep, and swine. Each animal has an Animal Identification Number (AIN), consisting of a 3-digit country code and a 12-digit randomly assigned number unique for each animal. RFID ear tags are an obvious choice for larger animals such as cattle, bison, camelids, and cervids. For horses, an implantable RFID tag has been recommended. For smaller animals, the department has not selected a tag format.

What is the process? Pets are tagged at the request of their owners. Owners pay $10–$15 to put their names into a database. The tags are injected under the skin or attached to the ear by veterinarians or by animal shelter personnel. When an animal is found, a reader wand can detect the registration number, which can be traced via the database to the owner.

Food animals are tagged at birth. Upon any movement, the PIN and AIN must be sent to the National Animal Records Repository for recording. NAIS does not require animals that would never leave a premises, such as a personal horse, to be tagged, although owners are encouraged to do so. The USDA recognizes that many ranches, farms, and marketing facilities will not wish to invest in the technology necessary to tag and monitor their stock, so it allows animals to be transported, without having been issued an AIN, to "tagging stations" where the tagging devices will be applied. Tagging stations may be specialized fixed facilities, or they may be a pre-existing facility such as a veterinary clinic, livestock marketing facility, or fairground.

What are the benefits? The main benefit of tagging pets is to enable those who find them to reunite them with their families. The stated purpose of tagging food animals is to enable authorities to respond to an outbreak within 48 hours, tracing the animal back to its sources and identifying all other animals at risk. Other benefits include documentation of origin (the system will replace the Country-of-Origin Labeling (COOL) program currently in place) and an increased ability to respond to bioterrorism incidents. Just as manufacturers merely adopting slap-and-ship compliance strategies deny themselves the opportunities to improve their internal processes, animal producers can treat NAIS as an additional cost, or they can work to leverage the technology. They can use it to improve their visibility on performance of sources, orders, feeds, locations, and so on. They can manage the cattle while on their premises in terms of temperature, weight, and location. They can use it to manage alliances and premium programs.

What are the drawbacks and issues? Pet tracking is not very controversial, but there are three separate databases in the United States, and a search may have to access all three of them to find the pet record. The system in Canada has run out of money and may be inaccessible.

In implementing NAIS, cost is an issue. The USDA has indicated that they will not pay for the tagging or the database, and producers are concerned that consumers will resist this increase in their costs.

Some people see the system as an assault on small operations, worrying that a house with a few chickens may have to register as a premises and pay to tag them.

What is the technology? Animal identification is best served with low frequency tags and readers because of its ability to penetrate liquids.

Some manufacturers combine sensing capabilities with RFID on their tags, so farmers can monitor the animal's health as well as its location.

5.2.1.10 Controlled Materials

What is tracked? Controlled materials such as drugs are tracked to comply with laws and regulations and to document and manage compliance. Hazardous materials pose a safety risk, which can be mitigated by knowing where they are and keeping records of where they have been and who has accessed them.

A new Japanese law makes businesses liable if their medical wastes are disposed of improperly, so a market has developed to enable them to track and trace these materials.

The United States Food and Drug Administration has issued guidelines for RFID to control drug counterfeiting. These guidelines are described in Chapter 7.

The United States military notes that explosives may be set off by radio frequency activity. They have published a specification called Hazardous Electromechanical Radiation to Ordinance (HERO) that certifies tags, readers, antennas, and power sources regarding their RF range and the minimum safe distance between the device and a shipment to avoid accidental detonation.

What is the process? We will describe a system for managing hazardous chemicals tested successfully at NASA's Dryden facility. A storage area is fitted with readers and chemical containers, ranging from metal drums to cardboard boxes with metal bladders, which are tagged prior to being placed in the storage area. Although metal and liquids are hostile to UHF tags, the facility was able to work with 915 MHz tags from Intermec, a tag called the Metal Mount Stick Tag especially designed for this application. One reader is mounted on the entry to the storage area and another is on the entrance to the covered shed. In addition, temperature sensors are mounted around the area and connected to the readers.

In Japan, the plastic and cardboard containers are tagged and readers are currently being tested to determine readout rates and precision, signal sensitivity, and other parameters.

What are the benefits? There are clear benefits to continuously monitoring hazardous materials. Immediate notification of tampering is valuable. So is direct access to documentation that may include pedigree, regulations, manuals, restrictions, diagrams, procedures, or contacts, all of which can improve emergency response. There is a value to having certainty that the right materials are in the right place and being used in the right application in the right amounts and concentrations. Identification of the items may also be an issue, as items may have fallen off a truck or pallet or their labels become unreadable. There may be issues of expiration or volatility that can be better managed with an automated system.

NASA engineers pointed out that storing chemicals at the correct temperature would extend their life and potency and cited temperature monitoring and control as one of the payoffs to the system.

What are the drawbacks and issues? The hazardous waste system's biggest drawback was cost. In fact, the system was not deployed because the center's 2005 budget was cut by about $41 million dollars to fund NASA's space shuttle program. NASA engineers were hopeful for 2006.

The monitoring of drugs throughout the supply chain also promises to be a costly endeavor. As noted in Chapter 7, the system will only be effective if tags are read whenever drugs change hands.

5.2.1.11 Library Books

What is tracked? Libraries today provide books, of course, and other materials as well. Their wares include magazines, videotapes, CDs, DVDs, video games, newspapers, and even toys and teddy bears.

What is the process? The physical retrofitting of library materials is generally simple. Existing bar codes are scanned into a conversion station and programmed RFID tags are produced. Choosing a conversion workflow process is less simple. Some libraries tag all of their new books as they are added to the collection. Others start with subsections of their collection in order to get accustomed to the system and field test operations. A combined self-check conversion machine would make it possible to bypass formal conversion altogether. The conversion can happen as patrons check out materials, and the most-used elements of the collection will receive the fastest conversion. After about a year, libraries run a report to see what remains to be tagged and do those.

The International Standard Book Number (ISBN) is an industry-supported numbering system for books. It identifies the title and edition of the book. The number libraries code into the tag is not the ISBN number, however. A library with 150 copies of *Harry Potter and the Goblet of Fire* needs a way to distinguish among the individual copies, and ISBN's are not serialized. Libraries use a unique, arbitrary identification number for each book, which keys back to the library's database to get its identity. This system has privacy protections, so no one without access to the library's internal computer system can scan the book to see what people are reading.

Patrons wishing to check out materials set the items on a desk. The system reads the tags and displays the titles on a screen. Patrons touch the screen, receive a receipt, and leave the library. When they return books, they receive a receipt indicating the items returned. Smart bins scan the returned books and sort them into appropriate bins.

What are the benefits? Libraries report faster checkout, less patron time standing in line, greater patron satisfaction. They appreciate the "wow" factor as patrons perform automatic checkout. Reshelving is more accurate, and staff can walk the aisles with wands to find books that are out of order. Taking inventory without RFID is time-consuming and costly. With RFID, it is a matter of waving the wands over the shelves. The University of Nevada's library reports saving $40,500 in replacement costs for the 500 items it found "lost" on their shelves. A major benefit is staff safety enhancement. Libraries spend hundreds of thousands of dollars on repetitive strain injuries. Grasping, stretching, reaching, and lifting are classified as "risky" operations, and staff members may perform these actions hundreds of times per hour.

What are the drawbacks and issues? The dominant negative issues are privacy, read accuracy, cost, and vandalism. There are arguments on both sides of the privacy question. While the ability to surreptitiously monitor what people are reading raises obvious privacy issues, some librarians argue that patrons would rather self-check out revealing books than confront a stranger with their choices. (Imagine checking out *Bankruptcy for Dummies, The Infertility Survival Handbook,* or *Gay Issues for Teens* under the watchful eye of your local librarian.) So they see RFID as improving, rather than reducing, privacy. Privacy advocates say the issue is the individuation itself. They cite this scenario: "You could be identified at a political rally, if not by name, then by the number of the book in your backpack. If you're arrested hours later, the book's RFID number would be enough to prove you'd been present at the rally."

Libraries do not have consistent numbering systems. This enhances patron privacy, but it reduces efficiency when interlibrary loans take place. As the Wal-Mart initiative goes forward, publishers may tag books with the ISBN number, which will reduce patron privacy.

Some libraries report a decline in tag reading accuracy when more than about five tags are in the read zone at one time.

What is the technology? Most library system vendors are ISO 15693–compliant, operating at 13.56 MHz. But there is no interoperability among different vendor systems because of proprietary protocols added to the tags. ISO 18000-3 is being adopted by library systems vendors. The identification numbers today are unique only to each library, but that may change in the future.

5.2.1.12 Parts in Airplanes, Military Vehicles, and Equipment

What is tracked? As described in Section 5.1.3, Boeing has mandated that maintenance-significant parts on their new airplane the 787 Dreamliner will be tagged by the supplier.

The United States military has put a premium on mobility, the ability to move its personnel, supplies, and equipment to remote places quickly. The role of RFID in this policy cannot be overstated: 35 percent of equipment was tagged in the Bosnia campaign in 1995; 85 percent was tagged in Afghanistan in 2001; and 100 percent in the Iraqi campaign.

What is the process? As envisioned by Boeing, as an airplane mechanic walks through a plane at an airport, he is watching his handheld screen. He's receiving status reports from various components, including their maintenance history and required repair schedule. Noticing that a part is nearly ready for replacement, he contacts the repair shop at the plane's next destination so mechanics there can ensure there is a new part on hand and install the part when the plane arrives.

The Department of Defense (DoD) has numerous RFID initiatives. These include active tags to support RTLS and movement of items into chaotic war zones. The DoD notes that $3 billion of the material shipped to troops in Desert Storm was never used. They attribute this to logistical system breakdowns. The DoD initiative is described in Chapter 6.

Maintenance-significant parts are processed as follows: when the container arrives at its destination, the active tag identifying the container is read, compared with the ASN, and the transaction is complete. As the container's contents are removed, the passive tag identifying each item is read, and the logistical database is updated. The item arrival information is used for work plan scheduling or inventory replenishment. The passive RFID tag is removed and discarded. An active tag with a Unique Identifier (UID) is affixed and updated when the item is used. The active tag reports status, time and telemetry (temperature, pressure, moisture, radioactivity, and so on) to the central system each time a unit is serviced or consumed.

What are the benefits? For Boeing, the benefits are cost savings for their customers. Airlines see maintenance as one of their largest expenditures and being able to improve management of the maintenance process by use of RFID tags is a competitive advantage. Airlines will see lower costs through streamlined access to information including pedigree, sensor-based telemetry messages, and on-the-spot notification of deadlines and maintenance checkpoints, as well as access to procedures, instructions, and manuals.

For the DoD, the ability to streamline maintenance is even more critical. Equipment is life-or-death for the soldier on the ground, and the ability of units to maintain and utilize their equipment is crucial.

What are the drawbacks and issues? For Boeing, the major drawback is the fact that the tags they wish to use are not available commercially. As mentioned in Section 5.1.3, Boeing public comments have acknowledged this fact, and Boeing expects the RFID industry to create these solutions.

Some have questioned whether having systems broadcasting the nature of the material in a hostile land is a good idea. Military spokespersons have said that the read range of the tags limits their exposure, and their methods of encryption provide adequate protection. They have also stated that the system architecture for passive tags, where only a UID is broadcasted, provides a level of protection.

What is the technology? DoD uses active tags at 433 MHz and 123 KHz. Their passive tags are chosen to comply with EPCglobal Gen-2 standards.

5.2.1.13 Track and Trace Conclusions

Track and trace applications raise issues ranging from privacy and cost to return on investment. The ability to trace individuals in any of these many contexts reduces our cherished freedom to come and go without accounting for our whereabouts. The widespread availability of our personal information has already claimed innumerable victims of mistaken identity, identity theft, and fraud. We are all subject to unwanted commercial messages because more and more people know how to find us. Privacy advocates are right to be concerned about increased ability of government, companies, and unscrupulous individuals to track the books we read, know the drugs we take, list the places we visit, know the tollbooths we pass through, and be able to find out the exact time and date we did so by merely launching an Internet search.

The cost of track and trace RFID systems is an entirely new cost in the economy. These are activities that were impossible a few years ago, and it is not clear who should pay for them. The fact that they may serve interests not directly chosen by the consumers raises questions of equity. Wal-Mart, Target, Best Buy, Tesco, and the other retailers clearly expect their suppliers to implement the technology and generate internal economies to pay for it. They will accept no increases in prices due to the cost of tagging. The DoD is more generous. They have allowed the cost of providing tags to be included in contract costs.

The advantages of being able to track and trace individual items throughout the product's lifecycle are significant, as well. We all have a stake in improving maintenance quality and timeliness, better management of prison populations, reduced incidence of injuries to librarians, improved management of hazardous materials, and reduced levels of theft and counterfeiting, and quicker response to disease outbreaks. The advantages of track and trace applications are significant.

The question of an ROI for an RFID system is discussed in Section 5.3 and also in Chapter 9.

5.2.2 Electronic Payment

Use of RFID for payment is the most widely used application for the technology. Systems such as Speedpass (see Section 5.1.6), the electronic toll payment systems seen on many highways throughout Europe and the United States, and fare collection systems used by millions every day on municipal railways and subways have become commonplace in the modern landscape.

Toll agencies around the world allow drivers to pay for tolls electronically. The customer opens an account and "loads" it with some amount of money. The account is maintained by the agency responsible for toll collection. The customer receives a tag with a unique identifying number. The tag is mounted on the driver's windshield or in back of the rear-view mirror. This positions it so it can be read by the readers located at tollbooths along the throughway. When the driver passes the reader, the tag broadcasts the account number, and the toll amount is subtracted from the account balance. The customer can attach their account to a credit card so that when it reaches some predetermined amount, money is transferred to refill it. Today, highway toll systems are separate from the systems that collect fares for subways. But agencies in Hong Kong, Singapore, and Japan are discussing combining them so that customers would have a single card and a single account for the tolls and subways. Many toll agencies have attempted to get merchants to accept the cards, hoping customers will use them instead of cash. None has so far succeeded; the latest to attempt is the London Transit System's Oyster card, which foundered on the sticky issue of who would pay for the necessary infrastructure.

RFID tags used to pay for goods and services are called *smart cards*. Smart cards have become popular in Europe and Asia, much more so than in the United States. Some smart cards, called *contactless smart cards*, work like credit cards, with the caveat that they hold more information than a simple magnetic strip can carry. In America, the power and success of the existing credit card authorization network has created a level of convenience unmatched in the rest of the world. Ironically, the level of convenience is so high that there is little incentive to move to smart cards. Abroad, with public transport more widely used and smart cards offering greater levels of convenience for payments, the devices are more widely adopted.

There seems to be a growth of interest in smart cards in the United States, however. American Express has released its ExpressPay, Master-Card has rolled out PayPass, Chase has released its Blink card, and Visa has begun enabling its cards as well (see Section 5.1.8).

On a related note, the European Central Bank is moving forward with plans to embed RFID tags as thin as a human hair into the fibers of Euro bank notes. The tags allow the currency to record information about each transaction in which it is passed. Governments and law enforcement agencies hail the technology as a means of preventing money laundering, black-market transactions, and even bribery demands for unmarked bills. However, consumers and privacy advocates fear that the technology will eliminate the anonymity that cash affords.

5.2.3 Access Control

RFID access control systems consist of antennas used to interrogate RFID tags embedded in access cards, electronics for data acquisition and control, a lock or some other physical security feature under the control of the system, network integration of the distributed electronics, and a centralized database that records the details of the use of access cards and the permissions of the population. After scanning the access card, the system determines whether the individual is authorized to pass and then unlocks the barrier. A record of the transaction is stored in the database. The systems can therefore report on the movement of individuals they are tracking. Aggregates of records can be used in logistics and cost analyses, building evacuation plans and generating required government reports. Most companies keep the record indefinitely. Many access control systems link with human resources databases, and some link with medical databases to allow first responders to scan a badge and call up relevant medical records during an emergency.

RFID systems offer benefits over card-swipe systems. Foremost is the instance when a number of people go through the door at the same time, (called *tailgating*). Groups of people rarely all swipe their cards individually, so many pass through, but few are recorded. RFID systems record them all.

Some vendors have begun offering systems that combine biometric data with RFID for use at military bases, nuclear and chemical plants, and other facilities that require multiple levels of security and control.

5.2.4 Telematics

In April 2004, the U.S. Federal Government announced "a new generation of RFID products aimed at bringing greater safety and new wireless applications to U.S. roads." The announcement heralded an award of $10.3 million to a group of manufacturers to establish a platform called DSRC (Dedicated Short-Range Communications). The manufacturers are Mark IV, Raytheon, Sirit, and TransCore, the companies that supply systems for RFID-based toll deployments. The award also set aside a 75-MHz block of radio bandwidth—the 5.9 GHz band—for this platform.

The award calls on the companies to develop technology systems for trial as part of the agency's goal to cut road fatalities in the United States by 50 percent within 10 years. The DSRC infrastructure is a prerequisite for introducing new roadway applications such as issuing alerts to drivers about impending collisions, rollovers, weather issues, or road hazards or warning the driver that their vehicle is going too fast for an upcoming curve. The technology could also be used for commercial applications such as downloading driving maps.

The 5.9 GHz system will support bandwidth higher than other RFID frequency ranges, and it will enable substantial differences in services. For example, either end of a link will be able to initiate a transaction. Vehicle-to-vehicle applications will be feasible. The system will consist of roadside monitors and sensors that can detect certain road conditions and situations and then transmit descriptions to DRSC transmitter/receivers installed in vehicles. A number of major automotive manufacturers are already studying the potential for such systems.

5.3 The Question of Return on Investment

The questions of cost and ROI arise in nearly every context: Where is the return on investment for RFID? The question is not simple. ROI is a tool for evaluating and comparing alternative investment opportunities. The ROI methodology quantifies the value of each potential investment over some fixed period of time and compares them. But ROI is famously ill-suited to evaluating new and disruptive technologies such as RFID. There is simply no certain way to quantify the results of something that will change the way a business performs. RFID is not a tactical investment; it is strategic, and calculating its ROI is nearly impossible.

A second reason calculating an ROI for RFID is difficult is that the technology is not a complete solution. It is part of a system, a tool that enables a raft of process changes and other applications, but by itself it does not generate much value. This reason is doubly important because the companies that implement slap-and-ship solutions will see only costs and virtually no benefits. To see a return on their investment, they will have to add or modify their software and revise their business processes.

Some observers have suggested that ROI is the wrong tool to evaluate RFID. They point instead to methods that value options, since RFID opens a number of options to a company, enabling a wide variety of value-creating investments and activities. This idea has not taken root.

5.4 Chapter Summary

Three forces are driving adoption of RFID technology: applicability to a very diverse set of problems, evolving standards, and declining costs. The range of applications is continually expanding. Some application types such as payment and access control are already well established. Some, such as RTLS, are emerging rapidly. And some, such as supply chain optimization, are the subject of intense activity. It would be easy to describe the retailer and DoD mandates as the drivers of this latter area, but surveys show that only about 20 percent of new RFID adopters cite mandates as their single most important reason. The rest cite the desire for business optimization or other specific goals. The mandates are important, but RFID has also clearly made its case as a productive addition to the world's technology portfolio.

As RFID professionals, however, we must not become sanguine about our prospects. The technology is easy to abuse, and in the past we have not been sufficiently sensitive to privacy issues. Consumers are justifiably wary of technologies that provide us little visible benefit but put our freedoms at risk and make our personal information available to those who wish us ill. We must be sensitive to these concerns as we design and implement these systems.

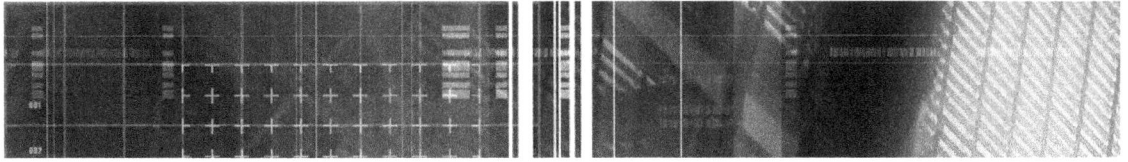

RFID at the United States Department of Defense

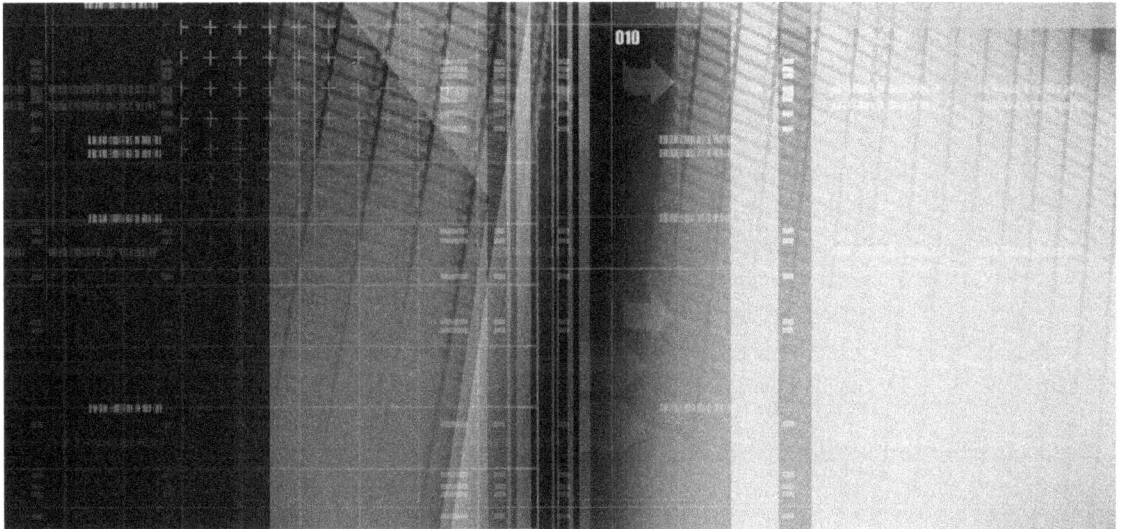

An adage from Napoleon rings true today: "An army marches on its stomach." The war fighter of any age, regardless of how valiant, is only as successful as his supply chain. Logistics is the foundation of combat power and the ability to exercise operational maneuvers. Without sound logistical support, the commander will find it difficult to control the tempo of operations and will be forced to pause at inopportune times in order to allow his logistical support to catch up.

Today, the American army, just like Napoleon's, depends on its supplies. The soldiers' vehicles, which provide speed and transport, must be fueled and maintained. Their ammunition quantities are greater because their tactics depend on massive firepower. Their complex weapons and communications gear need spare parts. These soldiers share with their Roman forebears the need for footwear, clothing, fuel, food and water, medical supplies, spare parts, and so on. But in addition, the modern soldier expects access to warm water, Sony Walkmans, DVDs of Madonna, mail from their families, and comfortable beds to bunk in. As of early 2005, the Defense Department had moved the following quantities of material to Afghanistan and Iraq:

- 1.2 million short tons of materials
- 62 million barrels of fuel
- 313,000 containers
- 11,280 high-explosive containers
- 140 million meals

The scale of activity described in these numbers is difficult to imagine. The ability to manage it rests on several concepts that have been developed over the past 12 years. Understanding the experience of the U.S. military with supply chains, logistics, and RFID technology is critical to anyone implementing it in their own company.

You are undoubtedly familiar with a range of technology innovations that began with the military and ultimately moved into civilian life. The communications of the Internet, the visual techniques used in special effects for movies, various materials such as Kevlar and ultra-strong fibers and ceramics, and freeze-dried food come readily to mind. It is clear that the military logistics problem shares many characteristics with its civilian counterpart. This chapter will detail how the United States military has found in RFID a solution to its oldest problem: getting the right product to the right customer at the right time at the least possible cost.

6.1 World War II Logistics

The World War II logistical model was formulated before the dawn of the Information Age. Operations were characterized by massive logistical efforts to amass material in the area of operations for further distribution to units needing material, accumulating sufficient quantities to last some predetermined number of days. This has been called the "just-in-case" logistics strategy. The method has some serious faults. First, just getting such a tremendous mass of material to the area of operations requires massive movement of material via land, sea, and air to transport such great quantities over such a distance in such a short amount of time. Priorities are difficult to manage, as immediate sustaining supplies compete for those destined for buildup. Once materials arrive, areas must be secured for the shipments to be disassembled and staged and for the items repackaged to be sent to the correct unit. The area itself becomes a tempting target for the enemy.

The just-in-case logistics strategy produced enormous amounts of waste, and it consumed enormous amounts of resources and energy. It gave rise to a well-documented underground economy as various personnel scrambled for the items, parts, and supplies that meant life and death for their comrades. With the coming of the Information Age, military logistics, would be, in the military's infelicitous jargon, *informationalized*.

6.2 Desert Storm Logistics

Desert Shield/Desert Storm saw several challenges to our logistics systems that were unprecedented. First, there was no time-phased force deployment data prepared. In other words, commanders rushed troops to the battleground with no time for detailed planning. Estimates had to be generated based on unprioritized requirements for units and material to be shipped to the area. Then this material had to be staged at embarkation points, loaded onto ships or aircraft, and transported. There were no systems for tracking material in transit. Once it was loaded onto a ship or airplane, there were no means for anyone to determine what had been shipped and what had not. Of the materials that had been shipped, no one could not tell which container, pallet, or bin contained it. Once the ship or airplane arrived, the logistical personnel in the theater had to figure out what was in each container and where it was supposed to go.

The material inside one container might be destined for several different commands, and each piece had to be identified and sorted separately. Of the 41,000 containers shipped to the area, 28,000 had to be opened on arrival, just to find out what was in them.

A second source of difficulty came as a result of the priority given to combat personnel at the expense of support personnel. This was required by the commander, to counter a feared invasion of Saudi Arabia. But the troops, once they arrived, would have no food, fuel, or ammunition.

A third source of difficulty was the lack of a common logistical system among the various branches of service. The Army, Air Force, Navy, and Marines each had their own systems for tracking, transporting, and delivering fuel, parts, ammunition, and personnel. There was no coordinated view of the joint operation's supply chain.

A fourth problem was there was no interface to commercial vendors. Items were often delivered without information to designate what they were or where they should be sent. These items were called "frustrated cargo." A significant number of items were simply addressed to "Operation Desert Shield." Overworked supply personnel at the stateside depot would load this cargo into a container and ship it to the theater, where even more overworked personnel in the theater would have to try to figure out what it was and what to do with it.

Underlying all these difficulties was the absence of supply chain information. The General Accounting Office (GAO) of the United States tells the anecdote of the colonel who ordered a critical part nearly a dozen times, since he was unable to tell from any available information whether in fact it was on its way to him or not. He took these actions to ensure that it would be available for his troops. This scenario, repeated thousands of times over, resulted in a staggering amount of waste. Over 42,000 containers were shipped by the Army, but by the end of the war, 8,000 of them remained unopened. They contained spare parts worth $2.7 billion. GAO observed that the Army itself, and in particular, commanders in the field, had virtually no ability to find out about assets in transit, so there was a tendency to reorder supplies that had not yet been received. The solution, a lesson for logistical operations everywhere, was an increase in what was called *visibility*. Visibility is the ability for authorized persons to see what parts, supplies, and equipment are en route from the factory to the foxhole (that is, over the total supply chain). Alan Estevez, Assistant Deputy Undersecretary of Defense for Supply Chain Integration, categorized visibility not as a goal for supply chain management, but as a tool to help with the following tasks.

- Reliably deliver the required item to the right location, in the correct quantity, when it is needed, and from the most appropriate source.
- Make tools and information available to the decision makers.
- Manage end-to-end capacities and available assets across the total supply chain to best support war fighter requirements.
- Enhance the ability of the supported combatant commander to exercise directive authority over logistics.

In response to GAO criticism, the Department of Defense created a visibility program, and called it Joint Total Asset Visibility (JTAV). The program, a work in progress at this point, consolidates requisition, inventory, and transportation data from 39 different data feeds. It makes this data available over the Internet to qualified users.

With active tags and a worldwide network leveraging the Internet, the Army would gain visibility on containers in transit. It would be unquestionably a great step forward. If material could be tracked reliably, the military would be able to schedule transportation resources—ships, planes, trucks, and personnel—more realistically. With greater confidence in the system that visibility would provide, commanders were less likely to over-order needed items. With less items being ordered just-in-case, more transportation resources would be available for other required items. With greater visibility, commanders could better exercise oversight and command judgment over the logistics they depended on.

6.3 Between the Wars

As the military conducted "lessons learned" and moved to improve its management of logistics, an additional constraint became important. The military required an overall assessment of their total supply chain footprint. Having a massive depot located in Saudi Arabia gave Al Qaeda a convenient starting point in its rhetorical war against the United States. To limit the size of stockpiles, much better tracking of material will be required to produce sufficient reliability in the total supply chain. For this to happen, all decision-makers need to be able to "see" the total supply chain over its entire length, knowing where any item of interest is and when it will arrive.

Several initiatives have helped improve in-transit visibility over the years. The agencies studied and adopted computerized tracking systems similar to those used by FedEx and UPS. These systems connected identification technologies with GPS satellite communications to enable precise location of items. In Desert Storm, identification for containers and pallets was based on bar coding. This was essentially an improved manual system that was faster and less prone to errors. But during this period, the DoD began using active RFID tags as well. These tags provided not only easier tracking, but also contained valuable information describing the contents of the container. Throughout the interwar years, the DoD studied and planned, but it took the terrorist attacks of September 11, 2001 to provide the impetus and funding to seriously upgrade the military logistics system.

6.4 Logistics in Afghanistan: Enduring Freedom (2001)

Enduring Freedom saw significant improvements in visibility. The Central Command had put together a system with more than 60 readers, 500 trackable trucks, and a reasonably robust joint logistics tracking system with satellite relays by February 2003. But there were still holes in visibility at the operational level. Visibility had improved markedly on items en route to the theater. But at the theater, the problems persisted. Containers holding items destined for multiple customers still had to be broken down into individual deliverables. These often did not have the necessary tracking information for visibility. Additionally, the systems were thrown together and personnel lacked training in their use. There were no standard operating procedures to provide consistent guidance on how to use the equipment and systems.

Visibility stopped at the distribution point, where supplies are disaggregated out to end users.

6.5 Logistics in Iraq I: Iraqi Freedom

Col. Glenn W. Walker, writing in *Army Logistician* in February 2005, said, "During Operation Iraqi Freedom, containers arrived in theater with RFID tags carefully mounted and full of data on what was inside the containers." However, there was no capability at the port to read the tags. No one at the port could "forward the containers to their correct destinations when they came off the commercial vessel, so all that labeling work was a wasted effort." Walker observed, "Adding technology without first implementing the right organizational and doctrinal changes only means that we know more quickly that we're in trouble—and we have no way to fix it."

Four months later, *RFID Journal* described the progress that had been made. Readers were installed at the Theater Distribution Center (TDC) in Kuwait. The report noted that the TDC had no electrical power and no computers until they brought in a trailer that could serve as an office with computers and a generator. The article described the process as each container was broken down. The contents destined for various units were sorted into pallets and containers bound for those units. A bar code was scanned, and the item identifier was written to the container's active tag. Each time a tag is read or a bar code is scanned, the in-transit visibility system is updated. To solve that problem and to gain visibility over the "last mile," the agency has turned to passive RFID. The reason for this decision included the lower cost of tags and the existence of a standard infrastructure—EPCglobal Gen-2. Agency spokespersons pointed out that "on a pallet full of a single type of item, a passive tag would be appropriate." However, a pallet that is going to be taken apart and restocked several times in the supply chain might need an active tag, so the data can be rewritten every time the pallet is altered.

The agency is working with its supplier, Savi Technology, to add batteries to passive tags so they can operate sensors to monitor telemetry data such as temperature and shock. The technology was tested in the shipping of food items.

But the tags were not enough. There also needed to be an information system to support operations from one end of the supply chain to the other. DoD needs to have visibility over its worldwide logistics operations, if it is to influence the battle through logistics. In Iraq, we may have seen the first evidence that RFID can deliver on the battlefield. Col. Mark Nixon, head of the Marine Corps' Logistics Vision and Strategy Center, said it was the first time battalion commanders on the ground could see and control the flow of replenishment supplies. RFID readers are set up at distribution centers and at checkpoints along main arteries into Iraq. When trucks pass the readers, the location of the goods they are carrying is updated in the DoD's in-transit visibility network database.

The U.S. military has proven to its own satisfaction the value of active tags to track pallets, containers and commercial sustainment shipments, all air pallets, prepositioning shipments, and vehicles and equipment being moved as part of an entire unit move.

Table 6-1 describes how different tags support different layers of the military supply chain.

So you can see that the DoD is an enthusiastic user of the range of RFID capabilities. Their needs are intense, but there are experts who say that global competition is just as intense, and the company that masters logistics, as the DoD has, will build sustainable competitive advantage.

Table 6-1

Military Supply
Chain Layers

Layer	Description	RFID Tag Class	Tag Characteristics	Supply Chain Area
5	Movement vehicle	4 (active tag)	Tag-to-tag communications Ad-hoc networking	Supply transportation, consolidation, distribution
4	Freight container	4 (active tag)	Tag-to-tag communications Ad-hoc networking	Supply transportation, consolidation, distribution
3	Unit load	2 or 3 (semipassive tag)	Rewritable passive or semipassive tag	Supply, Unique Identification Number (UID), transport, maintenance, aggregation/ consolidation, distribution
2	Transport unit	1 or 2	Identity tag Optional read-write user memory Optional packetized communications	Supply, UID, transport, maintenance, aggregation/ consolidation, distribution
1	Item package	1(identity tag)	Passive backscatter WORM tag	Supply, UID, transportation, maintenance, disposal
0	Item	1 (identity tag)	Passive backscatter WORM tag	Supply, UID maintenance, distribution

6.6 The DoD RFID Mandate

In January 2005, the DoD issued a mandate to its suppliers that items sold to them must be marked with a passive RFID tag. They have published, adopted, or amended three standards documents to accomplish this, which are listed in this section. This section summarizes the DoD mandate and is included for two purposes. First, if your company is a supplier for the DoD, you need to know this information. Second, the requirements are a good reference for your own project, even if you are not subject to this particular mandate.

6.6.1 DoD Standards

The first two documents constituting the RFID mandate were published by the DoD. The third is published by EPCglobal and is the same specification used by Wal-Mart, Target, and other retailers. The three documents are summarized in the following sections.

6.6.1.1 Military Marking for Shipment and Storage: Military Standard 129P; October 29, 2004

Revision P of this standard established significant new requirements for DoD contractors that ship packaged materials. They must now provide both linear and 2-D bar codes on military shipping labels. Code 39 bar codes will continue to be required on interior packages and on shipping containers. The standard also describes the requirements for usage of RFID tags.

6.6.1.2 Identification Marking of U.S. Military Property: Military Standard 130L; October 10, 2003

This standard describes the informational content of the label on shipments to the DoD. It includes requirements for labels that are intended to be read by either humans or machines. This standard does not discuss RFID directly, but the requirements for machine readable labeling are written broadly enough to be applicable.

6.6.1.3 EPC Tag Data Standards Version 1.1 Rev. 1.27: EPCglobal; April 1, 2004

This EPC standard is the final authority on how a tag is to be coded with information. All of the formats are defined in this document.

6.6.2 Phase I

The Department of Defense has set up three phases for its mandate.

Phase I began on January 1, 2005. It provided for tagging a particular set of items but only when they are going to two particular destinations. The items are listed as follows:

- **Class I subclass**   Packaged operational rations
- **Class II**   Clothing, individual equipment, and tools
- **Class VI**   Personal demand items
- **Class IX**   Weapon system repair parts and components

The destinations are Susquehanna, PA (DDSP) and San Joaquin, CA (DDJC).

The requirements apply to individual cases, cases that are part of a pallet of goods, and the pallets themselves. Individual items are not required to have labels until Phase III, on January 1, 2007.

6.6.3 Phase II

Phase II began on January 1, 2006. It continues the requirements of Phase I and adds several more. The additional items are listed here:

- **Class I**   Subsistence and comfort items
- **Class III**   Packaged petroleum, lubricants, oils, preservatives, chemicals, additives
- **Class IV**   Construction and barrier equipment
- **Class V**   Ammunition
- **Class VII**   Major end items
- **Class VIII**   Pharmaceuticals and medical materials

The additional destinations are listed in Table 6-2.

6.6.4 Phase III

The requirements of Phase III are that everything shipped to any location in the Department of Defense must be identified with an

Table 6-2

Military Command	Phase II Destinations
USMC	Albany, GA
Marine Corps Maintenance Depot	Barstow, CA
U.S. Army	Anniston, AL
Army Maintenance Depot	Corpus Christi, TX Red River, TX Tobyhanna, PA
USTRANSCOM	Charleston Air Force Base, SC
Air Mobility Command Terminal	Dover Air Force Base, Dover, DE Naval Air Station, Norfolk, VA Travis Air Force Base, Fairfield, CA
U.S. Air Force	Hill Air Force Base, Ogden, UT
Air Logistics Center	Tinker Air Force Base, Oklahoma City, OK Warner Robbins, GA
U.S. Navy	Cherry Point, NC
Naval Aviation Depot	Jacksonville, FL North Island, San Diego, CA
DLA	Albany, GA Anniston, AL
Defense Distribution Depot	Barstow, CA, Columbus, OH Cherry Point, NC, Corpus Christi, TX Hill AFB, Ogden, UT Tinker AFB, OK Norfolk, VA, Jacksonville, FL Puget Sound, WA, Red River, TX Richmond, VA, Tobyhanna, PA North Island, San Diego, CA Warner Robbins, GA

RFID tag. Individual items must have their own tags. Phase III begins January 1, 2007.

6.6.5 Tag Selection

Items must be tagged so that each item can be uniquely identified using only the information in the tag. This is interpreted to mean "license plate" tags.

The DoD's goal is to move to the use of EPCglobal Class 1 Gen-2 tags as they become available. In the interim, suppliers are urged to use Class 0 and Class 1 tags. Note that these are 64-bit tags. Ultimately, the Gen-2 tags, which contain 96 bits, will be required.

6.6.6 Operating Frequency

Tags are required to operate in the 860–960 MHz (UHF) frequency band. This is a new requirement. Previous DoD requirements specified 433–434 MHz.

6.6.7 Tag Readability Requirements

When passing through a portal, the tag must be readable from a distance of up to three meters while traveling at a speed of up to 10 miles per hour. This will occur, for example, when a forklift carries a pallet of goods through the portal.

On a conveyor, the tag must be readable from a distance of one meter while moving at a speed of up to 600 feet per minute. Pallets are not expected on conveyors, so this requirement applies to case-level tags.

6.6.8 Encoding Standards

DoD suppliers can choose one of two encoding standards.

If the supplier is a member of EPCglobal, they may use one of the following EAN.UCC codes:

SGTIN-64	SGTIN-96
GRAI-64	GRAI-96
GIAI-64	GIAI-96
SSCC-64	SSCC-96

If the supplier is not a member of EPCglobal, the DoD has defined an RFID encoding scheme based on the CAGE (Commercial and Government Entity) or NCAGE (NATO CAGE) code. The CAGE code is the unique identifier for the supplier. It is a 5-byte alphanumeric string assigned by the government. When the CAGE code is used in a 64-bit tag, only 30 bits are used, which means it is compressed. In a 96-bit tag, the CAGE code is padded on the left with space characters, to be 48 bits in length.

6.6.9 Costs

DoD suppliers are permitted to include the costs of compliance in their billing to the government. (Wal-Mart's suppliers are not. Wal-Mart believes that the suppliers will be able to utilize RFID to reduce their own costs more than enough to cover the costs of compliance.)

6.6.10 Exceptions to the Mandate

Bulk commodities, such as sand, water, coal, and animal feed, will not have to be tagged.

Explosives are exempt. Requirements for materials that could explode will be published after additional testing has been done.

6.6.11 Smart Label Placement

Placement of the smart label must enable opening of the package without cutting the tag. The smart label must not be covered with packing materials. Two tags on the same package must be at least 10 centimeters apart. Placement must uniquely associate the tag with the item.

6.6.12 Data

The precise information on the smart label for humans to read will be specified by each individual contract between DoD and supplier. The code uses industry standard formats and encoding schemes. The code must be a unique identifier for each item. The tag is encoded in a 96-bit format, so only a simple license plate is envisioned.

6.6.13 Advanced Shipment Notice (ASN)

An ASN or a Receiving Report must be sent from shipping entity to receiving entity when products identified by RFID tags are going to be delivered. The exact contents of the ASN will be specified in each contract. The ASN may be sent through an interactive web application or an established EDI connection or by using Secure FTP.

6.7 The Challenge

Read rates are proving to be a challenge in the military deployment of passive RFID tags. As has been the case in commercial applications, DoD has experienced read rates in the 80 percent range in tests at various locations. This has required human intervention to supplement the automatic methods of data entry. It also means that RFID is not a practical basis for payments. But in many cases, repositioning antennas, optimizing placement of tags on packages, restacking of cases and approach to readers have significantly enhanced read rates.

On the other hand, the military seems to have solved the two intractable problems of commercial implementations. First, they seem to have addressed issues of data quality and data synchronization. Those bottlenecks, so difficult in commercial implementations, do not bedevil the military. Second, they have a doctrine in Total Asset Visibility (TAV) that leverages RFID data in support of their mission, so, as the data is available, it results in improved performance.

Overall, the United States Department of Defense has found in RFID the solution to its problems of defining customer needs and meeting them. The underpinning is visibility, the ability to know where in the supply chain every element is located. Knowing that and leveraging the Internet to make that knowledge available to commanders everywhere enables them to reduce excess inventory, utilize assets according to the highest priorities, adopt best practices such as cross-docking from the commercial sector, and support commanders in the field with timely and correct shipment of material.

6.8 Chapter Summary

Military logistics can result in a particularly demanding set of supply chain problems, and the consequences of errors are uniquely intense. The United States Department of Defense has adopted RFID as its technology solution for supply chain improvement. They have issued a mandate similar to that of major retailers, but covering a broader swathe of technology. The DoD mandate embraces EPCglobal standards where they exist, but goes beyond them to meet requirements for active tags and semipassive tags. The mandate rolls out to only two destinations initially and then to a wider set of destinations in 2006 and to all military destinations in 2007.

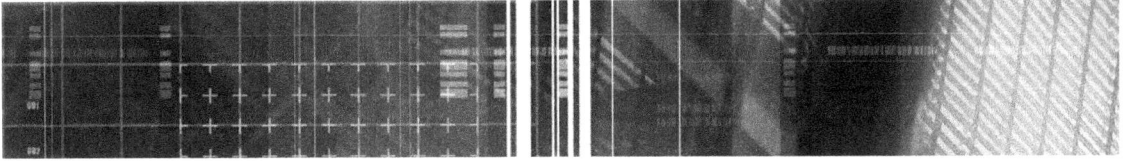

CHAPTER 7

The Pharmaceutical Industry

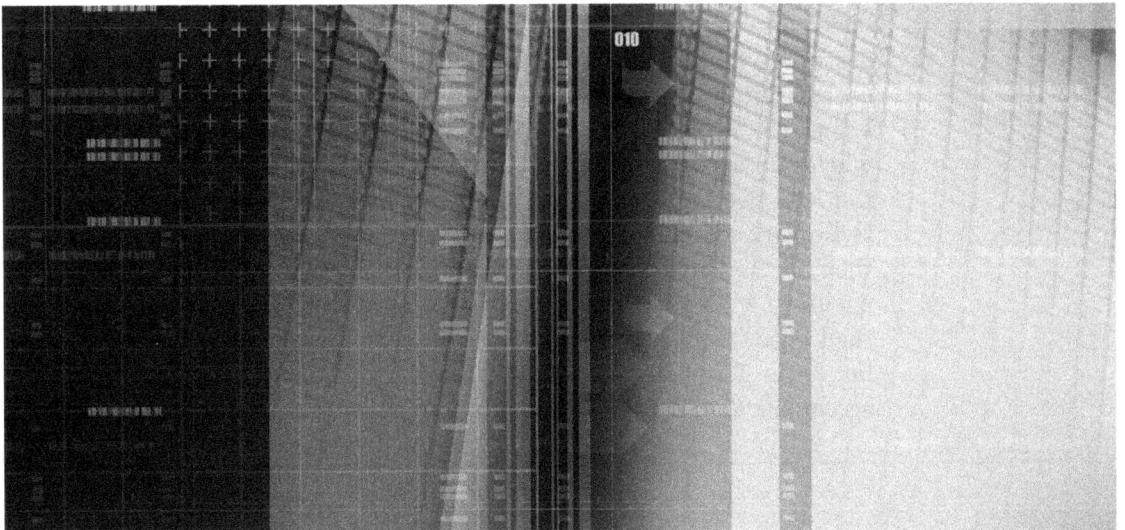

The pharmaceutical industry presents a special case for RFID. The reasons for adoption are the same as for any other manufacturer, but the needs in this industry are more urgent and demanding because of the products. Pharmaceutical companies say they are losing close to $2 billion annually due to overstock and expiration dates associated with their products. In addition, the estimates of loss attributed to counterfeiting of drugs exceed $30 billion annually.

As a result of these conditions, pharmaceutical companies are an especially fertile ground for RFID adoption. In 2006, pharmaceutical applications accounted for about 35 percent of the tags sold; by 2010, they could account for 60 percent. Specific drivers for the deployment of RFID in the pharmaceutical industry are

- Regulatory bodies (states, FDA) are putting pressure on them to do so.
- Counterfeiting is a particularly severe problem in the industry.
- The logistics of product returns are particularly significant.
- Inventory management issues are severe.
- Patient safety is impacted.
- Quality control can be enhanced.

7.1 The Role of the U.S. Federal Drug Administration

In 1987, Congress passed the Prescription Drug Marketing Act (PDMA), and it was signed into law by President Reagan in April 1988. The bill was enacted to ensure that prescription drug products purchased by consumers would be safe and effective and to avoid an unacceptable risk that counterfeit, adulterated, misbranded, subpotent, or expired drugs were being sold to the American public. The PDMA requires state licensing of wholesale distributors of prescription drugs. It recognizes a distinction between two legitimate classes of distributors. Primary distributors are those with a direct relationship with the manufacturer; there are five of them in the United States. Secondary distributors, equally legitimate, do not have a direct relationship with the manufacturer but may acquire their goods from authorized distributors or from manufacturers on an ad hoc basis. There are over 4,000 secondary distributors. The act requires secondary distributors to provide purchasers a statement, also called a pedigree, identifying each prior sale of the drug. The bill does not require primary distributors to provide such information. Since primary distributors are not required to furnish it, the secondary distributors who acquire goods from them cannot provide it.

Until recently, the FDA permitted secondary distributors to furnish a pedigree showing all transactions only back to the primary distributor. This workaround was temporary and was set to expire in December 2000. Responding to letters and petitions from industry, trade associations, and members of Congress, the agency delayed applicability of the requirement for pharmaceutical pedigrees until December 1, 2006. The agency also recommended that Congress revisit the exemption from pedigree requirements granted to primary distributors.

While technology did not exist in December 2000 that could provide full pedigrees for individual products, it is feasible with RFID today. Thus, the FDA has been particularly supportive of the industry's use of RFID systems. The FDA says they have "stepped up [their] efforts to improve the safety and security of the nation's drug supply by encouraging use of a state-of-the-art technology that tags product packaging electronically."

The agency has singled out RFID as the technology of choice to address their concerns regarding patient safety and security. They have encouraged companies to implement pilot programs to determine the best ways to use it. The FDA says that RFID makes it easier to ensure that drugs are authentic. It creates an electronic pedigree for every item in the drug supply chain and a record of every place the item has been, from manufacture to consumption. These pedigrees will improve patient safety and protect the public's health by allowing wholesalers and retailers to identify, quarantine, and report suspected counterfeit drugs. It will also enable them to conduct efficient, targeted recalls.

In November 2004, the FDA published a compliance policy guide for industry on implementing RFID studies and pilot programs. Specifically, the agency announced it would "exercise enforcement discretion" by which it meant that supply chain members who initiated RFID pilot programs would not be unduly penalized by the agency provided they complied with the following guidelines:

- A manufacturer, repackager, relabeler, distributor, retailer, or others acting at their direction will attach RFID tags only to immediate containers, secondary packaging, shipping containers, and/or pallets of drugs that are being placed into commerce. There is no limit to the number of tags or readers that may be used in the study.

- The drugs involved will be limited to prescription or over-the-counter finished products. The drugs involved will not include those approved under a Biologics License Application or protein drugs covered by a New Drug Application. The study need not have a predetermined time limit or endpoint, except that tag placement for the study will be completed by December 31, 2007.

- RFID will be used only for inventory control, tracking and tracing of products, verification of shipment and receipt of such products, or finished product authentication.

- RFID will not be used to fulfill existing FDA regulatory requirements (for example, fulfillment of labeling or Current Good Manufacturing Practice requirements; provision of chemistry, manufacture, and control information; storage of information in fulfillment of a regulatory requirement; or performance of label and product reconciliation).

- RFID will not be used in lieu of current labeling control systems to ensure correct labeling processes.

- The study may use passive, semiactive, or active tags.

- Information will be written to the tag at the time that the tag is manufactured (on read-only tags), after the tag is manufactured but before it is affixed to a drug's container (on read-write tags), or after the tag is affixed to a drug's container. The tags will contain a serial number (an electronic product code) that uniquely identifies the object to which the tag is attached and may also contain other information such as storage and handling conditions, information from the FDA approved label and labeling, lot number, and product expiration date.

- The tags will not contain or transmit information for the healthcare practitioner.

- The tags will not contain or transmit information for the consumer.

- The tags will not contain or transmit advertisements or information about product indications or off-label product uses.

- A seal containing a logo, an inventory control message unrelated to the product (for example, a message informing the custodian that the package contains an RFID tag), and/or a unique serial number may be placed over the RFID tag or elsewhere on a drug's immediate container, secondary packaging, and/or shipping container.

- The addition of the RFID tag and seal will not block, obscure, or alter any of the product's existing and approved label and labeling information.
- The RFID tag will not substitute for, replace, or interfere with a linear bar code required.
- Participants will "read" the tags as needed to identify the product and/or conduct the study.
- The tag readers will work radio frequencies of 13.56 MHz.

The agency guidelines go on to say that if a study is in compliance with all of the parameters listed here, the FDA will not initiate a regulatory action on the basis that the study fails to comply with any part of a particular list of regulatory or statutory requirements of the Federal Food, Drug, and Cosmetic Act when those requirements are triggered by the use of RFID in the study. The agency intends to limit its exercise of enforcement discretion to those regulatory issues specifically triggered by RFID (that is, triggered by the use of RFID readers, the addition of RFID tags, or the placement of seals).

The policy of enforcement discretion expires on December 31, 2007, although the agency may change it sooner or deviate from it in particular cases in order to protect the public health. The agency expects RFID technology to be in widespread commercial use in 2007.

7.2 Regulatory Pressure by States

Since February 2005, other pressures on the industry have emerged. The State of Florida now requires drug wholesalers to provide documented histories for the entire pedigree—that is, all the steps of the supply chain for all instances of a named set of drugs. Drugs without the requisite documentation are deemed "adulterated" and may not be sold. RFID is the least expensive way to collect this information. Florida's law is effective as of July 1, 2006. California has a similar law, effective in December 2006. Colorado, Nevada, and New Mexico are considering similar legislation. The pedigrees contain such information as date, quantity of the drug, and locations of manufacture and storage. It should also be mentioned that Wal-Mart has required that all pharmaceutical products be RFID tagged since mid-2004.

7.3 Drug Counterfeiting

Drug counterfeiting is a very costly problem for the pharmaceutical industry. Observers note that, as the product moves from manufacturer to distributor to retailer, there are many opportunities for counterfeits to enter the supply chain. Having each bottle tagged with its pedigree would enable retailers—and regulators—to account for each bottle and monitor its movement and history through the supply chain. RFID tags are currently costly to manufacture. The technology by itself will not prevent counterfeiting, but it is an important part of a general anticounterfeiting strategy.

7.4 Product Returns

When distributors or retailers return products to the pharmaceutical manufacturers for credit, the pharmaceutical companies will read the secure pedigrees stored on the tags, compare them with their databases, and give credit only where the product is legitimate and legitimately in the hands of the company requesting the credit.

7.5 Inventory Management

In the pharmaceutical industry, there are significant pressures and oversight on drug expiration dates. This puts unusual pressure on the pharmaceutical inventory management procedures. Because the products are often very expensive to manufacture, temporary misplacement of the product is particularly costly. Individual pills may cost $100 each, and a pallet of such pills misrouted or misplaced for a few days may have to be discarded or returned. These consequences are not nearly as severe in other industries.

7.6 Patient Safety and Quality Control

Tagged pharmaceuticals open a number of intriguing possibilities. First is the management of expiration dates: systems can detect possible use of expired drugs and guard against them. Second, just as systems can detect expiration dates, they can also relate drugs to patient records and guard against erroneous or conflicting prescriptions. Third, the system can check to make sure the correct drug in the correct dosage is administered and flag errors before they take place. Fourth, the system can also check the pedigree and ensure that at no point in its history was the container tampered with or did the temperature or moisture go out of specified ranges.

Tagging pharmaceuticals will enable regulators and diligent health professionals to monitor these issues as no other system can promise.

7.7 Product Recalls

As we have discussed in other contexts, systems that manage tagged goods enable recalls to be quicker, less disruptive, less costly, more accurate, and more targeted. Nearly every party in the supply chain benefits from this effect. Pharmaceuticals that are tainted, mislabeled, erroneously manufactured, or contaminated are often a threat to the very lives of the customers, so the urgency to improve the management of recalls is extreme.

7.8 Chapter Summary

The pharmaceutical industry has the incentive and the means to lead the way in adoption of RFID technology. No other industry will derive such a compelling array of benefits, including saving lives; preventing sickness; reducing theft, fraud, counterfeiting, and costs; and providing more responsive customer service and recalls of higher integrity. Pharmaceuticals are very highly regulated, and regulatory bodies in the United States and around the world are pressing manufacturers to utilize RFID to deliver these benefits.

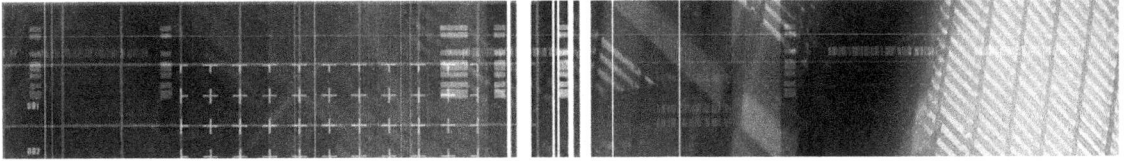

CHAPTER **8**

RFID in the Supply Chain

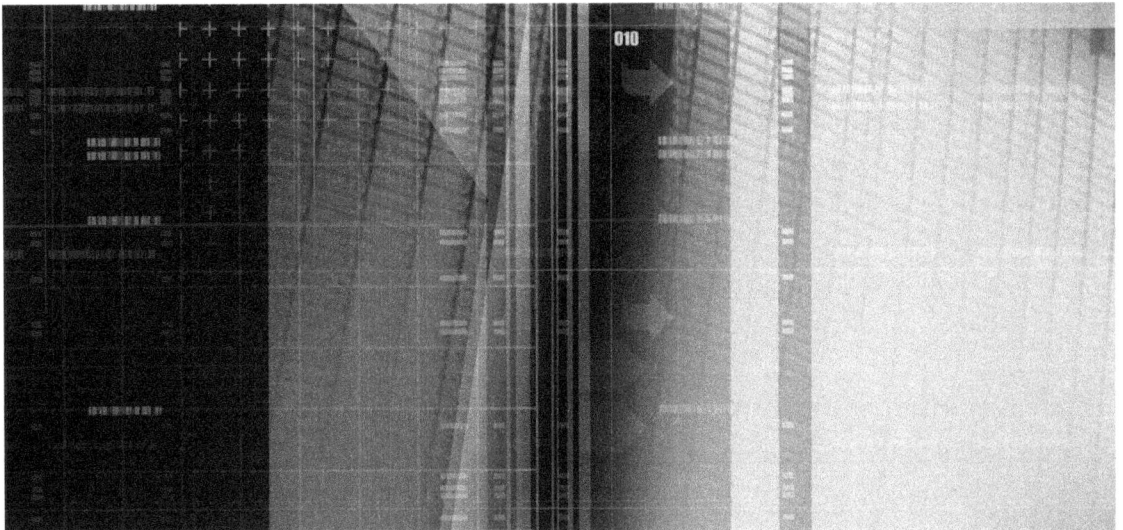

The supply chain consists of the organizations, equipment, processes, activities, and people that produce goods from raw materials and services and deliver them to end customers. It consists of raw-material suppliers, manufacturing facilities, distribution centers, warehouses, stores, knowledge workers, and customers.

The terminology is relational; each member participant in the supply chain (except the first and last) is in turn a customer to its predecessors and a supplier to its successors, as shown in Figure 8-1. Even when a "supplier" and the "customer" work for the same organization, it is useful to maintain this terminology. The goods in the supply chain consist of supplies, work-in-process inventory, finished goods, and knowledge. Again, the terminology is relational; each member's goods input can be viewed as "supplies," and its output can be viewed as "finished goods."

For the purpose of this book, we will limit our discussion of supply chain to physical products. We can't place a tag on a service or knowledge.

The supply chain involves the movement of supplies and products from one place to another. The business process that accomplishes this movement is called *logistics*. Logistics uses personnel, transportation equipment, warehouses, and warehouse equipment to move raw materials, components, subassemblies, and work in process from one location to the next. Logistics is a very big business. By one measure, logistics alone accounts for about 11 percent of the United States Gross Domestic Product.

The supply chain itself is bidirectional. It flows in two directions. Goods flow from a supplier to a customer; we call this *downstream* movement. But three commodities flow in the opposite direction; they flow *upstream*.

Figure 8-1
The supply chain

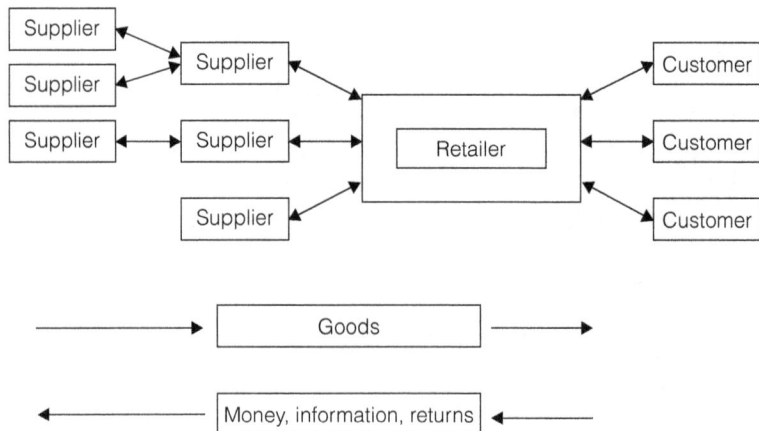

First, money flows upstream. Customers pay money to their suppliers for the goods they buy. Second, information flows upstream from customer to supplier. Each member receives demand information from their customers and in turn places orders for materials from their suppliers. The third commodity that flows upstream is goods that have been returned for one reason or another. The organizations, departments, equipment, and processes that move these returned goods upstream are called *reverse logistics.*

The supply chain has been subject to intense scrutiny, study, research, and attention in the past 50 years. Chapter 6 describes how the Army relies on its supply chain to enable it to be successful, and a retail organization is no less dependent. Like the Army, the retail organization will be only as successful and only as capable of serving its customers as is permitted by the effectiveness of its supply chain. Thus, it is not surprising that the leading retailers are driving RFID utilization so energetically. This chapter will explore these dynamics.

8.1 The Retailer in the Supply Chain

Retail has undergone massive change in the past 16 years. Of the 15 most profitable retailers in 1985, only 6 remained on the list just 10 years later. Retail space has increased more than 55 percent, and the competition has become far more fierce. The same products can be purchased in numerous stores at various price points, and retailers are under tremendous pressure because consumer choice has exploded. It is no longer enough for retailers to just buy the right products at the right cost and make them available on shelves. Retailers must also get them to the right place at the right time within very tight constraints of their operations budgets. The customer's retail "experience" is not just having a single product on the shelf. The experience is created by having the right mix of products so customers will come into the store to buy them. This requires careful planning, a sound strategy, and unprecedented coordination between merchandising, supply chain functions, and logistics. It requires the combination of demand analysis, supplier capability/coordination, logistics planning, and management to deliver goods to customers as inexpensively as possible. Those information nuggets are difficult to arrive at and they typically reside in different silos within the supply chain. Demand management information also has a very short period of time when it is relevant. The organization that can consistently and systematically assemble this information and translate it into meaningful knowledge in order to drive action will have a sustainable competitive advantage. This chapter will describe the role of RFID, technology, business processes, and organizations in the retail supply chain.

The role of the retailer in the supply chain is to identify and meet the demands of the end-customer (see Figure 8-2). The retailer generates sales and profitability by meeting these demands in a timely and cost-effective way. The retailer invests in facilities, personnel, systems, and inventory in order to provide services to the end-customer, who will come in and pay for the products. The demands of the end-customer are inherently dynamic, and to some extent, unpredictable. Nonetheless, the retailer must create an organization that can respond to them effectively and efficiently. The tension for the retailer arises from the fact that there are delays in the supply chain. So the retailer cannot respond to customer demands today; he must forecast what demand will be at some point in the future. Reducing or eliminating delays and the impact

Figure 8-2
Retail supply chain

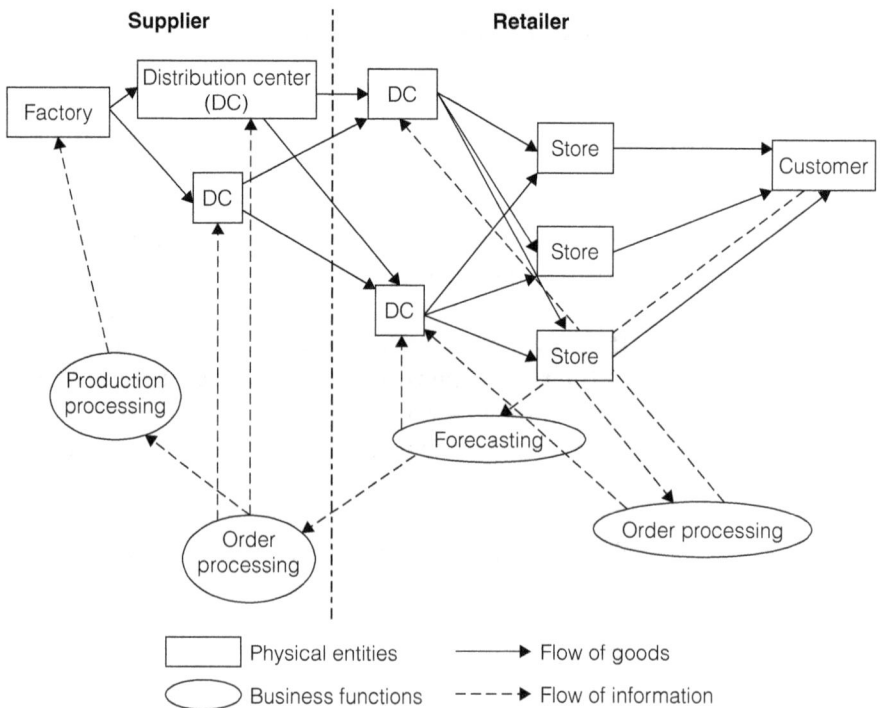

of such delays within the total supply chain is a principal objective of supply chain management.

8.1.1 Retail Purchasing and Distribution Functions

A retailer needs to know what to buy, how to buy, how much to buy, and when to buy. These questions are enormously complex for a large retailer and need to be considered in terms of retail strategy, forecast demand, logistics capacity, existing inventory, and goods in transit.

The purchasing process is driven by the business planning activity, which manages the portfolio of product categories. It leads to decisions on category roles, sales targets, and budgets. The retailer has corporate targets for revenue and profit, and the planning function decides how they can be met by categories, assortments, and SKUs at the store level. The store managers and merchandisers provide estimates at the store level, and the two groups agree on an assortment plan. The assortment plan is detailed to determine the purchasing budget for each SKU at the store level. To accomplish this planning activity, the participants must know:

- Sales and gross margins across product categories at all stores
- How to analyze performance metrics of each category in the merchandize mix

The purchasing budgets extend over a period of time, and they must be monitored on the basis of the differences between planned sales and real sales. Every month, the buyers calculate the Open To Buy (OTB) for each product they manage.

The fundamental OTB calculation is as follows:

Planned Purchases
– Inventory
– Stock already ordered
– In transit
= OTB

OTB is the dollar amount of merchandise that a buyer can order for a particular period for specific products. The planned and actual figures for sales, markdowns, and opening and closing stock levels are factored into the OTB calculation. The OTB is calculated at the item level at each store;

it translates into requirements for replenishment at each store. The distribution center (DC) servicing a group of stores makes the purchases, receives the goods, and transfers them to the stores based on store requirements. The purchase orders contain details for multiple items with requirements for partial shipments over a given time frame. The DC receives and holds the goods and then makes deliveries to stores in reaction to shelf and store inventory levels and buying trends. The DC in turn requests additional quantities of the item from the supplier to replenish its own stock.

The exact delivery quantities and timing are decided based on the following factors:

- Level of inventory already present
- Presentation minimum
- Store capacity
- DC capacity
- Delivery truck capacity
- Number of trucks available
- Distance between supplier's DC and retailer's DC
- Rate of product demand and demand variability
- Inventory holding costs for both supplier and retailer
- Transportation costs
- Lead time of supplier
- Other factors

As you can see, this is an extremely complex calculation. When you factor in the uncertainty and unavailability of much of this data, the task of ordering economic quantities becomes apparent. In addition, the conflicting needs of the two organizations make matters even worse. The retailer wants short supplier lead times, high fill rates, low inventory holding costs, and low transportation costs. The supplier wants lower inventory and transportation costs and full-truck loads. These two sets of wants are obviously in conflict with each other.

8.1.2 Challenges to Supply Chain Effectiveness

We have indicated that the decisions of what and how much to ship and when and where to ship product are very complex and dynamic.

This section describes some of the challenges of managing the volatility that resides in supply chains. Recall that the objective is twofold: minimizing inventory and other costs, and maintaining or improving customer service. Section 8.2. describes *Supply Chain Management (SCM),* the process and tools companies use to meet these challenges. These are the challenges to supply chain economic effectiveness:

- Demand variability
- Translating forecasts into orders
- Visibility of purchase volume
- Accurately assessing costs
- Vendor management
- Internal collaboration and cooperation
- Inventory management
- Retail information systems and organizational challenges

8.1.2.1 Demand Variability

The supply chain manager is challenged to accurately forecast demand at the store level and then to translate that demand to replenishment plans from the DC to the store and from the manufacturer to the DC. Modeling and building a reliable demand chain forecast is the key to matching demand with supply. Many planners simply extrapolate past shipment data and are not able to factor in specific drivers of sales. They do not have a unified view based on recent sales, current inventory levels, current inventory requirements, and lead times. They lack visibility on items in the pipeline, even within their own company. Apart from a few top-selling items or big promotional sales, buying patterns are random and hard to predict. However, the buying patterns become more predictable if you look at a group of stores served by a distribution center, overall company sales, and even item sales across all retailers. *Retail Logistics Quarterly*, a publication of McKinsey Corporation, studied retail forecast accuracy and found that the forecast error for single stores was 51 percent. The same measure for the DCs was 26 percent, and for the total company it was only 13 percent.

Supply chain members who serve the retailer are plagued by a phenomenon called the *bullwhip effect*, as shown in Figure 8-3.

"Bullwhip effect" is the name given to the process whereby even very small changes in perceived demand at the retailer level will cause

Figure 8-3
The bullwhip effect

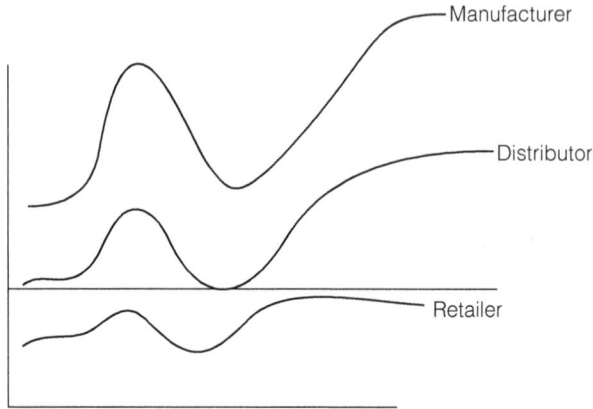

ever larger changes as you move up the supply chain. The bullwhip effect can be caused by

- Production problems causing delayed shipments. Demand will spike when the product suddenly becomes available, showing up in the data as a rise in demand.
- Quality problems causing extraordinary returns levels.
- Strikes, accidents, hurricanes, fires.
- Promotions or discounts.
- Competitors' actions.
- Changes in consumer tastes or preferences.
- Changes in season and weather.

Any or all of these cause the demand to fluctuate and lead planners to make changes to their order levels. The longer the lead times are, the more drastic the changes are. To the extent that their changes are unwarranted, inventories will increase or decrease unnecessarily.

When planners see customer orders unfilled, they tend to overreact, placing orders larger than were needed for the goods causing the backlog. Similarly, when planners see inventories growing, they tend to reduce their orders more than was necessary to bring inventory levels into balance. Anxieties are raised further the longer the lead time is for the products. As time goes on, planners, lacking in-transit visibility, are induced to place orders for more products, even though sufficient stocks may be in the pipelines. As planners are under pressure to reduce costs, they may batch orders. The batching of a group of orders with a single supplier may reduce costs by transport economies such as full truckloads and other supplier economies such as their own internal costs. But this occurs at the expense of retailer economics. In other words, the retailer may accept higher inventory costs in order to take advantage of temporary or special price reductions. During periods of shortage, customers may order more than they need in the hope that the partial shipments they receive will be enough to meet their needs.

Currently, managers who try to counter the bullwhip effect do so by placing smaller orders on a regular basis. In practice, this is difficult to do with most of the systems in place today. It is cumbersome to place frequent small orders when shipping costs for uneconomic batches are high. The effort involved may be justified for large, expensive items, but it is difficult to justify for the numerous smaller less costly items that many retailers manage.

The real problem is much larger than the issues of inventory and bullwhips. If the links in a supply chain are separate, each transition adds significant hurdles, expense, and time to the process, and often they add little value. Worse, they create barriers to the flow of information, as different organizations with different priorities grant or withhold access. Information synchronization also becomes a difficult challenge as data travels through different systems. Information becomes converted, misinterpreted, and altered, either by terminology and/or human error. Managers trying to counter the bullwhip effect by placing smaller orders more frequently could better solve their problem by giving suppliers access to point-of-sale data. The suppliers could plan their production around their own needs and still produce high levels of satisfaction.

RFID systems create a common pool of homogeneous information that planners throughout the total supply chain can access.

8.1.2.2 Translating Forecasts into Orders

Planners can translate forecasts into orders by following these steps. Of course, this assumes that the data elements are available, accurate, synchronized, and in a form that can be consistently interpreted and used.

1. Set time horizon for the forecast.
2. Forecast units required for the time horizon.
3. Subtract number of items in inventory.
4. Subtract number of items in transit to the DC.
5. Subtract number of items in transit to store.

6. Add in forecast sales between forecast and delivery.

7. Result: OTB.

8. Verify that OTB can be produced by supplier.

9. Determine lowest cost shipping method to DC. Verify availability.

10. Determine lowest cost DC space. Verify availability.

11. Determine lowest cost shipping method to store. Verify availability.

12. Place order.

Many planners are unable to assemble the data they need to follow this methodology. As a result, they simply place orders based on past shipment data. But by following that approach, they cannot factor in the miscellaneous data values required, and they do not have a unified view based on sales, inventory requirements, logistics capacities, and lead times to adhere to the above process. Planners often lack the tools to forecast at the DC level based on forecasted requirements at the store levels. This leads to excessive use of rules of thumb without any real analytical basis for their decisions.

8.1.2.3 Visibility of Purchase Volume

Lack of visibility on total spending information by supplier can arise from issues related to data quality and inadequate functionality of systems to capture purchase information. As odd as it sounds, retailers who purchase numerous products from a particular supplier are often unable to combine these separate data sets into a complete picture of their true purchase volumes. Lacking this information, they may not be able to negotiate the best discounts, terms, and customer service levels.

8.1.2.4 Accurately Assessing Costs

The pricing quotation of a given supplier may not reflect the real cost of the items to the buyer. It may not consider transportation costs, inventory carrying costs, duties or tariffs, or terms for delivery. It also may not consider how costs will vary depending on levels of commitment purchasers are able to make.

Information sharing will help suppliers and retailers have a common basis for negotiations. The parties can then cooperatively evaluate options on price and terms based on which party takes responsibility for various risks and cost items such as shipping and warehouse management.

Information sharing also allows for more subtle contracts that provide compensation for costs that were either previously hidden or not understood.

For the retailer, information about how the cost per unit varies with changes in volume of sale enables the retailer to redefine category plans. The retailer can then be in a position to benchmark suppliers and negotiate effectively on the basis of real data connecting cost, quality, and volumes. Retailers can establish processes and systems to gather this data and carry out analysis to support their negotiations. With superior quality and currency of data and the tools to derive and share key information from this data, retailers will be able to negotiate on the basis of hard facts.

8.1.2.5 Vendor Management

Vendor deviation from terms regarding order fulfillment has a cascading effect on every downstream element of the supply chain. It can lead to higher costs, misaligned inventories, and lost sales. The retailer needs to define key metrics for vendor performance and negotiate contracts specifying them. Continuous monitoring of vendor performance on the basis of defined metrics and factoring limitations of vendors on lead times into the planning process is difficult, but it enables retailers to mitigate risks in vendor relationships.

RFID systems can self-report shipment dates and receipt dates and compare them to schedules. RFID systems can also be designed to pay upon receipt automatically, if (and only if) no errors have been detected. In short, RFID can make vendor performance visible and transparent and open the way for incentives and penalties to motivate suppliers.

8.1.2.6 Internal Collaboration and Cooperation

With the increase in the number of specialized roles within a single company, opportunities to collaborate and plan collectively within an organization tend to get ignored. Internal collaboration between specialists such as planning and purchase order management must be instituted. In addition, external collaboration between the purchasing function and suppliers must also be instituted by sharing information on sales and inventory levels at store and DC levels and, by sharing the basis of reordering, will pay dividends. Organizations could structure incentives as well as processes and systems to support collaboration and joint planning.

Having every process owner at every step in the total supply chain using a common data set would drastically improve the efficiency of commerce.

8.1.2.7 Inventory Management

Effective inventory management is the central challenge in the supply chain. Too much of any item ties up working capital and capacity and creates inventory carrying costs and risks of obsolescence or spoilage. Insufficient stock, out-of-stocks, or late availability results in opportunity costs in terms of resource utilization, lost sales, and diminished customer good will.

The retailer also is faced with costs of accurately pinpointing where items are physically located. These include high labor costs involved in searching, identifying, and sorting items and variability in throughput of items, product returns, and more.

One obstacle to translating demand forecasts to effective inventory management is a quality data set. This is discussed fully in Chapter 13. Several leading retailers have identified RFID system implementation as the motivation and occasion to make data quality a reality because

- RFID systems report the location of every item in the supply chain automatically.
- Once installed, they reduce the cost and effort of capturing this information.
- When combined with modern Internet capabilities, these systems make the data available to anyone with access permissions.
- They reduce the time involved in processing the data into usable information.
- They alert users to conditions requiring immediate attention.

8.1.2.8 Retail Information Systems and Organizational Challenges

Retail IT (information technology) systems have generally grown up to meet specific requirements of specific business functions. Thus they often have little data sharing between them, and they tend to have poor data quality arising from inconsistency across systems. This inconsistency makes it difficult to share data among internal departments and even more difficult to share with external entities. Solving this problem will cause one or more departments to make costly and disruptive changes to their internal systems and procedures for the good of the enterprise. Convincing them to do this can be a challenge. Even small changes to systems that people depend on to do their jobs can be disruptive. Wholesale replacement of entire systems can be very disruptive for long periods of time.

Among companies, there is a second set of dynamics. Retailers have hoarded their Point-of-Sale (POS) data for years. This strategy enables them to play competitive suppliers against one another, under the assumption that this enables them to get the best pricing. In the short run this may be true, but, as we have seen, overall costs will be higher as a result of this approach.

A third set of dynamics works against sharing of data. When account managers and salespeople are compensated based on the number of units they ship, this creates incentives for behaviors that benefit themselves at the expense of the larger organization, and they resist changes in the system that would enable it to run more efficiently.

Even when the incentives are not so explicit, individuals who have tasks to do may resist making changes. A manager who needs to get 100 orders out this week will not support changes to his information system, and she may not be as careful as she could be to make sure her people enter newly required information in the right column.

Table 9-1 lists the various challenges to effective supply chain management and how RFID technology contributes to the solution.

8.2 Supply Chain Management (SCM)

Supply Chain Management (SCM) is the set of activities undertaken to increase the throughput while simultaneously reducing both inventory and operating expense. Throughput is defined as the rate at which sales to end-customers occur. Supply chain managers will always need to trade efficiency against service levels or find ways to improve one without diminishing the other.

A few markets support a build-to-order model. Customers in these markets are willing to wait while their products are built for them. But most markets build to stock and then service customers from inventory. The costs of creating and carrying inventory are substantial, so the answer to the question, "How much should we build?" is the subject of intense scrutiny. Building in lot sizes too small inflates the costs. Building in lot sizes too large translates into unnecessary inventory. Building too little creates shortages. Building too much creates surpluses and costs. Various surveys agree that popular products are out of stock about 7 percent of the time.

Table 8-1

Supply Chain
Management
Challenge and
RFID System
Solution

SCM Challenge	RFID Systems Contribution to Solution
Demand variability	RFID systems can provide functionality to aggregate stores' demands and to vary time horizons and employ different modeling techniques. Data collected through RFID systems can be captured throughout the total supply chain in real time. This data can then be utilized to significantly improve forecasting, demand management, inventory, and customer satisfaction.
Translate forecast to orders	RFID systems enable planners to consider goods-in-transit and on-order when making their forecasts. Supply chain–wide visibility also enables planners to factor in availability of inputs to their suppliers and logistics capacities.
Visibility of purchase volume	RFID systems create a base of data and employ a standardized naming system for manufacturers so ERP systems can accumulate deliveries by vendor.
Assessing costs accurately	RFID systems create a record of every time a product changes location. Systems can assign a cost to each transaction and accumulate the costs.
Vendor management	The record of product movement contains date and time information, which can be part of a report comparing contracted levels with actual performance.
Internal collaboration	The record of product movement creates a common data source for different departments to use, eliminating discrepancies in the data.
Inventory management	Decisions of what to buy, where to put it, how to stock it, and when to receive it all are improved by using RFID-generated data to assess capacities, loading patterns, costs, and opportunities.
Retail information systems	Having a common data source eliminates inconsistent data. Incentives can be created to motivate departments to make the switch.

8.2.1 Five Drivers of Supply Chain Performance

SCM is concerned with five variables:

- Production
- Inventory

- Location

- Transportation

- Information

These are the performance drivers for any supply chain. They are subject to the control and decisions of the supply chain manager, and their characteristics will determine the managers' ability to reduce inventory and operating expense while increasing throughput. The characteristics are described in the following sections.

8.2.1.1 Production

The production capability of the supply chain consists of its factories and its warehouses and gives the chain its ability to create and store products. Over a longer time horizon, the configuration of the factories and warehouses is subject to management decision. Over a shorter time horizon, it is fixed and it constrains the manager's decisions. Factories and warehouses must balance the twin drivers of responsiveness and efficiency. The most efficient factories perform within their walls the entire range of operations to make a single product. But because these factories cannot easily respond to changes in demand, they reduce the manager's ability to provide high service levels in a climate of changing demand. The alternative is to design factories that do just a few operations, such as making a select set of parts. These supply chains can then assemble the parts into many different kinds of products. These factories are not as efficient, and costs will be higher, but the flexibility may be worth it. A second arena for the wrestling match between responsiveness and efficiency is the capacity of the factory. The most efficient factory will be running at or near capacity, but such a factory cannot respond to sudden increases in demand. A higher service level demands some spare capacity that be brought into service when needed, but this means that the company is paying for capacity it is not using.

Similarly, warehouses must be designed to balance responsiveness against efficiency. Warehouses can store products in one of three ways. Traditionally, all of a given type of product is stored together. This is efficient and easy to understand. A second approach is to store all the different products related to the needs of a certain type of customer or job together. This permits more efficient picking and packing but usually requires more storage space than storing all of a given type together. The third approach is called *cross docking*. Cross docking is the practice of moving materials directly—across the dock—from inbound to outbound trucks without putting it away first. This increases demands on the information infrastructure, but it reduces the costs of warehousing products considerably.

Over a short time frame, the decisions made about warehouses and factories cannot be changed easily. They represent sizable investments. Over a longer time frame, the supply chain managers provide input to help make decisions about how to design and configure their facilities. Then they must live with their (or their predecessors') decisions.

In the short term, supply chain managers decide what items to order, the quantities to be delivered, the delivery destination, and the date of delivery.

8.2.1.2 Inventory

Inventory is distributed throughout the supply chain. It includes raw materials, work-in-process, and finished goods held by various supply chain members. Managers must balance responsiveness and efficiency against cost and flexibility in their decisions about what inventory and how much of it to hold. High levels of inventory enable high levels of customer service. However, purchasing and holding high levels of inventory is costly, and it increases the risk of loss, damage, or obsolescence. High levels of inventory also incur high storage and handling costs and even transportation costs, as excess inventory must be moved around to areas where it is needed, instead of purchasing from a supplier who might be closer. The manager's decision of how much inventory of each item to hold is one of the key areas of supply chain management. It determines a large portion of the total supply chain's cost structure. The inventory strategy determines the target inventory level at any point in time. It also determines the production decisions that translate into orders for specific items to be manufactured and delivered to specific locations.

8.2.1.3 Location

Location decisions determine where various activities take place. They determine the site of the factories, warehouses, and stores as well as which activities will take place in each facility. These location decisions over the long term are subject to management control. Over the short term, like the production decisions, factory and warehouse location decisions are fixed, and their parameters guide and constrain the manager's ability to respond efficiently to customers' needs.

In the short term, managers decide where to locate the items they have purchased. They must evaluate transportation availability and cost, both inbound and outbound. They evaluate demand histories and trends. They evaluate storage capacities in various locations, and they may need to evaluate handling capabilities as well. Their decisions are likely to be more efficient if they have visibility on what other shipments are bound for and scheduled from the locations they are considering. Opportunities for cross docking play a role in these location decisions, as well.

8.2.1.4 Transportation

The movement of goods between different facilities is another area for trading off responsiveness against efficiency. Using ships to move goods has a low cost structure but is slow and limited. Using air freight is very responsive, but it is very expensive. Trucks are also quick and flexible and also relatively expensive. Supply chain managers, as always, must balance their needs for responsiveness against the costs.

8.2.1.5 Information

Information is used for three purposes in the supply chain. First, it is used to coordinate the daily activities of the other four drivers of the supply chain: production, inventory, location, and transportation. The managers use the information to decide on 1) inventory levels, 2) production schedules, 3) transportation methods and routes, and 4) stocking locations.

The second purpose is forecasting and planning, which is the basis for monthly and quarterly production schedules and timetables.

Third, information is the basis for what we have called "longer-term" decisions such as whether to build new factories and where to locate them.

If the information is accurate, planners can make good decisions about production, inventory levels, stocking location, and transportation. If it is not, planners cannot make good decisions based on current data and must rely on past data, trends, feelings, and guesses. They will tend to compensate for poor information with too much production and too much inventory. This drives up costs in all areas and leads ultimately to markdowns and degraded financial performance. Poor information can also drive down customer service levels, as customers will be unable to find the products they need at a location and time that they require.

Managers need to decide how much information to share across the total supply chain. Open sharing of critical information promotes efficiency *and* responsiveness, but providing good information to business partners can be expensive. It also often runs against long-standing corporate practices. Sharing bad information can create the opposite effect, driving up total costs without providing increased value.

8.2.2 Challenges in Supply Chain Management (SCM)

SCM is the business process we use to balance our responsiveness and our costs. The battle is waged around the five drivers of supply chain performance just discussed. From about 1955 through the 1980s, companies sought to improve supply chain performance by improving their production planning and control systems. This is still true in many companies. There were many different methods introduced in this time period, but Materials Requirement Planning (MRP) got the most management attention. MRP promised very large benefits. It grew out of earlier systems called Bill of Material Processors (BOMP).

8.2.3 Computerized Supply Chain Management Solutions

Before computers, stock control was conducted on an order point/order quantity basis. For example, in one factory, airplanes were planned 18 months before the aircraft was due. Orders for parts and assemblies were launched into the shops according to week number. Three months before the start of assembly, the expediters checked stock to make sure parts were available. The parts that were in stock were marshaled into kits, and those that weren't went on the shortage list. Parts for the next order were chased down until the shortages were cleared. Orders went out to suppliers with due dates in the past. One factory routinely cleared about a quarter-million shortages a month and had about 250 people whose job it was to expedite the orders by clearing the shortages. One person typed the parts list twice every month: once by part number and once again sorted by department. Rumor has it she never made a mistake!

8.2.3.1 Bill of Material Processor (BOMP)

BOMP provided the very first real database facilities. It could look at the usage data that was typed every month and then re-sort it into lists of where each of the parts was used. It could "explode" an assembly into its components and print the lists.

8.2.3.2 Materials Requirements Planning (MRP)

MRP software built upon the BOMP algorithms and added the capability to convert a production plan for finished goods into a plan for producing or purchasing components. The MRP system could explode each top-level item or assembly in the production schedule and generate demand levels for that each component. This explosion could go any number of levels deep, meaning each assembly could be made of subassemblies, in turn consisting of individual items and subassemblies, and so forth. The promise of MRP was to improve the visibility of dependencies. Thus companies could plan procurement based on their actual needs, and they could respond quickly to changes in demand. In practice, however, the system introduced a good deal of instability. Each run produced wildly different results due to normal fluctuations in demand and supply. A second problem was that the systems increased inventory instead of reducing it. Every fluctuation upwards increased supply orders, which could not be easily reduced, so inventory stock went up instead of down, as had been promised. Also, MRP could not consider capacities, so it produced a schedule assuming that capacity was infinite. It required manual adjustment to make the schedule work in the real world. In the end, MRP produced output that people could not trust. It ended up being a glorified and expensive typewriter.

8.2.3.3 Manufacturing Resource Planning (MRP II)

To overcome the problems of MRP, new packages called MRP II were introduced in the early 1970s. These packages added the idea of a master production schedule and the ability to constrain the schedule according to resources available. It also expanded the scope of the inputs so engineering, marketing, finance, and human resources departments could have input.

The master schedule was derived from a company's business plan and told, in the aggregate, the schedule of finished goods the company must produce.

The MRP II package took the master plan as an input and produced a schedule of requirements for the subassemblies, component parts, and raw materials needed to produce the finished items in the specified time frame.

MRP II also added the capability to access an inventory master part file. The inventory file contained the part number, lot size, supplier, lead times, and other information. This enhanced feature enabled MRP II to use and report order quantity net of existing inventory and to tell when the items needed to be ordered. In theory, this would solve the problems of MRP. In practice, it highlighted the difficulty of having accurate inventory, which is rarely achieved, even today. Any practitioner knows that a perfect inventory count is nearly impossible, and it becomes obsolete very quickly.

MRP II also accessed an open purchase order file and an open shop order file, both of which were very difficult to maintain accurately or consistently in an automated system. Most of these were done manually or with autonomous systems with limited file sharing capabilities. To get this data into the MRP II system, people had to do a lot of retyping of other systems' reports.

For these reasons, and because the package had no way to account for priorities, MRP II never worked very well in practice. It could not resolve conflicts in a useful way.

8.2.3.4 Enterprise Resource Planning (ERP)

MRP II has given way to Enterprise Resource Planning (ERP) systems, which are used widely around the world. The market leader, SAP, claims 19,000 customers in 120 countries, adding up to 12 million users at about 65,000 sites. ERP is fundamentally a database, a way to input, store, standardize, retrieve, and use data that would otherwise be inaccessible. The scope of the ERP product enables easier access to data formerly stored in separate systems. The fact is that modern businesses are complex, and trying to plan requires access to a huge amount of data. ERP systems make available to planners large amounts of data that were not available in the past.

ERP systems also improve the quality of the data available. Recall that data within an early inventory system was generally inaccessible to other packages. This is still true today in many cases. Data in order entry systems, financial systems, human resources systems, and logistics systems can also be inaccessible to the planners. ERP advocates argue that only by centralizing on a single system can the most accurate, up-to-date information be available to planners. They contend that with enough data and the proper methods of analysis, reasonable projections of future outcomes become feasible.

Since ERP is at its essence a data management tool, it follows that any improvement in the way data is obtained, organized, and employed will have a significant improvement in planning and performance. For example, the algorithms for capacity-bound MRP are available, but the application has rarely been implemented successfully because of the lack of real-time data needed for a meaningful solution.

As you can see, the problem of getting timely and accurate data has plagued planners for years. Even with advances in the sophistication of the databases, the problem of acquiring and handling data has remained a challenge for the systems. Advances in data capture, such as bar codes, have expanded the reach of ERP to the better quality and greater quantities of the data needed for effective planning. However, most ERP systems depend on batch operations to process the data collected from various devices such as bar code readers. In many situations, updates occur only once per day. This falls far short of the continuous stream that planning systems really need. Further, much of the data is not granular; it does not provide unique identification for individual objects. A survey by APICS showed that 55 percent of respondents cited "improved inventory accuracy" as the main goal of implementing an RFID system. Even today, the achievement of better planning for the supply chain is often thwarted by lack of high quality, accurate, consistent, real-time data.

8.2.4 Improving the Supply Chain

Modern thought suggests that supply chain inefficiencies are the result of organizational, personal, and parochial mindsets and behaviors that inhibit synchronization. Synchronizing the supply chain will require three complementary activities:

- Data synchronization
- Processes to enable collaboration
- Automatic data capture and distribution

8.2.4.1 Data Synchronization

It is not possible for supply chain members to collaborate closely unless they have a common understanding and definitions for their data.

Without clear, unambiguous agreement on the meaning of codes, formats, units, time periods, and so forth, it will be very difficult to synchronize their systems. Fortunately, EPCglobal has published a standard, and they have embraced EAN and UCC data structures that have already gained wide acceptance. These standards are described in Chapter 2.

The quality of data in most organizations is also suspect; see Chapter 13.

8.2.4.2 Collaborative Processes

We say throughout this book that the payoff for a technology implementation such as RFID lies in the new business processes it enables. We describe in this section how some companies get their trading partners to synchronize their activities. One methodology for doing this was published by the Voluntary Interindustry Commerce Standards Association (VICS).

VICS has published a set of standards and processes that it calls Collaborative Planning, Forecasting, and Replenishment (CPFR). VICS claims that CPFR "combines the intelligence of multiple trading partners in the planning and fulfillment of customer demand." Using CPFR, trading partners have a roadmap to institutionalize sharing their plans for future events and reporting deviations or changes to the plan. In this way, partners have time to react, adjusting their schedules and their inventories as appropriate.

VICS notes that "an effective CPFR program builds upon a firm foundation of synchronized product data and electronic commerce messaging standards" such as those offered by EPCglobal.

VICS recognizes that different companies and different industries will implement various levels of collaboration. The organization lists the following levels:

- Front-end agreement, with executive sponsors, dispute resolution, confidentiality agreements, and success criteria
- Joint business plans for promotions, inventory policy changes, store openings/closings, and product changes
- Sales forecast collaboration
- Order forecast collaboration and order generation and delivery execution

Another advanced form or collaboration is called Vendor Managed Inventory (VMI). VMI is a business practice in which the supplier creates the purchase orders based on the demand information provided by the retailer. In this model instead of the customer managing his inventory and deciding how much to fulfill and when, the supplier does it. VMI provides improved visibility across the supply chain pipeline that helps suppliers improve production planning, reduce inventory, improve inventory turnover, and improve stock availability. With information available at a more detailed level, it allows the manufacturer to be more customer-specific in its planning. VMI is used where the end-customer's demand for products is relatively stable. VMI proponents claim:

- Reduced inventory. Suppliers can control the lead times better than the retailer can. Also, since the supplier takes greater responsibility, there is a lower need for safety stock.
- Reduced stock-outs. Suppliers take responsibility for product availability, reducing stock-outs for this customer.
- Reduced forecasting and purchasing activities.
- Increased sales due to less stock-out situations.
- Improved customer satisfaction, since products are on the shelf more often.
- Reduced errors.

VMI has challenges and limitations in practice as well:

- Supplier companies may not have the mechanisms, technologies, or processes to leverage the customer-specific information for production planning, so many continue to manufacture to stock and suffer the costs of poor inventory practices.
- Insufficient levels of system integration may result in poor utilization of the data.
- Retailers continue to raise expectations of performance.
- Sales forces resist losing control.

At its heart, VMI is just one way to get better information to the suppliers so they can act on it more quickly. It removes the time and expense of the forecasting and purchasing function from the retailer. If the data is actually utilized, two trading partners can operate as a single entity, improving the speed, lowering the cost, and increasing the responsiveness of the system.

8.2.4.3 Automatic Data Capture and Distribution

RFID in this context is part of the required infrastructure to enable synchronization of the supply chain; it is an automating technology. It allows the entire flood of items coursing through the supply chain to be accurately and quickly identified and tracked with little or no human action.

This introduces several efficiencies that enable collaborative processes such as VMI and CPFR. This is due to the timeliness, accuracy, and depth of data that RFID can produce. In high velocity items with high throughputs throughout the supply chain, this automation is critical. Retailers are deriving efficiencies and profitability through their large scale purchasing, distribution, and sales. A well functioning RFID system increases and enriches the data available to managers and planners and can also relieve data collection bottlenecks that might slow down the logistics. This is being done today with bar code systems, but a system relying on humans pointing readers at bar codes is too slow and does not scale up very well.

RFID scales up easily to large quantities and high velocities for items that move through the supply chain. It prevents numerous kinds of errors and thus improves order fill rates. An ASN can be generated automatically and instantly upon dispatch of a shipment, which enables logistics entities to plan better and thus improve their own performance.

8.2.5 RFID Impact on Supply Chain Management Activities

Managing the supply chain consists of the following well-defined activities:

- Forecasting and replenishment
- Inventory management
- Placement and picking in stores and DCs

8.2.5.1 Forecasting and Replenishment

The RFID-enabled supply chain can get real-time information from all interfaces in the total supply chain accurately and quickly. This enables companies to aggregate their forecasts automatically and make them more accurate at the store, DC, and supplier levels. It also provides greater visibility, allowing reorder points for replenishment of items to be set more accurately. In addition, the retailer can optimize the amount of stock he has on hand and on order, which reduces the average lead times of the items, in turn reducing safety stock and inventory costs, increasing service level, and decreasing the number of stock-outs.

Overall response time of the supplier also decreases, and the improved forecasting allows greater flexibility and reduces turnaround time in searching for and acquiring transportation equipment. The supplier can better plan their shipments, minimizing partial truckloads and enabling less expensive logistics. It also enables the supplier to reduce nonrequired inventory.

8.2.5.2 Inventory Management

Stores and warehouses can achieve better use of shelf and floor space if the forecasts are better. This is because the quantities more closely match actual requirements, and it is easier and faster to track and locate tagged items. Better visibility leads to lower shrinkage. Many times goods are in the store but not on the shelves because existing retail inventory systems lack the precision required to identify SKU-level inventory locations within the store.

8.2.5.3 Store Level and DC Level Efficiencies

At the DC level, better forecasts enable more efficient placement and picking strategies. This will decrease lead times, safety stock, and costs. At the store level, because of better inventory management, logistics become more effective and responsive to customer demand. Customer returns can be tracked all the way to the supplier, and the cost of servicing returns can be reduced.

8.3 The Adaptive Network

Section 8.1 opened by describing the pressures on retailers as competitive dynamics increase the speed, size, and importance of the supply chain decisions. It described the various systems and processes that businesses have tried over the years to meet the competing demands of greater throughput and customer service on one side and lower costs on the other. Let's now look forward and describe the successful supply chain of the very near future.

The advent of the Internet has been chronicled numerous times, but the underlying forces are much bigger than just the Internet. The steam engine launched the Industrial Age, but the forces it unleashed were far bigger than just the engine. In fact, the Industrial Age outgrew James Watts' little engine in just a few years.

The Internet gives us a global network that is easily accessible, always on, and virtually free. This means that companies routinely stay connected. They can acquire and use information far more quickly than was possible just a few years ago. The skills to leverage this connection and this information enable some companies (think Wal-Mart and Dell) to become more agile and more profitable than their competitors.

If competitors are still working in batch mode, they will be slower to react, decide, execute, and learn. Ultimately, they respond more slowly and incur higher costs overall.

Case Study: Retail Innovation at Tesco

Tesco is Great Britain's largest retailer, with over 4,000 hypermarkets, supermarkets, and convenience stores in the U.K., Ireland, Central Europe, and Asia. The company completed its RFID pilots and went on to purchase 4,000 readers and 16,000 antennas for delivery in 2005.

Tesco announced its radio bar code and secure supply chain programs in May 2000, with the goals of increasing product availability, cutting operating costs, and improving customer service.

Tesco's RFID rollout is taking a different approach than that of Wal-Mart and other United States–based retailers. First, the company is not requiring suppliers to provide the tags. Second, the company is not tagging pallets and cases. The retailer's IT director, Colin Cobain, was quoted as saying, "The technology isn't yet able to read all cases on a pallet. So what? Don't roll those out then."

The company has identified several areas where RFID tags provide immediate value. One application involves putting RFID tags on DVDs and electronic games. Fewer items are stolen, and workers are more productive because they know exactly where to find products. The project will produce a payback when tags can be applied automatically.

Tesco's major RFID project consists of putting tags on containers on wheeled dollies, which are used to deliver high value goods from the DCs to the stores. "The accuracy of the stock records is much better," Cobain said. "We're also getting an infrastructure which we can then use for multiple future uses. We've got this bigger picture in our heads, and this is an important part of that."

Tesco has long been a practitioner of what it calls *lean provisioning* (LP), a technique that runs counter to the supply chain management tenets many other companies live by. Most SCM professionals struggle to develop economies and savings based on the lower costs of large batches. Large batches, they claim, optimize the use of logistics resources, producing the lowest unit cost. Large orders optimize the use of ordering and management resources: it takes as much time to process a small order as a large one. LP differs from both of these assumptions.

LP begins by connecting the POS terminal to the DC decision point. This makes the end-customer the regulator of the supply chain. Contrary to accepted SCM practices, Tesco sends a truck from the DC on a "milk run" to stores every few hours, replacing the high-velocity items that were sold since the last visit. At the stores, the items are wheeled directly onto the selling floor on wheeled dollies that take the place of sales racks. Empty dollies are loaded onto the truck for the return trip. This innovation removes several touch points on the items. Previously, employees moved items from pallets to roll cages in the DC and from the roll cages to the trucks; they drove the products to the stores and then loaded them onto dollies and rolled them to the store shelves. Tesco found that half its cost in operating this supply chain was the labor in the store required just to fill the shelves. The final innovation was to turn the DC into a cross dock rather than a warehouse, with goods from suppliers moving quickly from their trucks to the store-bound trucks across the dock.

The results of LP at Tesco are impressive. Counter to what traditional supply chain managers would expect, the number of miles and the logistics costs to service the stores have gone down. The number of times the item is touched between factory and customer has gone from 150 to only 50. The total throughput time from the filling line at the supplier to the customer leaving the store with his product has declined from 20 days to 5. The number of inventory stocking points has been reduced from five to two, and the supplier's distribution center for the items has disappeared. Safety stock is maintained only at the DC, and it is much smaller than before.

Tesco has combined LP and RFID with other innovations. They have stores in four different formats, all of which share a core of high-velocity products. They have a loyalty card for the entire chain, and nearly 80 percent of their purchases are made in conjunction with the cards. This means the company can identify which customers purchase which products at which (kinds of) stores. It also means they can stock each store with products that appeal to its particular geography and demographic. Finally, the company has launched a web-based home shopping service. They have been able to deliver superior service levels by utilizing existing stores and using existing store personnel to pick web orders when store traffic is low.

Companies that can leverage a common data source and work together can achieve supply chain performance that is more profitable to all of them. They can constantly adjust their behaviors to meet their objectives because they have the information and the financial incentives. When demand at an RFID-enabled store unexpectedly spikes or falls, the supplier can be notified instantly and move to adjust his shipments. That retailer-supplier combination has two competitive advantages: first, it has lower costs because it has not had to pay for safety stock, and second, it can respond instantly to sudden unanticipated changes.

As companies stay connected to one another, they routinely collect far more data, and they can quantify and assess the performance of one another. They can also identify the costs of errors because all the activities are online and in databases. The suppliers themselves can see their performance, and that alone creates an incentive to improve it.

Ultimately, the information will be available to all the members of a supply chain. At that point, it will cease to be a chain; it will be a network, with all the members sharing and utilizing the same sources of data, focused on the same goals, and motivated to do their part to make the network more efficient to the benefit of all its members.

8.4 Chapter Summary

The supply chain is the collection of organizations and processes that create and deliver products for companies and consumers. Supply chain management works to reduce its cost while improving its responsiveness. Various measures have been used over the years to do this; as pressures on retailers have increased, MRP has given way to ERP systems, and new paradigms of organization and technology have emerged. A key bottleneck has always been the availability of quality data in a useful time frame. RFID systems are a fundamental tool in this regard, as RFID systems reduce the cost of collecting the data, improve the quality of the data, and make it available in real time. Management of the supply chain is an ordered set of steps, each of which has information inputs and outputs. RFID systems can automate, improve, and speed many of those steps.

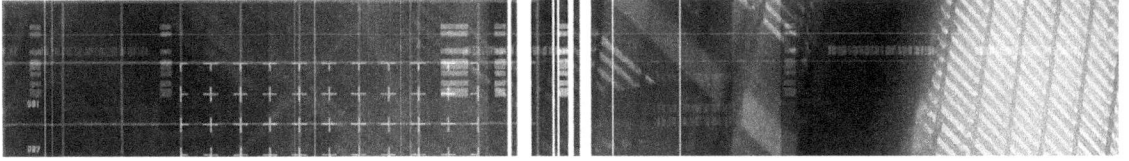

PART **III**

Your Project

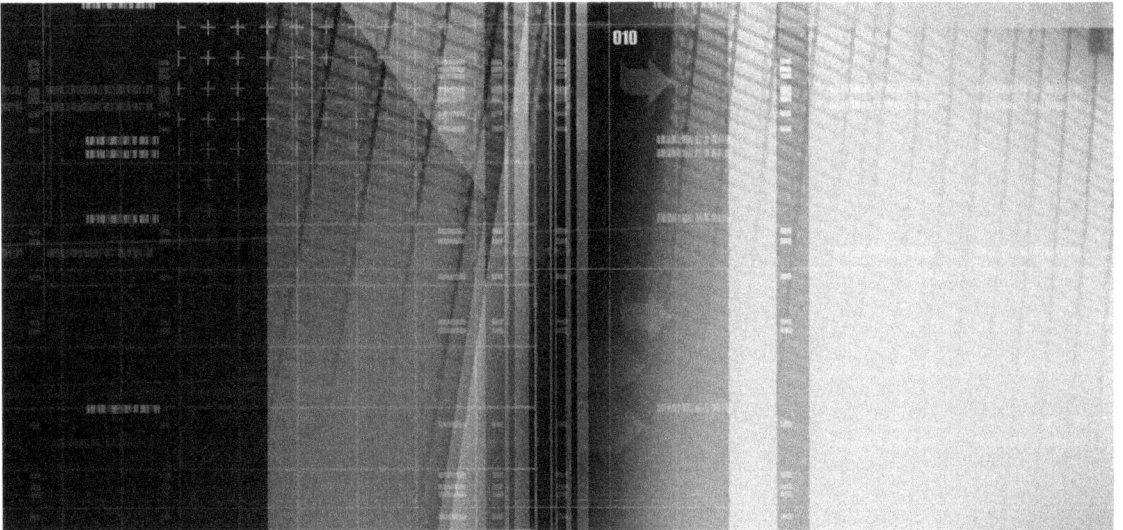

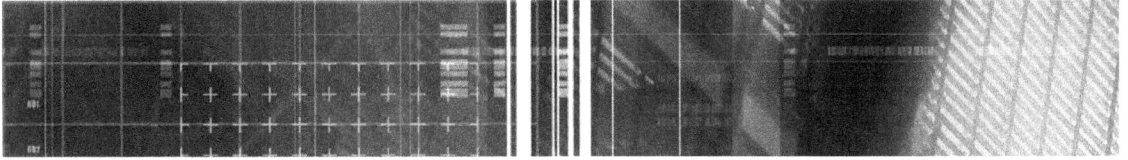

Business Justification for RFID

The business justification process lays out the costs, risks, and benefits of any project so executives can rationally make investment decisions. When all investment decisions are subject to the process as the basis of comparison, it optimizes the performance of the company, providing resources to the projects that offer the greatest net benefits. Business justification reduces the complexity of any investment to a single number, so management can compare many very different types of projects and decide where to invest their capital. The business justification process enables you to provide objective data about the benefits and costs of your project.

Even if your company has received a mandate from one of its customers or trading partners or is required to comply with a legal or regulatory requirement, you should still go through the business justification process. In addition to the benefits just mentioned, it will also help you select the right application, set milestones, prepare for the project, and focus on achieving the desired outcome.

The end result of the business justification process for your project is twofold. First, it enables you to decide which application(s) to pursue. This means you must identify and evaluate some number of applications and come up with a narrowed-down set of choices. Second, it prepares your company so your executives can allocate the appropriate types and quantities of resources and establish clear expectations. RFID has very broad applicability and can support a very large number of potential applications. Your job is to focus on the few that will give the maximum return for the investment and meet any regulatory or trading partner requirements.

Before we look at the steps of the business justification process in detail, we need to introduce a concept that is fundamental to financial analysis. This is the concept of *present value (PV)*. The present value concept is the basis of nearly all financial decisions. Present value quantifies the value of a stream of benefits and costs on the basis of the sensible notion that one would rather have a dollar today than a dollar a year from now. The present value of a dollar a year from now is something less than one dollar, and it can be calculated. Present value analysis reduces the complexity of gains and losses over time to a single number. The *present value (PV)* is the dollar value today of a stream of costs and benefits that stretch into the future. Since different investments have different time frames associated with their costs and revenues or savings, calculating the PV enables management to compare one proposed investment with another. For example, an investment with costs over three years and payoff over ten years can be compared with an investment with costs over one year and payoff over three years by calculating the PV of each, and management can decide which is superior.

To calculate a present value of a payment ("C" in the following formula) in the year t, you multiply it by a factor where i is the annual interest rate in the year in question, and t is the ordinal of the year (where the current year has a t of 0, the next year has a t of 1, and so forth):

$$PV = C \times (1 + i)^{-t}$$

The present value of the benefit or cost stream is equal to the sum of the present values for each of the years. We will use a time period of five years for our analysis and an interest rate of 5 percent.

For example, if you expected to benefit (or cost) $1,000 per year for five years, and the interest rate is estimated at 5 percent, the present value would be calculated as follows:

Year	t	Benefit	Present Value	Comments
2006	0	$1,000	$1,000	A thousand dollars in the current year is worth $1,000 today.
2007	1	$1,000	$ 952	A thousand dollars next year is worth $952 today.
2008	2	$1,000	$ 907	A thousand dollars two years from now is worth $907 today.
2009	3	$1,000	$ 864	
2010	4	$1,000	$ 822	
Totals		$5,000	$4,545	

The present value of the revenue or cost stream is the sum of each year's present value. In this case, the present value of a stream of payments of $1,000 per year for five years is $4,545.

Now we'll describe how to execute the business justification process. It requires your team to execute four steps:

1. Enumerate potential applications.
2. Build business cases.
3. Determine priorities.
4. Create milestones.

9.1 Enumerate Potential Applications

RFID can impact the following general areas of your business: sales, manufacturing, distribution and logistics, administration, payment, and security. Within each of these areas, there are numerous categories that can be targeted. Within the categories, you can identify specific areas where you will have the most impact. For example, within the area of security, Tesco has targeted DVD theft and has begun a project to tag all DVDs in their system. This is an application within the security area. Select just a few applications that have a high value, and justify your project based on those. In Section 9.3, we'll show you how to put values on your applications and use them to rank and prioritize them.

9.2 Build Business Cases

Build a business case for each application. Business cases look at several considerations in order to support a recommendation. We recommend you use these three:

- Benefit
- Cost
- Risk

Some analysts include complexity in the list of factors to consider, but we believe that complexity is best expressed as one of your risk factors.

9.2.1 Benefit

The benefit section of your justification document is where you quantify the financial value of the application. Financial value is generated by cost reduction, process improvement, and revenue enhancement. Strictly speaking, there are two kinds of benefits: hard and soft. Hard benefits are those to which you can ascribe an exact dollar amount. For example, removing a step from the workflow gives rise to an exact dollar amount because you can determine the cost for a specific element of a process.

Soft benefits are just as real, but it is difficult to calculate the exact dollar benefit realized from them. For example, better customer service is difficult to evaluate precisely, but it is known to be a good thing. Any quantification you can provide for even the softest of benefits is good, but your prioritization should consider only the hard benefits and costs. You should list and quantify the soft benefits whenever possible in your justification document and report on them throughout the project and in the project's final write-up.

9.2.1.1 Estimating Soft Benefits (and Soft Costs)

We have defined soft benefits (and soft costs, in the next section) as items that confer real benefit and cost real dollars but are difficult to value precisely. One way to get a number you can work with is to convene a panel of subject-matter experts (SMEs) from several departments of your company. Gather your panel in a room, brew a pot of coffee, and don't let them out until you all have agreed on a value for each item. Your SME team should include a representative from the affected department and someone from finance, operations, quality, customer service, sales, and IT. The team can discuss the issue and then ultimately agree to sign off on a number for you to use in your justification.

9.2.1.2 Cost Avoidance

One of the key values of an RFID implementation is the reduction in costs. The project may reduce the following costs: errors, shrinkage, inventory, handling, excess assets, misplaced assets, and others. You can determine the current cost of any of these elements, and then you can estimate the RFID impact. For example, if you believe you can reduce shrinkage by 30 percent, you can find out what the cost of shrinkage is now and calculate the value of the benefit. Similarly, if you believe you can cut the incidence of errors or searching for misplaced documents, you can determine a cost figure, calculate the benefit over the foreseeable future, and then compute its present value.

Figure 9-1 illustrates how to present the value of the cost avoidance benefit and calculate its present value.

Figure 9-1
Cost avoidance benefit

Cost Avoidance	Factors	$ 2,006	2007	2008	2009	2010	Comments
Cost of errors	10,000,000	$ 66,667	$ 66,667	$ 66,667	$ 66,667	$ 66,667	one third of 2% of sales
Cost of shrinkage		$ 132,000	$ 132,000	$ 132,000	$ 132,000	$ 132,000	half of 4% of sales
Cost of inventory		$ 50,000	$ 50,000	$ 50,000	$ 50,000	$ 50,000	10% of 5% of sales
Cost of handling		$ 10,000	$ 10,000	$ 10,000	$ 10,000	$ 10,000	10% of 1% of sales
Excess assets	5,000,000	$ 37,500	$ 37,500	$ 37,500	$ 37,500	$ 37,500	3/4 of 1% of assets
Misplaced assets	30,000	$ 15,000	$ 15,000	$ 15,000	$ 15,000	$ 15,000	half of lost asset value
		$ 311,167	$ 311,167	$ 311,167	$ 311,167	$ 311,167	$ **1,555,833**
Year		$ -	1	2	3	4	
Interest Rate		$ 0	0.05	0.05	0.05	0.05	
Total Benefit		$ 311,167	$ 296,349	$ 282,237	$ 268,797	$ 255,998	each year
Present Value		$ 311,167	$ 607,516	$ 889,753	$ 1,158,551	$ **1,414,548**	cumulative PV

This example shows the cost avoidance impact of an RFID project on a company or business unit with annual sales of $10 million. We have chosen to use a five-year projection period. We have estimated that the company has $5 million in excess assets and $30,000 in assets lost every year.

In this example, we have discovered that errors cost the company 2 percent of sales, and we believe we can eliminate a third of them. To be conservative, we have estimated no sales growth over the five years. Shrinkage costs 4 percent of sales, and we estimate we can eliminate half of that shrinkage. Inventory costs 5 percent of sales, and we estimate we can cut it by 10 percent. All of these are annual numbers. We take credit for these benefits for our full five-year horizon. This generates a total of $1,555,833 in cost avoidance, which has a present value of $1,414,548 if you assume a 5 percent interest rate.

9.2.1.3 Process Improvement

It may be that replacing humans reading bar codes with RFID tagging will reduce the time involved in, for example, receiving a pallet into inventory. More importantly, in the following example, it enables us to introduce a new process called cross-docking. In the following example diagrams, the project does not just reduce the time, it eliminates steps from the workflow altogether. To document the value of this type of benefit, you will need to utilize two tools of business process re-engineering: process flow diagrams and use cases. These will show exactly what the process consists of before your project and then its steps after your project. Some steps of the process will take less time, some will take more, some will be removed, and some steps will have to be added.

To document process improvement, follow these steps:

1. Document the process "as-is." The company might not yet have business-flow diagrams for its processes, so you may have to create them. Figure 9-2 is an example of a business-flow diagram for a hypothetical workflow. Each business-flow diagram is accompanied by a use case, giving details about each step in the diagram. Figure 9-3 is its accompanying use case statement.

In Figure 9-2, the actors are Office Personnel (OP), Dock Personnel (DP), and Warehouse Personnel (WP).

The document shown in Figure 9-3 is a use case, which accompanies and explains each step in the Business Flow Document (BFD). The use

Figure 9-2
Business flow
diagram example

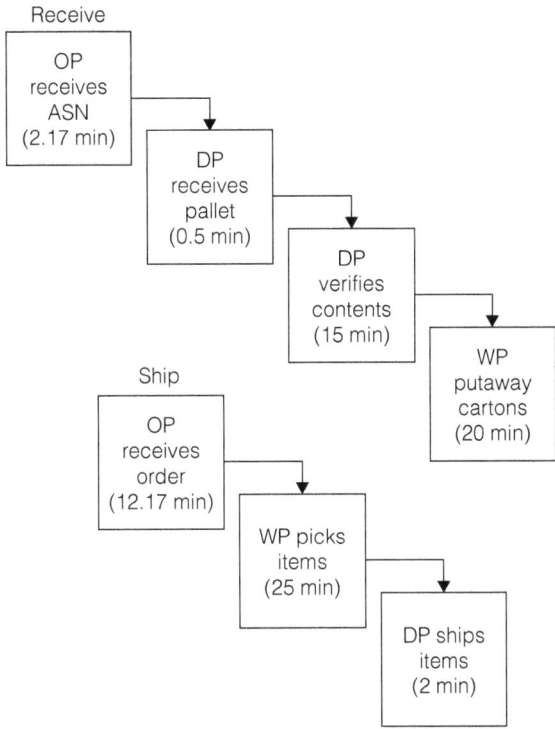

Receive

OP
receives
ASN
(2.17 min)

DP
receives
pallet
(0.5 min)

DP
verifies
contents
(15 min)

WP
putaway
cartons
(20 min)

Ship

OP
receives
order
(12.17 min)

WP picks
items
(25 min)

DP ships
items
(2 min)

case documents exactly how much time each step takes and how it is performed. Note the list of exception conditions at the bottom. Exceptions are events that disrupt the workflow. You may wish to augment the analysis by estimating the frequency of each exception and adding its cost to your total.

This analysis shows that the process of receiving a pallet into inventory, putting away its contents, picking the items, and shipping them takes 77 minutes at a cost of $102.44. Contrast this with the revised process in Figure 9-6.

2. Examine the process for RFID touch-points. Touch-points are points in the process where your RFID project will impact it. Figure 9-4 shows the business flow diagram with RFID touch-points marked.

3. Revise the business flow diagram and use case for the process as it will work once the RFID system has been installed, as shown in Figure 9-5.

Figure 9-3

Example use case

Use Case 101 As Is		
Use Case Name	DC receives and ships a pallet	**Use Case Number**
Actors	Dock Personnel(DP), Warehouse Personnel(WP), Office Personnel (OP), WMS System (System)	
Begin Use Case	ASN Arrives	
End Use Case	Pallet is sent to customer; Inventory is updated, invoice sent to customer	
Assumptions	No exceptions occur	
Description	DC receives ASN, and pallet, Receives it into inventory, then receives order order for the same item.	
Estimated Time/Cost-to-Complete	76.83333	$102.44
warehouse hourly factor		$80

Steps	Minutes	
OP Receive ASN		
1 OP receive ASN by fax (10 seconds)	0.17	$0.22
2 OP file ASN (2 minutes)	2	$2.67
DP Receive Pallet		
3 Truck offload pallet onto Dock		
4 DP move pallet to Sorting Area (30 seconds)	0.5	$0.67
DP Verify Contents		
5 DP retrieve ASN from OP (5 minutes)*	5	$6.67
6 DP depalletize contents (5 minutes)	5	$6.67
7 DP verify contents (5 minutes)*	5	$6.67
8 System updates inventory (0 seconds)		$0.00
9 System update order processing, approves for payment (0 seconds)		$0.00
WP Putaway contents		
10 WP repalletize contents (5 minutes)	5	$6.67
11 WP move pallet to correct shelf (15 minutes)*	15	$20.00
DC Receive Order		
12 OP receive order by fax (10 seconds)	0.17	$0.22
13 OP enter order items into WMS (10 minutes)*	10.00	$13.33
14 WMS print Pick List (2 minutes)	2	$2.67
WP Pick Items		$0.00
15 Pick each item (7 minutes)*	7	$9.33
16 WP Deliver each item to staging area (5 minutes)	5	$6.67
17 WP Pack items on Pallet (10 minutes)	10	$13.33
18 WP Verify items (3 minutes)	3	$4.00
DP Ship Items		
19 DP Load pallet onto truck (1 minute)*	1	$1.33
20 DP Mark Order 'Shipped' (1 minute)	1	$1.33
Totals	76.83333	$102.44

*Exception Conditions / *Remedy (time)*
5 Can't find ASN / *Return Shipment (15 min)*
7 Items on pallet not ordered / *Return Shipment (15 min)*
7 Items missing from order / *Notify shipper; adjust records (7 min)*
11 Putaway Error / *Restock item on correct shelf (10 min)*
13 Out of Stock / *Backorder item (5 min)*
15 Can't find item / *Backorder item (5 min)*
19 No room on truck / *Re-do paperwork and ship next avail (15 min)*

As you can see by comparing Figures 9-4 and 9-5, we have dramatically reduced the time involved in this process by eliminating steps and reducing the manual labor associated with others. We have also introduced a new actor, the RFID system (RFID). The real drama, however, is in the details. Figure 9-6 shows how to document the savings from using cross-docking instead of the normal putaway-pick cycle. Note how clearly the steps removed from the previous diagram are illustrated. The use case illustrated in Figure 9-6 shows the new business process taking 9.5 minutes compared with 76 minutes before. This is an almost 86 percent reduction in time spent, and it translates into 86 percent reduction in logistics costs for this shipment. This reduction

Figure 9-4
Business flow
diagram with RFID
touch-points marked

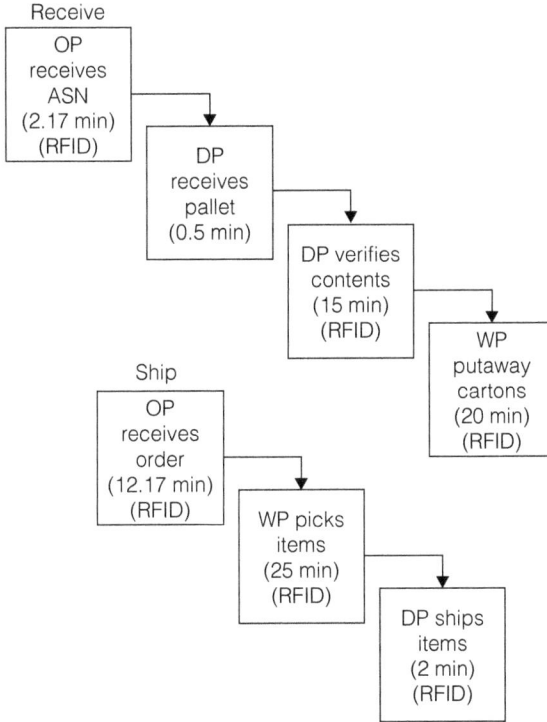

Receive

OP
receives
ASN
(2.17 min)
(RFID)

DP
receives
pallet
(0.5 min)

DP verifies
contents
(15 min)
(RFID)

WP
putaway
cartons
(20 min)
(RFID)

Ship

OP
receives
order
(12.17 min)
(RFID)

WP picks
items
(25 min)
(RFID)

DP ships
items
(2 min)
(RFID)

Figure 9-5
Business flow
diagram revised

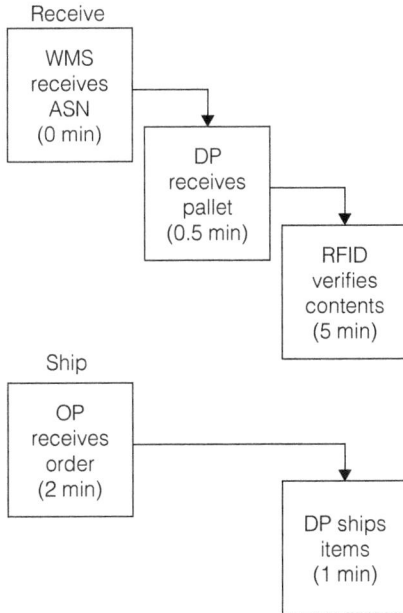

Receive

WMS
receives
ASN
(0 min)

DP
receives
pallet
(0.5 min)

RFID
verifies
contents
(5 min)

Ship

OP
receives
order
(2 min)

DP ships
items
(1 min)

Figure 9-6

Use case cross-dock
with RFID

Use Case Name	DC receives and ships a pallet		Use Case Number	
Actors	Dock Personnel(DP), Warehouse Personnel(WP), Office Personnel (OP)			
Begin Use Case	ASN Arrives			
End Use Case	Pallet is sent to customer; Inventory is updated, invoice sent to customer			
Assumptions	No exceptions			
Description	DC receives ASN, and pallet, Receives it into inventory, then receives order for the same goods			
Estimated Time-to-Complete			9.5 minutes	
			Warehouse Hourly Factor	
Steps			Minutes	
OP Receive ASN				
1 WMS receive ASN			0	
2 OP file ASN (2 minutes)			0	
DP Receive Pallet				
3 Truck offload pallet onto Dock			0	
4 DP move pallet to sorting area (30 seconds)			0.5	
DP Verify Contents				
5 DP retrieve ASN from OP (5 minutes)*			0	
6 DP depalletize contents (5 minutes)			0	
7 DP drive pallet through portal; System verifies contents (5 minutes)			5	
8 System updates inventory (0 seconds)				
9 System update order processing, approves for payment (0 seconds)				
DC Receives Order				
12 OP match order to ASN			2.00	
DC Ship Items				
19 DP Load pallet onto truck (1 minute)*			1	
20 DP mark order 'Shipped' (1 minute)			1	
Total minutes / cost			9.50	
* Exception Condition / *Remedy*				
5 Can't find ASN/ *Return Shipment (15 min)*				
7 Items on pallet not ordered. / *Return shipment(15 min)*				
7 Items missing from order. / *Notify shipper; adjust records(7 min)*				
11 Putaway error. / *Restock item on correct shelf.(10 min)*				
~~13 Out of Stock. / Backorder item (5 min)~~				
~~15 Can't find item. / Backorder item.(5 min)~~				
19 No room on truck. / *Re-do paperwork and ship next avail. (15 min)*				

is accomplished in two completely separate phases. First, it replaces manual steps with the system performing the tasks of identification, reconciliation, comparison, and updating of vital records. Merely by driving the pallet through the portal, all records are updated.

Second, it enables cross-docking on a larger scale. Cross-docking can be implemented in a number of ways, but it can be exercised much more often with the level of visibility that the RFID system provides. Cross-docking reduces costs in a number of ways besides the reduction in work time shown here. For example, it reduces the amount of inventory the organization must carry, since fewer cartons are stored on the shelf. It reduces the time the company holds the cartons. It may increase stocking levels (without increasing inventory!), and it improves responsiveness. These are soft benefits, not easily quantified for these business justification processes, but you should make the effort to include them in your document.

Notice also that two costly exception items have been removed. If you can determine the frequency of these occurrences, you can capture this value for your justification as well.

Figure 9-7
Process improvement

Process Improvement	Factors	$	2,006	2007	2008	2009	2010	Comments
OP Receive ASN		$	2.89	$ 2.89	$ 2.89	$ 2.89	$ 2.89	
DP Receive Pallet		$	-	$ -	$ -	$ -	$ -	
DP Verify Contents		$	13.33	$ 13.33	$ 13.33	$ 13.33	$ 13.33	
WP Putaway Cartons		$	6.67	$ 6.67	$ 6.67	$ 6.67	$ 6.67	
OP Receive Order		$	0.44	$ 0.44	$ 0.44	$ 0.44	$ 0.44	
WP Pick Items		$	33.33	$ 33.33	$ 33.33	$ 33.33	$ 33.33	
DP Ship Items		$	-	$ -	$ -	$ -	$ -	
Total		$	56.66	$ 56.66	$ 56.66	$ 56.66	$ 56.66	
								start at 20/day, 200
Number of Pallets Cross-docked	20 X 200		4000	5000	6250	7812.5	9766	days / year
Benefit		$	226,640	$ 283,300	$ 354,125	$ 442,656	$ 553,320	
Year		$	-	1	2	3	4	
Interest Rate			5.00%	5.00%	5.00%	5.00%	5.00%	
		$	226,640	$ 269,810	$ 321,202	$ 382,383	$ 455,218	each year
Present Value		$	226,640	$ 496,450	$ 817,651	$ 1,200,034	$ 1,655,252	cumulative PV

Figure 9-7 shows how the addition of RFID and cross-docking generates a benefit stream with a present value of $1.65 million. Figure 9-7 also illustrates how to capture the process avoidance benefit for your business justification document. This table displays the *difference* between the two use cases, which is the actual time and dollar value of the project.

9.2.1.4 Revenue Enhancement

The third business impact is revenue enhancement. Many people are surprised that a technical project like RFID can deliver revenue enhancement, but early adopters are reporting that it is the case. For example, most consumer products are out of stock about 7 percent of the time. This means that some revenue enhancement can be captured if out-of-stocks (OOS) can be reduced. You are cautioned that using industry averages like this 7 percent figure can be misleading in a business justification document. It is far better to gather the actual statistics for your products and your organization. In Figure 9-8, we will use 7 percent OOS, and estimate that we can reduce it by 30 percent.

A second potential contributor to revenue enhancement is more difficult to forecast but nonetheless real: brand enhancement. RFID enables a variety of information-based services that can distinguish the brand in the marketplace. For instance, a food item RFID tag can contain safety information, dates of packing, assurance of meeting temperature requirements, or recommended usage information. Hazardous materials can carry cautions, usage notes, reference materials, or directions for use. An example is discussed in Section 5.1.3. In that example, Boeing is using RFID to improve the processes of its customers for its maintenance-sensitive parts.

If you are going to include brand enhancement in your justification for your project, you will need to describe the mechanisms that will deliver it.

A third potential revenue enhancement application is the management of promotions. Manufacturers have begun to notice that promotional displays and materials often languish in the back room of stores for some or all of the promotion period. They have also noticed that promoted items are often out of stock, as inventory replenishment orders are not coordinated

Figure 9-8
Revenue
enhancement

Revenue Enhancement	Factors	2006	2007	2008	2009	2010	Comments
Annual Sales of Tagged Products (.000)		$ 500,000	$ 1,000,000	$ 2,000,000	$ 4,000,000	$ 8,000,000	
Pro Forma Out of Stock	7%	$ 35,000	$ 70,000	$ 140,000	$ 280,000	$ 560,000	7% of products are OOS RFID can reduce OOS
Reduce OOS by 30%	30%	$ 10,500	$ 21,000	$ 42,000	$ 84,000	$ 168,000	by 30% 45% of the additional in-stock will result in
Benefit: Increased Sales at 45%	45%	$ 4,725	$ 9,450	$ 18,900	$ 37,800	$ 75,600	additional sales
Year		0	1	2	3	4	
Interest Rate		5.00%	5.00%	5.00%	5.00%	5.00%	
Total Benefit		$ 4,725	$ 9,000	$ 17,143	$ 32,653	$ 62,198	
Present Value		$ 4,725	$ 13,725	$ 30,868	$ 63,521	$ 125,717	

Figure 9-9
Summary of
project benefits

Benefit Element	Present Value
Cost Avoidance	$ 1,414,548
Process Improvement	$ 1,655,252
Revenue Enhancements	$ 125,717
Total	$ 3,195,518

with promotions. Walgreen's RFID implementation focused on promotion management as its main application, and Proctor & Gamble has spoken publicly about the value of this particular application.

For each benefit, estimate its value over the time horizon. See Figure 9-8 for an example.

The next step is to summarize the project benefits in a single chart so you can sum them and generate a number you can use to calculate the ROI; see Figure 9-9 for an example.

9.2.2 Cost

The business justification process requires you to predict the cost of your project. The major costs to report are

- RFID hardware
- Application software
- Middleware
- Installation and configuration services
- Testing
- Training
- Maintenance
- Tags
- Business process change
- Downtime in operations to accommodate installation

Let's look at each one of these in detail.

9.2.2.1 RFID Hardware

Readers and antennas are a one-time expense and a highly visible element of the cost of any RFID system. As your installation grows, you will need to add controllers. In addition, you may need to budget for additional computers and network gear such as routers, bridges, and wireless access ports. You will also need additional data storage capacity. These are one-time costs and can be amortized over the life of your project. They may also support applications other than RFID, or other RFID applications. So their costs might be allocated over several projects. In this case, your justification could show just your allocation of the costs. Be explicit about your reasoning.

Figure 9-10 illustrates how to calculate and display the reader cost for a single distribution center. We assume we are equipping two docks the first year, three the second, and so on. We are also equipping one storage room in 2007, two in 2008, and so on.

Readers for one distribution center	Factors	2006	2007	2008	2009	2010	Comments
Receiving docks		2	3	6	6	7	24
Storage Rooms		0	1	2	3	3	9
Readers							
Handheld		2	3	6	6	7	one handheld per dock
Fixed Readers		2	4	8	9	10	one fixed reader per dock, plus storage
Antennas per fixed reader		4	4	4	4	4	
Antennas		8	16	32	36	40	
Damaged Handhelds replaced		1	1	1	1	1	
Damaged Fixed readers replaced		1	1	1	1	1	
Damaged antennas replaced		2	2	2	4	4	
Cost per handheld		$1,500	$1,313	$1,148	$1,005	$879	Costs expected to decline
Cost per fixed reader		$2,500	$2,188	$1,914	$1,675	$1,465	
Cost per antenna		$225	$197	$172	$151	$132	
Readers and Antennas		$14,250	$19,731	$31,123	$29,812	$28,957	
Installation and Configuration	$500	$1,000	$2,000	$4,000	$4,500	$5,000	
Wiring Cost per dock		$1,200	$ 1,320	$ 1,452	$ 1,597	$ 1,757	
Wiring Cost		$ 2,400	$ 5,280	$ 11,616	$ 14,375	$ 17,569	
Maintenance at 20%	20%	$ 2,850	$ 6,796	$ 13,021	$ 18,983	$ 24,775	
Miscellaneous Costs	$400	$800	$1,600	$3,200	$3,600	$4,000	Racks, extra cabling etc. based on fixed reader count
Total Readers' Cost		$21,300	$35,408	$62,959	$71,269	$80,301	$271,237
Year		0	1	2	3	4	
Interest Rate		5%	5%	5%	5%	5%	
		$ 21,300	$ 33,721	$ 57,106	$ 61,565	$ 66,064	each year
Present Value		$ 21,300	$ 55,021	$ 112,127	$ 173,693	$ **239,757**	cumulative PV

Figure 9-10
Reader cost for one distribution center

9.2.2.2 Application Software

Each application will have to be supported by a software package. In your business justification process, try to identify a couple of candidates for each application and estimate the cost of their purchase; see Figure 9-11. These purchases are a one-time investment and can be amortized, but software maintenance and upgrades are ongoing expenses. If existing software can be used or upgraded, show those costs here.

9.2.2.3 Middleware

RFID middleware is an important purchase expense. It is the key to using your system and integrating the data stream into other business applications. For business justification purposes, you will need a number you can use. Figure 9-12 shows how to find it and illustrate it. You should be able to contact a couple of vendors and get ballpark figures. Middleware is software, so it will incur annual maintenance costs, and upgrade costs

Application Software	Factors		2006	2007	2008	2009	2010	Comments / Totals
Software Purchase			$65,000					
Installation & Configuration			$35,000					
Maintenance	20%	$	13,000	$ 13,650	$ 14,333	$ 15,049	$ 15,802	$ 71,833
Internal Support Cost	15%	$	9,750	$ 10,238	$ 10,749	$ 11,287	$ 11,851	
Total Application Software Cost			$122,750	$23,888	$25,082	$26,336	$27,653	$ 225,708
Year			0	1	2	3	4	
Interest Rate			0.05	0.05	0.05	0.05	0.05	
Present Value		$	122,750	$ 22,750	$ 22,750	$ 22,750	$ 22,750	each year
Present Value Cumulative		$	122,750	$ 145,500	$ 168,250	$ 191,000	$ 213,750	cumulative PV

Figure 9-11
Cost of application software

RFID Middleware	Factors		2006	2007	2008	2009	2010	Comments
Software Purchase			**$55,000**					
Installation & Configuration			$27,500					
Maintenance	20%	$	**11,000**	$ 11,550	$ 12,128	$ 12,734	$ 13,371	
Total Middleware			$93,500	$11,550	$12,128	$12,734	$13,371	$ 143,282
Year			0	1	2	3	4	
Interest Rate			0.05	0.05	0.05	0.05	0.05	
		$	93,500	$ 11,000	$ 11,000	$ 11,000	$ 11,000	each year
Present Value		$	93,500	$ 104,500	$ 115,500	$ 126,500	$ **137,500**	cumulative PV

Figure 9-12
Cost of middleware

are likely to occur as well. In this example, upgrades are included in the maintenance figures.

9.2.2.4 Installation and Configuration Services

There are many installation and configuration services that your project might need. You will need expertise in installing and configuring

- Readers and antennas
- Application software
- Middleware
- Cables and power connections

This book describes how to perform these functions, but if you are just beginning to work with any of these technologies, you budget for someone to work with you to supplement your skills.

Budgeting for these services is easier if the scope of the project is very small and grows iteratively based on a successful trial.

We have included these costs in their respective areas, so installation and configuration of the readers and antennas, for example, is included in Figure 9-10.

9.2.2.5 Testing

You may choose to do your own testing, or you may choose to hire an outside testing service. Regardless, there will be costs associated with the testing required. You'll need to run RF tests every place you plan to install a reader. In addition, you will need to test tags and SKUs. Testing protocols are listed and described in Chapter 12.

9.2.2.6 Training

There are several ways to acquire training services, each with its drawbacks and advantages. You might engage vendors to provide it. Alternatively, you might have vendors provide *knowledge transfer* and have your own people provide the training. Knowledge transfer is an informal process. It consists of customer personnel working with vendor personnel to learn what they do and absorb whatever information they can about the product. Training is an area where many companies pinch their pennies, and it sometimes hurts them. The ability of your people to run the software and operate the equipment is critical to your success, and paying the extra dollars to get it right could be an excellent investment. Training, as described in Section 11.6, takes place in a number of venues. You must provide training in the following areas:

- Configuring and maintaining the readers, antennas, and reader network
- Configuring and maintaining any new application software
- All new work processes

The vendors' installation teams can provide the installation and maintenance of the first two items. Some vendors offer formal courses; others rely on informal knowledge transfer. Whichever you choose, the end result should be a documented class that personnel from your company can teach. That way, you will not need to pay the vendor to provide training when you expand your operations or your personnel change.

Training on new work processes will be provided by your team. Again, it is important to end up with a documented class that personnel from your company can teach in the future.

The cost of training will be a price negotiated with your vendors. To that, you must add the personnel cost of creating, updating, and delivering the training thereafter. We recommend that training be repeated in two-year intervals due to technology advances and staff turn over. Figure 9-13 illustrates how to calculate and display the cost of training.

9.2.2.7 Maintenance

The cost of maintaining the RFID application is significant and includes spare antennas, readers, controllers, sensors, and actuators. It also includes maintenance contracts on the equipment and the middleware and application software. Maintenance generally costs from 15 percent to 20 percent of the original cost of an item. You may include costs of internal application support in this item, as well.

Training	Factors	2006	2007	2008	2009	2010	Comments
Reader Maintenance and Configuration		$3,500		$3,500			
Application Software Configuration		$5,000			$ 2,000		
Work Processes	$	4,000	$ 2,500		$ 2,500		
Total Training		$12,500	$2,500	$3,500	$4,500	$0 $	23,000
Year		0	1	2	3	4	
Interest Rate		0.05	0.05	0.05	0.05	0.05	
	$	12,500 $	2,381 $	3,175 $	3,887 $	-	each year
Present Value	$	12,500 $	14,881 $	18,056 $	21,943 $	**21,943**	cumulative PV

Figure 9-13
Cost of training

Pallet tagging for one distribution center		2006	2007	2008	2009	2010	Comments
Cost of tags							
Pallets per year		38,400	57,600	76,800	96,000	115,200	
Defective Tags Percent		5%	4%	3%	2%	2%	
Defective Tags		1,920	2,304	2,304	1,920	2,304	
Tags Required		40,320	59,904	79,104	97,920	117,504	
Cost per Tag	$	0.25 $	0.15 $	0.10 $	0.05 $	0.03	
Labor Cost for tagging		33,600	49,920	65,920	81,600	97,920	
Cost of tags		10,080	8,986	7,910	4,896	3,525	
Total Tag Cost		43,680	58,906	73,830	86,496	101,445	
Tagging Cost per Pallet		1.14	1.02	0.96	0.90	0.88	
* Labor at $25.00 / hour	$25.00						
Average tags per minute	0.5						Two min per tag
Year		0	1	2	3	4	
Interest Rate		0.05	0.05	0.05	0.05	0.05	
	$	43,680 $	56,101 $	66,966 $	74,718 $	83,459	each year
Present Value	$	43,680 $	99,781 $	166,747 $	241,465 $	**324,925**	cumulative PV

Figure 9-14
Cost of tags

Maintenance costs have been included in various sections of this model. Thus, for example, the maintenance of the readers and antennas is included in Figure 9-10.

9.2.2.8 Tags

The cost of tags is probably the largest single cost in the project. This is particularly true if you are using UHF passive tags that remain with the product or the packaging as it goes to the consumer and are destroyed or consumed, not recycled. Figure 9-14 shows a tag failure rate of 5 percent and declining, but as of early 2006, manufacturing techniques give you approximately a 5 percent failure rate for tags, but this will come down in the next few years, as will the cost of the tags themselves.

9.2.2.9 Business Process Change

Business process change is a difficult cost to estimate, but you should consider the following elements:

- Cost of workers and consultants studying and documenting as-is processes
- Cost of workers and consultants designing and building new processes

- Cost of training workers on new processes (don't double-count this item in the training section)
- Loss of productivity as new process becomes effective

In some ways, this is a soft cost, since it is so difficult to quantify. Section 9.2.1.1 describes one method for estimating the value of soft benefits and soft costs.

As shown in Figure 9-15, this entire cost occurs during the first year, so we show this as a one-time cost.

9.2.2.10 Downtime

During installation, you may impact the ability of the facility to perform its work, and this will be a cost to the company. Your company may provide a "standard cost" for each hour of facility usage. In our model shown in Figure 9-16, we used $80 per hour. We thus assign a value to each hour

Business Process Change	Factors	2006	2007	2008	2009	2010	Comments
Studying As Is Processes		$4,000					
Designing New Processes		$5,000					
Training on New Processes	$	2,500					
Initial Loss of Productivity	$	15,000					
Total Business Process Change		$26,500					
Year		0					
Interest Rate		0.05					
Present Value	$	26,500	$ -	$ -	$ -	$ -	only first year
Present Value Cumulative	$	26,500	$ 26,500	$ 26,500	$ 26,500	$ 26,500	cumulative PV

Figure 9-15
Cost of business process change

Downtime for Installation	Factors	2006	2007	2008	2009	2010	Comments
Receiving docks		2	3	6	6	7	24
Storage Rooms		0	1	2	3	3	9
Dock downtime for installation		$1,200	$1,800	$3,600	$3,600	$4,200	4 hours, $150 per hour per dock
Store-room downtime for installation		$0	$400	$800	$1,200	$1,200	4 hours, $100 per hour per store room
Total Cost of Downtime		$1,200	$2,200	$4,400	$4,800	$5,400	
Year		0	1	2	3	4	
Interest Rate		0.05	0.05	0.05	0.05	0.05	
Present Value	$	1,200	$ 2,095	$ 3,991	$ 4,146	$ 4,443	one time cost for each dock and storeroom
	$	1,200	$ 3,295	$ 7,286	$ 11,433	$ 15,875	cumulative PV

Figure 9-16
Cost of downtime

Figure 9-17
Cost summary

Cost Element	Present Value
Readers for one distribution center	$ 239,757
Application Software	$ 213,750
RFID Middleware	$ 137,500
Training	$ 21,943
Tags	$ 324,925
Business Process Change	$ 26,500
Downtime for Installation	$ 15,875
Total	$ 980,249

of the lost time and estimate the time. If you do not have a standard cost, see Section 9.2.1.1 for guidance.

9.2.2.11 Summarizing Costs

For each cost, we have estimated its amount over the five-year time span. Then we used the present value calculation to reduce each cost stream to a single number. Those numbers are displayed in Figure 9-17.

Costs become more difficult to estimate as the project grows larger. One way to manage the costs is to keep each phase of the project small enough to give you confidence in your numbers. Start as small as possible and iteratively grow each application. Start with a single product, one or two portals within a single distribution center, and just one or two stores. If you are tagging to meet a mandate, start with as constrained a project as possible to meet your requirements, and then build on your success by adding similar items. Complete each iteration with a rigorous "lessons learned" session and apply them. Make changes to your project ruthlessly until you have a reliable working model. In this way, your exposure to runaway costs is minimized. Also, as described in Chapter 10, plan and document rigorously before executing.

Many times, however, senior management will want to see the entire project cost before they approve it, so you will probably have to estimate the entire project and get approval for the larger number before you begin.

9.2.3 Risk

Any technology project carries risks, and an important part of the project manager's job is to manage these risks. A statement of risk is an important part of the business justification process. Consider these risks:

- Company reputation might suffer.
- The technology might not work as advertised.
- The project might take longer than expected.

- Unforeseen costs might arise.

- Workers might be unable to execute as required.

- External data sources might be insufficiently accurate.

- Successful pilot projects might not scale.

- Interfacing with existing operating systems may be more costly and time-consuming than originally estimated.

The preceding list is a representative sample of risks you may document. For each risk, describe steps you plan to take to mitigate and manage it. As described in Chapter 10, track these risks and mitigations as part of your project management process.

For business justification purposes, you can assign a risk factor to each application. Evaluate each risk by assigning it a weight and a value. Weight each according to its importance, and value it according to its severity. Rate importance on a scale of 1 (low) to 5 (high), and the risk itself as a percentage. Work with your team members, including vendors, to assign factors. As shown in Figure 9-18, multiply the percentage by the risk value and get a risk factor. Divide the total by the sum of the importance factors to express a risk factor for the project. As we observe here, the risk factor is 22 percent, which is high. But 40 percent of the risk comes from the external data sources risk, which can be managed. Cost and schedule risk are important, but they also can be managed.

One factor surrounding risk is particularly important. Adjust your risk factor by considering to what extent you have access to a model, an application as similar as possible to the one you are contemplating. If you have someone on your team who has done exactly this application (the same types of RF interference, the same opaque materials, configuration of readers and antennas, vendors, applications, network standards, degree of external data source problems, and so on), or an industry-standard document describing how to do it, your risk is considerably less.

Figure 9-18
Risk factors

	A Importance (1-5)	B Risk Value	C Risk Factor A*B	D Comments
Risk				
Company Reputation	5	3%	0.15	*ultra important, not likely*
Technology Risk	3	10%	0.30	*important, can be managed*
Schedule Risk	2	30%	0.60	*less important, harder to mange*
Cost risk	2	30%	0.60	
Unable to execute	3	10%	0.30	*important, not likely*
External data sources	4	75%	3.00	*very imporant, very likely*
Pilots not scale	4	3%	0.12	*very important, not likely*
Total	23		5.07	*divide by 23*
Risk			**22.04%**	

Take the risk factors, multiply them, and normalize the number to a percent scale. For example, the technology risk, risk that the technology might not work as advertised may be relatively important (5) but low (10 percent). So you might assign it a weight of 3 out of 5, but a value of only 10 percent, as shown in Figure 9-18. This generates a risk factor of .30 for the technology risk. Dividing the total of your risk factors by the sum of the importance factors normalizes the risk factor for the project and gives you a factor of 22 percent.

9.3 Determine Priorities

The prioritization process consists of evaluating all the applications that you reviewed and ranking them in order of attractiveness. This means quantifying the factors of benefit, cost, and risk. You can achieve a guideline for prioritization by calculating a prioritization factor, the risk-adjusted ROI.

The risk adjusted ROI is calculated as follows:

1. Calculate the present value of the cost stream. Use the same method described in the introduction to this chapter.

2. For each application, divide the benefit by its cost to come up with a percentage. That percentage is the raw ROI.

3. Multiply each ROI number by the inverse of the risk factor. Thus, if your risk factor is 22 percent, multiply the ROI by 78 percent. This is the risk-adjusted ROI (RA ROI), as shown in Figure 9-19.

4. Rank the investments according to the risk-adjusted ROI.

You can see from the chart in Figure 9-19 that Case 101 is the superior choice to begin your project. It has the highest risk-adjusted ROI. The other projects offer good ROI as well, but note that a good portion of the benefit of Case 101 lies in the benefits of cross-docking. This illustrates that the payoff from the RFID comes from the improved business processes, not the technology itself. In fact, the cost avoidance from each pallet that can be cross-docked is so great, that the company should do everything it can to maximize the number of pallets that are cross-docked. Other steps include closer coordination with shippers and customers and faster flow of information so your suppliers can assemble pallets specifically for your customers, increasing again the number of pallets that can be cross-docked.

Figure 9-19
Risk-adjusted ROI

	Case Name	Benefit	Risk	Cost	RA ROI	Benefit Assumption
101	DC receive and ship a pallet with crossdock	$ 3,809,023	22%	$ 1,065,870	279%	warehouse cost is $80/hour and we'll be able to crossdock 40 pallets per day
105	Improve maintenance of vehicles	$ 3,333,333	10%	$ 1,423,147	211%	add 6 months' life to 100-vehicle fleet
103	Reduce shrinkage of DVDs	$ 1,656,000	10%	$ 957,110	156%	each month about $30,000 in cost
102	Reduce stockouts of paper goods	$ 1,700,000	20%	$ 921,110	148%	Paper is out of stock about 10% of the time, and our paper revenue is $34 million.
104	Track and Trace laptop computers	$ 22,080	5%	$ 20,500	102%	we spend about 2 man days per month keeping track of who has which of 600 laptops

Figure 9-20
Milestone chart

Business Process	RFID Pallets with Cross Docking	
Milestone		
1	Description	Pilot for 2 dock doors
	Time	8 weeks
2	Description	Expand to include 2 more dock doors
	Time	8 weeks
3	Description	Install, configure and integrate middleware
	Time	5 weeks
4	Description	Expand to include 8 more dock doors
	Time	8 weeks
5	Description	Repeat pilot in second Distribution Center
	Time	12 weeks
6	Description	Expand to the rest of second Distribution Center
	Time	12 weeks
7	Description	Pilot third distribution center
	Time	4 weeks
8	Description	Pilot third DC
	Time	Expand to rest of third DC

9.4 Create Milestones

The last step of the business justification process is to identify project milestones you expect to achieve on each application. For each milestone, give its description and duration.

Figure 9-20 shows a milestone chart for Business Case 101, assuming your project is to install systems in three distribution centers; each center has 24 docks.

9.5 The Controversy Around ROI for RFID

Throughout 2005 and 2006, the press carried articles about ROI for RFID. The following is a summary of points that various authors have made regarding the return on investment for RFID:

- There is no discernible return on the RFID investment.
- The technology is too new to support substantial investment.
- The benefits all flow to the retailers while manufacturers bear the costs of RFID implementations.
- Most of the benefits of RFID can be achieved by other, less expensive means.
- The logic of ROI calculations cannot establish a value for a disruptive technology like RFID. Instead, one author wrote, you should value the *options* that RFID enables and use those figures to determine the return on your RFID investment.
- Some writers have theorized that many modern manufacturers are so optimized in their operations that the benefits of RFID will not be substantial. Earlier in this book, we said that the most advanced operations are the ones to benefit the most from an RFID installation. Companies whose level of performance is low will face obstacles such as poor data quality, suboptimal ordering and stocking policies, and a culture of secretiveness that will reduce the gains available.
- Nearly all experts agree that slap-and-ship strategies are simply an additional cost, with no significant ROI. They all agree that any benefit from these RFIDs arise from being able to use it to generate useful information and then to act on it. Several authors, however, urge readers to begin with slap-and-ship applications and then build on them to start to generate useful information.

Let's look at each of these points in detail.

9.5.1 No Discernible Return

When people take a narrow look at the technology, it is difficult to find a return on the investment. The return comes from using it to generate data that enables productive changes in the business processes.

Business processes are made more productive by removing or automating steps and by executing the steps more quickly and more reliably. Even greater productivity improvements can be gained by replacing the processes with new ones that reduce critical cycle times for the entire enterprise.

It is not often discussed today, but a new idea took root in manufacturing thinking in the early 1980s, and a review of its history is useful. Before this idea took root, the drive for improving manufacturing productivity focused on determining formulas for the Economic Order Quantity (EOQ). This was important because it balanced the advantages of small quantities against the costs of numerous setups. Setups took hours to perform and took the machines and facilities out of production. Reducing this down time was the most manageable way to reduce costs. Longer production runs reduced the costs of setups but drove inventory costs higher.

In the early 1980s, the introduction of Single-Minute-Exchange-of-Dies (SMED) techniques allowed manufacturers to slash setup times. This redefined a core assumption on which EOQ-based production planning was based. It is unlikely that ROI models at the time were successful in capturing the value of this new technique, but in retrospect it enabled an entirely new approach to production control (lean manufacturing, pull/kanban-based production, and so on). Companies that embraced it were able to build substantial competitive advantages until their competitors either embraced it also or went out of business. Similarly, RFID will transform supply chain management by enabling new levels of cooperation and coordination, if the companies use it to build new processes. This generates massive return, but it is unlikely to be captured in ROI models.

9.5.2 Technology Too New

Some experts say that RFID technology is not new but many of its current applications are. The claim arises as an excuse to cover situations where vendors were insufficiently prepared for their pilots, practiced hit-or-miss tag positioning techniques, or failed to deal with ambient radio interference at the read sites. It also is heard when the tags have a high failure rate or are incapable of functioning in a particular environment. New tags for particular situations are constantly being developed.

There is a certain amount of discovery going on in the RFID marketplace right now. For example, companies are questioning whether placing readers at all the portals of a distribution center is the optimal strategy.

Instead, the same benefits may be gained in some circumstances by placing mobile readers on the forklifts and trucks that move goods around and putting tags at the portals so the system can still record the goods' locations, at a fraction of the cost.

The existence of experimentation and uncertainty should not stop companies from adopting and learning the techniques and systems of RFID. Companies that put it off too long will find themselves at a disadvantage as their more adept competitors begin to see the results of their successful implementation.

9.5.3 Benefits to Retailers; Costs to Manufacturers

The idea that RFID would provide benefits to retailers but costs to manufacturers was first advanced by Douglas MacDonald, who was then with the A.T. Kearney Company. He reported that retailers would see reduced inventory, generating a one-time cash savings of 5 percent of total inventory. They would see an annual reduction of 7.5 percent of warehouse staff and expenditures. In addition, they would benefit about $700,000 per $1 billion in sales from reduction in out-of-stock conditions. Their cost, the report said, would be about $400,000 per distribution center and $100,000 per store. Most importantly, this cost would be nonrecurring.

Manufacturers, he went on to say, would have the same equipment costs, but they would also have the ongoing cost of the tags, which the retailers would not bear. Manufacturers could see better tracking and inventory visibility, enhanced labor efficiency, and improved fulfillment.

The other set of manufacturers' benefits depends on retailers providing manufacturers more detailed information regarding out-of-stock items, inventory, and unsold items. This requires a higher level of information sharing and process collaboration than is currently practiced.

For both these reasons, the benefits would accrue primarily to the retailers, while the manufacturers would be stuck with the costs, according to this argument.

Since this argument was made, many manufacturers have held off large-scale implementation. But the public successes of such major manufacturers as Kimberly-Clark, International Paper, and Proctor & Gamble have made it clear that the technology can deliver profitability if it is accompanied by strong management and a determination to utilize the data RFID produces.

9.5.4 The Benefits Are Achievable by Other Means

It is probably true that any single benefit promised by RFID technology can be delivered in some other way. It is not true, however, that any system can provide all the benefits with a single investment. RFID promises to enhance revenue by reducing out-of-stocks, improve security by reducing shrink, speed up critical cycles, and lay a foundation for very productive collaboration. In addition, RFID generally improves visibility, which often leads to unexpected gains. Also, getting the company ready for RFID may involve data cleansing and process re-engineering, which should provide a whole other set of cost reductions.

The business justification process can give you data to deal with this issue as you compare it with the investing in multiple applications.

9.5.5 ROI Methodology Cannot Represent Disruptive Technology

The idea that ROI methodology cannot represent disruptive technology was advanced by Michael Witty of *Manufacturing Insights* magazine. Witty observed that the ROI model failed to show adequate returns, which led companies to adopt low-cost slap-and-ship strategies that limited their ability to capitalize on the technology in the future.

In order to avoid another round of throwaway solutions, Witty suggested that companies abandon the ROI model and adopt instead an options-valuation model. This would require them to look beyond the technology and enumerate a number of possible applications. They would then assign a value to each of these options, using an options-valuation technique such as the Black-Scholes model. Other writers have observed that if the executive team is not comfortable with Black-Scholes, a simple decision-tree model will suffice.

As we have shown in this chapter, the ROI methodology can capture some portion of the justification case for RFID. In any event, it is a useful guidepost. The ROI model is very flexible in what you include as costs or benefits, so you can reflect the gains and costs of any application, as we have shown. Many corporations are wedded to the ROI model, and their executives understand it.

9.5.6 Manufacturers Too Optimized to Reap RFID's Benefits

This argument was made by MacDonald when he observed that most manufacturers are "well beyond the basics" when it comes to supply chain efficiency, and there might not be much left to gain from these benefits.

A parallel argument describes adoption of smart cards in the United States. Experts observe that smart cards are in wide use throughout Europe and Asia but are struggling for a foothold in the U.S. The reason, they say, is that the United States has an effective network for credit card transactions based on existing technology. So any additional convenience for the customer by adopting a smart card is small. Similarly, MacDonald seemed to argue that most manufacturers have highly effective systems, so any incremental benefits from RFID would be limited.

Many writers are less sanguine about the condition of most manufacturers' supply-chain management systems. They point out the high incidence of out-of-stocks, shrinkage, loss, spoilage, counterfeiting, fraudulent returns, and so on as evidence that MacDonald is too optimistic. There is, they say, substantial room for improvement. When companies as sophisticated as Alcoa and DuPont report serious revenue shortfalls due to disruptions from hurricanes, it seems to suggest that there is room to improve the logistics systems' performance of even the most sophisticated companies.

In Chapter 10, you will learn to evaluate the level of performance of your company's supply chain. If your company has robust, effective systems that hold inventories, costs, errors, and losses to low levels and foster productive levels of coordination, it may be that RFID might provide a limited benefit to your company. Experts suggest that most companies are not so fortunate.

9.5.7 Slap and Ship

Slap-and-ship is fully discussed in Chapter 11.

9.6 Chapter Summary

The process of business justification investigates the costs and benefits of any project in a methodical and disciplined manner. It reduces complex expectations, benefits, and costs to a single defensible number that enables any project to be compared with any other project. Even if you are going to recommend the project based on a customer mandate or regulatory requirement, the business justification process should be executed in full. It will enable you to target your project at the most productive applications and set the stage to manage the project with the least amount of risk.

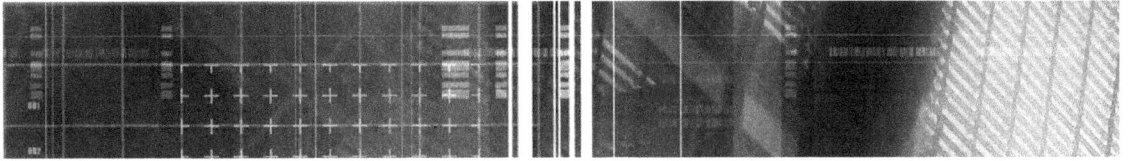

CHAPTER 10

Plan Your Project

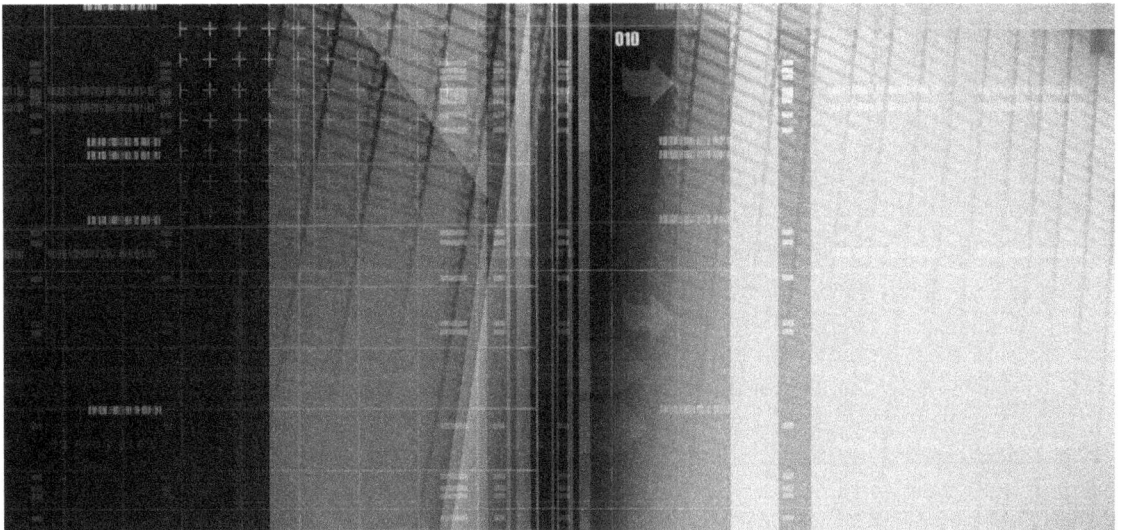

In general, good project management is simple to explain and devilishly difficult to put into practice. The requirements are clear and straightforward, but each company is different and the circumstances around such simple advice as "get executive support" or "select the reader that can read all your tags" are different in each case. Nonetheless, the rules are tried and true for all significant projects, and it's worth taking the time to get them right before moving on to equipment purchases and configuration.

Project management is an iterative process, and this outline reflects that. This project has four phases. During Phase 1, you make assumptions and estimates that will be confirmed, refined, expanded, validated, and revised during subsequent phases. Each document will be reviewed and potentially revised for each phase of the project. This project planning methodology enables you to revise your assumptions as more precise information is developed. Good project planning may take up to 40 percent of the entire project timeline. Often senior executives are impatient to start seeing readers installed and tags flowing by on the conveyor belt right away and may have little patience for the time spent in planning. But time spent getting the planning right will pay off in an enhanced chance for a successful project. The mechanism is this: first document what you're going to do, and then do it!

It is said that the great director Alfred Hitchcock rarely visited the sets on which his motion pictures were being shot. His planning and documentation were so thorough that the cameramen, technicians, actors, and actresses could make his great movies from the documents and he did not have to be there. Your documentation should be similarly complete, clear, and sufficient.

Four types of documents must be produced, and the document preparation process is built to include stakeholders to make sure their needs are met so you can benefit from their insights and contributions.

For any phase of the project, the first type of document is a list, and the process of creating a list is called *enumeration*. The list document names the elements to be documented. It may include short descriptive information that has been gathered as the enumeration was conducted, or it may include full-scale descriptions. For example, the Objectives Document is a list and includes all we need to say about the objectives. Considerable investigation may be required to create the list. For example, we enumerate (list) the workflows that will be impacted by first the pilot, then the technical integration, and then the rollout. The list document may include a map or a diagram of how the listed elements are related to one another. It may also include a statement as to why an item was included in the list.

The System Requirements List (SRL) enumerates everything that will be needed to complete the project.

The second type of document is a "description document." This document contains a section for each item listed in its corresponding enumeration. Once you have an enumeration approved, you can begin to document its descriptions. Description documents may contain a list of activities to be performed as you prepare the enumerated items for their role in your project. Most description documents are created for the use of the project team. The team will create and use description documents to plan and execute the activities. Some description documents, such as the Reader Onsite Documentation, are finished products of the project.

The third type of document contains protocols that describe how to perform the activities required. We have provided several protocols in this chapter, Chapter 11, and Chapter 12. You may wish to include a copy of the protocol for each document at the beginning of that document, for easy reference. As the project progresses, you will probably add your own steps, notes, cautions, and clarifications to the protocols.

The fourth type of document is a schedule that arranges the various activities, including document preparation, along a timeline so you can see when activities need to begin and end. The Project Plan is a schedule.

This chapter arranges the project into four phases. Business justification precedes Phase 1 and must be completed and approved before launching the project; it was covered in the previous chapter. Phase 1 of the project is the Preliminary Planning Phase. This is where you build on the work you did during business justification to create the recommendations and documents that will frame the rest of your project. Phase 1 includes assembling and briefing the Executive Steering Committee, assembling the initial project team, setting project objectives, and developing a testing strategy. Phase 1 will encompass some level of planning, but that should not be confused with the detailed planning that the later phases will require. Detailed planning will be done during Step 1 of each successive phase.

Figure 10-1 shows the phases and steps of the RFID project plan.

Phases 2, 3, and 4 are each broken down into two steps. Step 1 is the detailed planning step. It encompasses interviews, research, investigation, and above all, documentation. The success of this project depends on careful documentation so that activities are planned and necessary information is collected, available, and accessible when needed. Step 2 is the execution step. This is where the planning and documentation pay off: execution is where the visible work gets done. At the end of each phase, you will review your activities with team members, relevant stakeholders,

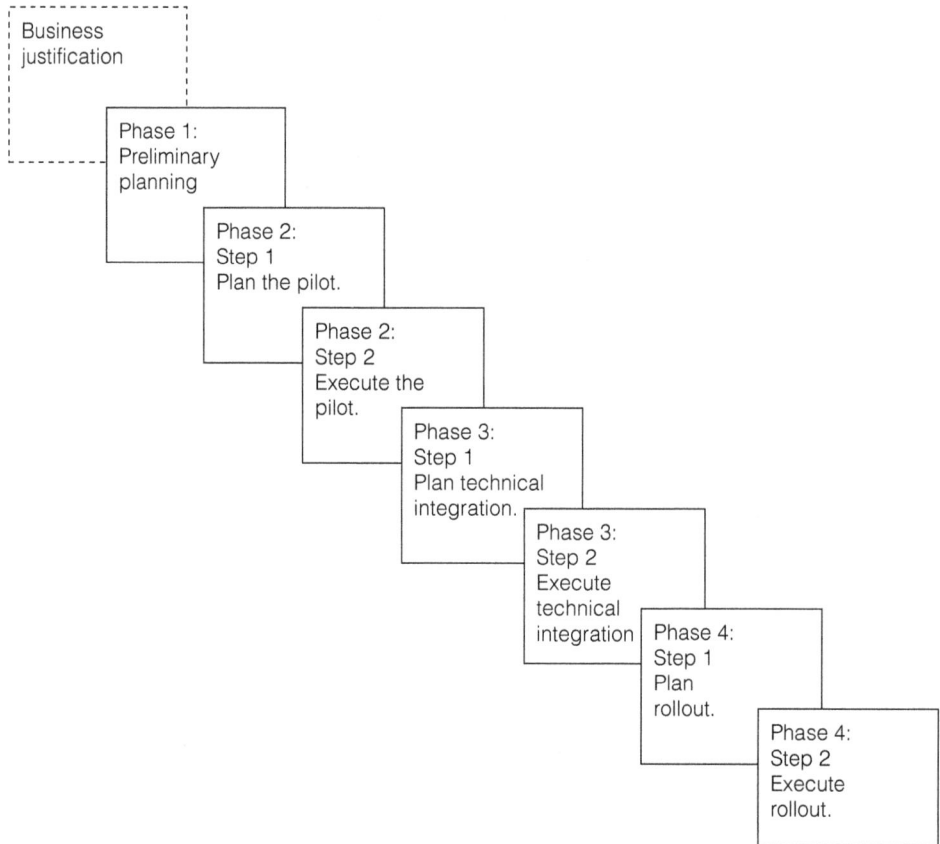

Figure 10-1
Project plan overview

and ultimately with your Executive Steering Committee (ESC). You will assemble preliminary information for the next step so your presentation to the ESC can include both the evaluation and your recommendations for the phase to follow.

Phase 2 consists of assembling your core team and then planning and executing the pilot. We've provided a detailed list of the skills you will need on the team. Phase 2 begins by arranging for those skills to be available for you. It then lays out the analysis, planning and preparation, and documentation you will need to accomplish before you can begin execution. Planning considers facilities, products, and readers. It also considers systems, workflows, data flows, and network impacts. It prepares for testing, selection, procurement, installation, and testing again.

NOTE *As we use the term* reader *in this chapter, we also include* encoders *and* printers.

During the second step of Phase 2, you will execute the pilot. This is where you test the ability of your team to deliver a functioning system. Then you will conduct a review of the pilot and compile your lessons learned.

Phase 3 will interface the reader network (actually, only one or two readers) to the enterprise network and connect the tag data to a single application, both in a scalable fashion.

Phase 4 is the final rollout. This is the production part of the project. If your previous phases were executed properly, rollout will scale the system as necessary to accomplish the project objectives and deliver business value.

Before you begin, review the implementation notes in Section 10.9. These notes are valuable tips that other project managers have learned the hard way and will keep you from making many common mistakes.

In the actual project, you may find yourself going through these steps several times, as you approach a new application or a new facility. In each case, follow the essential methodology of plan, pilot, rollout, and evaluate. When you start in a new facility, test the RF environment, and then plan, pilot, rollout, and evaluate. Wal-Mart has reportedly been able to RFID-enable an entire store in less than six hours, executing each of these steps each time.

10.1 Phase 1: Preliminary Planning

The purpose of Phase 1 is to build on the business justification process to create a set of documents and recommendations describing what it will take to achieve a successful RFID project for this company. The activities prescribed here will assemble a team and generate documents and information to support the recommendations you will make. It will also show you how to determine project costs so you can get approval for the project.

During Phase 1, you will need to provide the following deliverables:

- Designation of an Executive Steering Committee (list)
- Designation of an Executive Sponsor
- An RFID Assessment Team (list)
- An Objectives Document (list)
- A Project Scope Document (list)

- A Preliminary Budget Document (list)
- Touch-points and current performance levels (list)
- A testing strategy recommendation (description)
- An initial project plan (schedule)
- A selection of an operating frequency for your project (description)
- Initial recommendations (description)
- Executive support and sign-off

10.1.1 Executive Steering Committee (ESC)

The Executive Steering Committee generally consists of three to five executives (I have seen it work well with two) who manage and approve the project's final budget. The executives who have approval authority for projects will designate the members of the ESC. It is advisable to give the ESC members an overview of RFID technology and applications at the beginning of your project. How you interface with this committee depends on the protocols the company has for tracking major projects. In most cases, the committee receives regular (weekly or monthly) summary reports and notification of any important changes in the project. They also receive regular (monthly or quarterly) briefings from the Project Manager or Executive Sponsor. All contact with the ESC should be coordinated through your Executive Sponsor.

10.1.2 Executive Sponsor (ES)

The Executive Sponsor is a single executive the project reports to. The ES has day-to-day supervision of the project, helps the project manager make decisions, mediates issues with other departments, arranges for needed support, and keeps the project integrated with the rest of the company's operations.

10.1.3 RFID Assessment Team

A key element of your project's success is the membership of the RFID team. In Phase 1, you need to work with a group of people already interested and motivated by what RFID can offer the business. Find individuals who are excited by the project and capable of conceptualizing the changes that will need to be made and how the company can benefit. They should be leaders who can evangelize the project to their own departments and beyond. The team will prepare the initial justifications and documents as described next. These members will, in all likelihood, form the core of your team going forward. When you assemble your core team in Phase 2, you will add members who have particular skills and experiences.

10.1.4 Project Objectives

The Objectives Document must be as specific as possible. Accenture calls this process "Value Targeting," and they work with clients to categorize and identify the most promising types of applications and benefits for their particular company. One poll identified the following objectives, along with the percentage of respondents selecting each objective as the most important:

Improve inventory accuracy	55 percent
Trading partner requirement	12 percent
Increase inventory turns	10 percent
Reduce out-of-stocks	9 percent
Enhance supplier relationship	9 percent
Improve fill rates	4 percent

Intermec lists the following benefits companies have derived from the technology. You may wish to consider some of these as objective statements for your project. Or, consider each for the "out of scope" section discussed later.

- Smoother-running business environments
- Knowing precisely where inventory is
- Increased throughput and efficiency
- Shorter order cycles
- Faster shipping
- Better inventory management
- Reduced labor costs by reducing the workforce needed for bar coding, tracking, inventory management, and other

- Increased revenue
- Higher profits
- Better customer service
- Greater customer satisfaction
- Reduce stock-outs
- Reduce shrink

Other companies have identified increased sales as a goal of their project. The consulting firm A.T. Kearney predicted manufacturers would see sales gains in the range of 0.7 percent, based on improved product availability attributable to RFID systems. Another consulting firm, Accenture, predicted sales gains between 1 and 2 percent. Both noted that these incremental sales are much more profitable than revenues gained from promotions, new channels, or product introductions.

If the purpose of the project is simply to comply with a customer's mandate, and your management has decided not to pursue business advantage through RFID, this drives a much simpler project. If the purpose is to gain competitive advantage and improve business performance, the payoff will be higher. But so will the costs, and the effort will be greater. About 30 percent of Wal-Mart's suppliers reportedly took the mandate as an opportunity to upgrade their business performance. In the Objectives Document, include a general explanation of how your team will view this Objective in this project. Detailed explanations will be included in the Benefits Document.

If you are planning a closed, asset tracking application, the following objectives list may be helpful:

- Improve asset utilization
- Reduce time spent locating lost assets
- Reduce cost of assets
- Improve maintenance of assets
- Reduce loss, theft, or misrecording of assets

10.1.5 Project Scope Document

The Project Scope Document is a description of the project's boundaries. Will this project be inter-company, intra-company, company-wide, division-wide, include only one or two facilities, or confined to a single facility?

Will it focus on a single problem to solve, or will it scale to an enterprise installation? Will it encompass receiving and shipping, routing and put away, or only shipping? Is the outbound side of the system limited to just producing readable tags, or will it provide Advance Shipping Notices and update your Warehouse Management System, your ERP system, your billing system and any others? Each of these applications will raise data quality, software upgrade, and integration issues that your project will have to address.

Is it an open system, interfacing with entities outside your company, or is it closed? Is there an inbound side to the project? What systems will be fed RFID data? Will this project encompass only a few of the products or all of them? Will it replace or augment an existing bar code application? Will you be tagging pallets or cases, or will you be tagging to the item level? Will the tags have to be read by any entities outside your project? Will the project have to read tags provided by anybody else? Will the project only write tags once, or will it have to update tags at various steps in the work processes?

Nearly every project is threatened with what is called "scope creep." This means that there is pressure, once the project starts, to increase its scope. The scope creep is rarely accompanied by increased budget or timelines. (One solution is to start with a very limited pilot. We have suggested something like tagging only cross-docked pallets that are not broken down). But your best defense is a strong scope document that has been approved by senior management early in the project.

The Project Scope Document should list assumptions and dependencies on which the project depends and identify the gaps that other departments need to fill in order for you to be successful.

The scope document should contain a clear list, as complete as possible, of elements that are "out of scope." It should also list prerequisites that may impact your project, such as the need for high quality data, synchronized data, network infrastructure, storage, power, application software purchases, and upgrades, etc. Consider also features or services that you could be asked to provide but will not consider in this project. Getting agreement and sign-off on this list at this time makes a successful project much more likely.

The project scope should also include a cut-off time for the project to end.

10.1.6 Preliminary Budget Document

The preliminary budget is a list of items that will require expenditures and estimated amounts. It will give you a rough-order-of-magnitude estimate of the cost of the project. The Preliminary Budget Document is the beginning of a budget for the project. Cost items to include are

- Salaries of the team members for the duration of the project.
- Fees for consultants.
- A headcount of permanent additions to the company, if necessary.
- Equipment costs; you may rely on your vendors and project partners for some of this information.
- Estimates for any software, consulting, and integration services you will require.
- Network upgrades.
- Additional software.
- Additional servers and storage.
- Any additional IT manpower.
- Planning for equipment upgrades as the project progresses.
- Travel and lodging.

Present the budget broken down by phase of the project.

10.1.7 Touch-points

Touch-points are those points in your company where the RFID system will touch existing workflows, facilities, and systems. Each touch-point must be examined to determine the impact it will have on your project and the impact your project will have on it. This is an opportunity to assess current performance and determine how RFID technology can make it better. Just listing each touch-point gives you an indication of the scope of your project. Successful projects use this as an opportunity to get the team to work together and to reach out to various stakeholders. The touch-points are organized into three areas:

- Workflows
- Facilities
- Systems and networks

Workflow touch-points are documented in process maps you may have, showing the relevant workflows. Your team members will need to work with the stakeholders and managers to gather performance statistics and suggestions about ways to improve the processes using RFID. Work with your assessment team to enumerate as many process touch-points as you can. The result will be the initial Workflow List and will be substantially augmented as the project proceeds. Include in this list document the reason each process was included.

The second area for touch-points is the facilities. If one is available, get a map showing your company's various facilities. Create a document consisting of an initial list of the facilities that will be involved in this project. Facilities will include factories, warehouses, distribution centers, equipment yards, and office buildings. The Initial Facilities List should list each of the facilities, as well as the following information for each facility to the extent that it is known:

- The reason the facility was included
- The location of the facility
- If and how the facility connects to the corporate network
- Contact information for the manager responsible and any other relevant personnel
- What workflows are executed at the facility
- An estimate of the number of readers projected for the facility
- Any construction you may need for the facility
- Any required upgrades to the facility's communication capabilities
- Any known issues

This Initial Facilities List document will be updated prior to rollout. You will need to choose and recommend a single facility for the pilot, so consider the preceding questions as you are preparing this document. For the pilot, we recommend that you select a facility in the same city as your own base of operations. This will enable you to be very thorough in your testing and assessments at a lower cost for travel and logistics. In selecting a facility, keep in mind that in the technical integration phase you will need to commit to some number of business value statements at this location. Make sure your facility selection will support them; see Section 10.4.2.6.

The third area for touch-points is your network and systems infrastructures. Make a preliminary list of each network component, computer system, and application that will be touched by your project. Determine where RFID's data flows will impact network performance, system performance, and storage.

10.1.8 Testing Strategy Recommendation

Your project will need a disciplined way to test spaces, readers, antennas, and tags. As the project manager for this RFID project, you will need to make a recommendation regarding testing strategy for the project, so you need to tell how the project will test:

- Readers and tags, so you can select equipment that will work in your various configurations and situations.
- Facilities and reader locations, so you can document and remove sources of RF interference before they become insurmountable problems.
- Products and packages, so you can determine where and how to attach tags to maximize reliability and read rates.

Chapter 12 provides detailed testing protocols for these scenarios. Each time you consider a new reader or tag, you will need a way to test it and see how it performs with your products without disrupting existing production. As you move into rollout, you will be installing in some number of facilities. You will need a disciplined protocol for assessing the level of RF interference in any location where you plan to put a reader. Each time you decide to tag a new package, product, or pallet, you will need to test various tags and tag placement strategies to determine how and where to place the tag so you can get the best read rates. Some companies look into the future and see an ongoing requirement, as new packages and products are constantly being released. Others see a more stable future, with a smaller number of package types and lower rates of change. The companies with an ongoing need should invest in their own test range; Chapter 12 describes how to build one. Other companies will want their own test range just for the project and will then move to an outsourcing model for the testing of their new packages as they come out. Still other companies outsource the function throughout the project. Chapter 12 lists several companies that provide this service.

10.1.9 Initial Project Plan

The Initial Project Plan may be a simple Excel Spreadsheet. It should lay out the phases and tasks. The Initial Project Plan will show overall timelines and the structure of the project. You will develop the detailed Project Plan for the pilot in Phase 2.

Testing Services

As described in Chapter 12, you will need to simulate various elements of your solution and tune its performance. Many service companies do this by trial and error. They set up a system in a spare room and tune the reader and antenna until it "works." This is potentially a very time-consuming and costly method. Its results are of dubious reliability, and when problems arise, there is little data to help find a solution. However, Odin Technologies and several other vendors claim to design and test systems in a scientific manner. Odin claims a deep understanding of the physics of radio waves. Another resource is the University of Wisconsin, which has set up a laboratory including an anechoic chamber, a room guaranteed to be completely free of electromagnetic interference. This environment allows their customers to simulate their system in a controlled environment for testing. This means a company can begin to tune their system in an interference-free environment and then introduce various external sources of interference to see how your system handles motors, fluorescent lights, external radio waves, and so on. Sun offers its customers use of a laboratory with a 600-foot-per-second conveyor belt and other warehouse equipment and facilities. The University of Kansas also offers commercial companies a testing laboratory. The testing should proceed from the sanitized facility in an orderly fashion to your own installation, in steps that are precisely defined.

10.1.10 Operating Frequency for the System

As you plan your system, you will need to make a decision about what frequency you will use for your system. The operating frequency of the system controls many of the subsequent decisions, so the decision is very important. If it is not clear from the mandate or industry practice (see Chapter 5), consult with industry experts, complete your investigation, and come up with a sound recommendation. Chapter 1 describes the performance characteristics of each of the available RFID frequencies.

The frequency decision is the most basic technical decision you will make, and it will enable and/or constrain many future options. Frequencies differ in read ranges, costs of readers and tags, ability to penetrate opaque materials, software support, ability to interchange tags with trading partners, and likelihood of future development. Where you have a choice, you will need to consider all these factors. Your executives may not know enough to care which frequency you select, but they will care about the consequences. Make this basic decision carefully, at the beginning of the project. If you are not certain of your selection at this point, narrow the choices as much as possible and make this decision before the start of the pilot. Given the critical nature of this decision and its long-term cost impacts, you may want to seek professional advice when making this decision.

10.1.11 Recommendations

The recommendations should describe what actions you recommend based on the work done thus far. If you recommend that the project go forward, your Objectives Document should constitute the criteria for project success. Be certain that your objectives and scope are consistent!

Recommend the scope for your pilot project. Your recommendation will establish the profile for your pilot. You should recommend the facility where the pilot should take place and which pallets, products, and cases are to be tagged during the pilot. List the workflows to be created or changed. Discuss any facilities issues to be dealt with during the pilot (issues of construction, cabling, power, RF interference, and equipment relocation). This will drive networking, integration, and development issues. Last, you should recommend a methodology for determining which RFID tags and equipment are to be purchased for the pilot. Chapter 12 describes protocols for testing your products and readers to find the ones that work best for you. As part of your discussion of the pilot, you should list any assumptions to be tested during the pilot and its success criteria.

10.1.12 Executive Approval

Your presentation to the Executive Committee should include as handouts the documents you have prepared:

- Objectives Project Scope
- Preliminary budget
- Initial Processes List
- Initial Facilities List
- Initial Systems List

- Testing strategy recommendation
- Initial Project Plan
- Recommendations

During your presentation, review the objectives, scope, and preliminary budget. Describe the Processes List, review the ROI projections. Explain your recommendations and request approval for the project.

10.2 Phase 2: Step 1, Plan the Pilot

One of the key reasons the RFID industry is underperforming global market expectations is lackluster performance by several major vendors in the pilot phase. Poor results for pilots result in project delays and terminations. Today, relatively few projects are approved past the pilot phase. Tragically, the vendors, and some managers, view the pilot as the time to develop their information base, engage for the first time with the project, and make their mistakes. The results are predictable. They begin the pilot with key decisions unmade, poorly analyzed, or undocumented; they bring the wrong equipment; and they spend too much time with trial-and-error methods of installation and tag placement. They try to do the work without the information they need to do a satisfactory installation. The installation takes too long, it costs too much, and the system performs poorly. People then blame the "immature" technology. We strongly urge you to allocate at least 25 percent of the project to the pilot. Take the time and allocate the resources to doing it correctly, as we have documented. The pilot is not the time to make your mistakes. It is the time to demonstrate your ability and professionalism. This section describes the work that needs to be done *before* the pilot begins.

The plan for the pilot will include most of the activities of the project itself. The set of workflows to be dealt with is limited, the range of packages to be tagged is smaller, the project will be limited to just a single facility and one or two readers, and the system interfaces will be deferred until Phase 3. But the RF issues will be just as demanding, and the analysis should be just as deep.

This phase involves three sets of activities:

- Assemble your core team
- Create the Initial Business Infrastructure Document
- Expand and revise the document set created in Phase 1

10.2.1 Assemble RFID Project Team and Staff

There are three parts to the team you will need to deliver a successful project: a core team of dedicated professionals, a staff to do a lot of the project work, and a project team of people who are interested or will be impacted by the project. Team members should be kept informed about the project, consistently involved in decisions, and solicited for knowledge, insight, and guidance as needed.

First, assemble your core team. Review the skills set listed in the next section, and make sure you have sources available for all these resources. For the project to be successful, you will need access to this knowledge and experiences.

10.2.2 Core Team

The core team consists of individuals who are interested in the technology, can represent and articulate the concerns of their departments, and can evangelize the project back to the important departments. Core team members may come from operations, sales, security, finance, information technology, and perhaps some others. Review the project objectives and make sure the impacted departments are represented on the core team.

The core team will need access to all the following skills. Where possible, try to staff this from people within your own company. Identify the people, the departments, and their projected level of involvement. You should check their availability. If you have headcount, you may be able to hire the expertise you will need. If your company is committed to RFID, it will be beneficial in the long run to have this expertise in the technology within your company. Be open to providing training to promising and enthusiastic individuals who may wish to be on your team. Identify any training resources you envision using. Source the following list of skills and knowledge you will need for your team:

- An experienced RFID project manager
- Current equipment knowledge
- Current best practices in your field
- Deep knowledge of the technology, science, and physics of radio communication, including antenna selection, placement, and tuning

- Deep knowledge of the state of the art in printing, tag placement, and absorption characteristics of various materials
- Sources of equipment and tags
- Standards and mandate requirements
- Your company's internal processes
- Your company's IT networks and standards
- Construction methods and techniques

10.2.2.1 Project Manager

The RFID project manager will have planning, coordination, and communication responsibilities for the project. His tasks will include detailed planning of the project, identification of risks and developing mitigating strategies, developing and executing a communications plan, building a work plan, defining work tasks, assigning personnel, building a schedule, estimating and tracking costs, and providing leadership for the project. The project manager is the key person for the success of the project.

10.2.2.2 Current Equipment Knowledge

It would be beneficial if your team has access to hands-on, real experience with a variety of readers, printer/encoders, antennas, tags (see Chapter 3), and test equipment. Within the standards, as we have said, there is considerable room for variance and an almost unlimited list of potential pitfalls. Someone who knows the quirks and capabilities of various manufacturers and models can be very helpful. If you plan to do your own testing, someone who knows how to work with signal generators and spectrum analyzers will be helpful as well.

10.2.2.3 Current Best Practices in Your Field

It is likely that someone has already attempted the application you will be building. This is not to say that they will have worked with exactly your combination of radio interference, metals, liquids, sodium or graphite, size issues with fitting labels on your packages, or distance issues in terms of getting the read zones right on that particular wrapping station. Nonetheless, a set of best practices is emerging and someone who knows them can save your project a good deal of wasted time rediscovering what works.

10.2.2.4 Technology of Radio Device Performance

The ability to design, source, create, and connect a well-functioning read site is invaluable. You will need someone who knows how to read contour diagrams; work with limited cable lengths; select antennas, connectors, baluns, and chokes; and position and tune the antennas. This skill is relevant if your project contemplates reading tags. If you are just setting up a tag-and-ship operation, it is not as important.

10.2.2.5 Printing, Tag Placement, Absorption Characteristics of Various Materials

Knowledge of how to select, configure, and position tags is important for most RFID projects. As we have said, absorption and reflection of radio waves is an important variable to manage, and many projects are struggling with it today. Determining optimum tag placement on your package scan can be outsourced if you do not anticipate continually determining tag placement strategies for your packages.

If you contemplate working with only a few packages, you can acquire these skills by outsourcing the site preparation activity.

10.2.2.6 Source(s) of Equipment and Tags

Sourcing decisions are always tricky, and there are numerous companies offering choices. Having someone on your team with recent experience with relevant vendors will be useful. Engage your supply chain management function to assist you in qualifying suppliers and in negotiating the terms and conditions of all hardware and software agreements.

10.2.2.7 Knowledge of Regulations, Standards, and Mandate Requirements

You will need access to someone who can leverage relevant standards to help you install a system that will work well with other parts of your enterprise, as well as with your trading partners and business partners. If you are responding to a mandate, it will be helpful if someone on your team understands relevant mandates and what issues have impacted other installations.

10.2.2.8 Knowledge of Your Internal Business Processes That Will Be Affected

Your team will have to be able to design RFID-intelligent solutions for some of your existing business processes. Even a simple tag-and-ship installation will create new processes for dealing with packages that need to be tagged and shipped. It will introduce new equipment, supplies, and activities into the packing process that may include selecting the right label stock, dealing with bad tags, and interfacing to the host computer. If your project is more ambitious, the impacts will be even more substantial.

The assembly of human resources for your RFID project is more challenging than other IT projects. The base of information needed can be very wide, and the technology base is wider. Your team will have to understand not just databases, networks, and server architecture, but also radio communication engineering. Even in the traditional IT realm, the challenges of potentially huge amounts of data pouring into the servers may raise challenges.

10.2.2.9 IT Networks and Standards

You need someone on your team who knows how your company's networks are organized and the standards your company uses. If your corporate standard is Microsoft BizTalk, J2EE, SOAP, this knowledge will help you make sure your system can interface properly. Also, someone will need to decide how to configure firewalls to enable your readers to access the application programs and databases.

10.2.2.10 Construction Methods and Techniques

You will need someone on your team who is knowledgeable about construction matters. You will be making facilities recommendations, and someone who knows what it takes to move a wall, erect a shield, or run a power line to a location in the ceiling will be valuable.

10.2.2.11 Sourcing the Skills You Will Need

When the resources you need are not available within your company, augment your core team from your vendor population. Do not assume that these people will be available to your project at the time and to the level you will need them; get commitments for the resources, level of effort, and times you will need. You should secure commitments as soon as you know how many hours you will need and when, and list alternative sources for these skills.

10.2.2.12 Core Team Meetings

The core team will meet weekly with the project manager to review milestone achievements, identify all ongoing and new tasks for the coming week, and update the risks log, expenditures, and issues log. At each meeting, distribute final-draft copies of project documents to core team members for comment and to ensure that all members know what decisions have been made. Post completed documents on the project website.

10.2.3 Project Staff

The project will need staff as follows:

- **Documents coordinator** This person is responsible for making sure all documents are produced and contain all project knowledge, activity gathering is completed, and results are recorded in the correct documents. The documents coordinator must be informed whenever activities by one document team impact other teams' activities. For example, reader, software, systems, facilities, and other teams may generate activities for procurement document, or tag characteristics may drive considerations for the Data Description Document (DDD) team.
- **Project coordinator** This is the person who schedules and coordinates all visits and events. Anything that involves three or more people should be registered with the project coordinator, who is responsible for logistics, travel, facilities, and so on.
- **Participant coordinator** This person works with supply chain partners who might be interested in the project. The participant coordinator keeps them informed, actively solicits their involvement, and arranges (with the project coordinator) for briefings.
- **User feedback coordinator** This person solicits, documents, and expresses user issues and concerns to the project. Consider posting user quotations on the website.
- **Executive coordinator** This person coordinates executive review meetings and keeps executives informed about project progress.
- **Webmaster** This person is needed if, as we recommend, your project has a website. The webmaster will update the website with new information and notifications as it develops.

10.2.4 Project Team

You should also make a list of all the stakeholders in the project. Each of these persons should be contacted and briefed, and their concerns and expectations should be noted. These persons should be invited to join the team and to attend regular team meetings.

At this point in the project, you should review the support the project will require from various departments and outside vendors. Certainly, it will need vendors for the equipment and tags. The IT department can probably provide network support, software resources, servers, and storage hardware resources. The internal operations department should be able to provide facilities resources, so you do not need to coordinate any needed construction, installation of power lines, movement of conveyor systems, and so on. You will need to provide support for project management, RF site analysis, case analysis, testing, equipment selection, deployment, requirements definitions, and so on.

Create a team roster that lists the organization and the person or peoples' names providing the resource. As you develop the project plan, you will be able to schedule their time and their deliverables. Include on the roster contact information for each person and their general area of responsibility. You may wish to post this roster on the project website.

At this point, you should have the project coordinator contact vendors and request information about their capabilities. The author of this book will provide a list of companies providing a variety of RFID-related products and services upon request. Send an e-mail to dbrown@rfidrunner.com. Send the companies that look promising a Request for Information (RFI). The RFI should describe your project and ask the vendor to document what resources they can provide and what experience, expertise, and special resources they can offer your project. Based on the RFI responses, narrow your list of candidates to a manageable number and submit Requests for Quote (RFQs) for the products and services you will need. Rank the two or three top candidates in each category and ask them in for presentations. Select a primary and a secondary for each vendor type you will need to help you develop detailed costs, lead times, and timelines.

For all cases where you will use the company's internal expertise, you can identify the departments and have the project coordinator interview the persons and get commitments for their projected level of involvement. If you have headcount available, you may be able to hire expertise.

Begin the project by scheduling a core team kickoff meeting. The meeting should include introductions, distribution of the team roster, a discussion of general roles and responsibilities, timelines, and a review of project resources, project scope, and objectives. Assign responsibility for each document and discuss any issues. Schedule milestones and deliveries. Responsibility for the System Requirements List (SRL) may be divided so different sections of it are different people's responsibility.

Lay out activities for the project as a whole and for the coming month, including deliverables and specific responsibilities. You should also schedule a meeting of the extended team to cover the same issues at a higher level of generality.

Your presentation to the kickoff meeting should include a review of the project's general structure. Underline the project's dedication to the principle of "plan and document first, then execute." Agree on a time for weekly progress meetings and for project team meetings.

10.2.5 Pilot Documents

Many of the documents created during Phase 1 can be expanded to include more detail and provide more useful information for your pilot. In addition, several new documents will need to be created. The following is a list of the documents you will need to complete before you can start your pilot:

- Business Infrastructure Document (BID)
- Pilot Objectives Document (OD)
- Pilot System Requirements List (SRL)
- Pilot Workflows Description Document (WDD)
- Pilot Facilities Description Document (FDD)
- Pilot Data Description Document (DDD)
- Pilot Software Description Document (SDD)
- Pilot Systems and Storage Description Document (SDD)
- Products Description Document (PDD)
- Pilot SKU Description Document (SKUDD)
- Pilot Network Impacts Description Document (NIDD)
- Pilot Human Resources Description Document (HRDD)
- Pilot Tag Selection Document (TSD)
- Pilot Reader Selection and Deployment Document (RSDD)
- Pilot Procurement Document (PPD)
- Pilot training documents
- Onsite documentation
- Project Plan (PP)
- Budget

10.2.5.1 Business Infrastructure Document (BID)

After the kickoff meeting, start to collect information you will need about the existing business infrastructure. This will enable you to begin the Business Infrastructure Document. These issues should be investigated before beginning the pilot but with a focus on how they impact the pilot. The document will be extensively expanded during technical integration. These are the issues that must be investigated, documented, and resolved:

- Methods of identification, numbering, tracking and cross referencing for products, packages, locations, persons, fixed assets, inventory items, work-in-process. Review whether they comply with relevant standards. If not, consider recommending that they be changed or recommend a work-around.
- How the reader network will interface with enterprise network.
- How data volumes and flows will impact enterprise network.
- The data quality level in relevant systems.
- Current level of supply chain performance.

Your project will require detailed descriptions of how the company manages code numbers for parts, employees, facilities, equipment, products, locations, departments, documents, orders, invoices, memos, books, vehicles, vendors, suppliers, departments, and shelves in the warehouse. This is beyond the scope of the pilot, but it will certainly affect your project. You should understand how the various departments assign and coordinate identification numbers. We recommend that you begin to assemble this information as you prepare for the pilot and do a full-scale survey as you prepare for the complete rollout. For all of these encoding activities, there may be international or industry standards in place. EPCglobal has standards for locations, products, companies, returnable containers, and fixed assets. If your company's system does not use these standards, you will find it difficult to synchronize data with trading partners. At some point you may recommend that the system be changed. Look for anecdotes where various features of the numbering system created unusual values or caused problems or extra costs.

Many companies deal with nonstandard numbering system discrepancies by incorporating translation tables into the processes, rather than dealing with the level of upheaval and expense that introducing new identification systems may create.

The pilot has been designed to keep reader/network interface issues out of scope. However, if your project contemplates hundreds or thousands of readers distributed across various facilities, you will have to develop a plan for interfacing them to the corporate network. We have discussed the role of RFID middleware (see Section 3.9.1). Full description of this interface may be beyond the scope of your pilot, but it will be dealt with during technical integration. You may initiate discussions with the network planners early so that you can have this decided in time for that phase.

The pilot is probably small and will generate relatively little data. But when fully operational, RFID systems can produce enormous amounts of data, taxing existing networks. Preparing the network for expected data floods is outside the scope of the pilot, but could impact your ultimate rollout. We recommend alerting network planners so they can begin developing strategies of their own to be able to support you.

Your project can do everything right and still fail to deliver a substantial return on investment if the data resources of the company are of poor quality. Various surveys say that up to 90 percent of business records have some error in them. The RFID system needs quality data in order to create the records and provide the information that actually will improve business performance. Data quality is a project in its own right, but we have provided an overview of what is needed in Chapter 13.

One key question that should be answered carefully at the beginning is, "What is your current supply chain performance level?" If your supply chain is excellent, which means that it is efficient and makes full use of the data available, then RFID is a logical next step. If the chain is inefficient and it does not use the data available well, then other, less expensive measures may produce gains. For example:

- Does the company's order management process use actual demand, logistical, and availability data? Many companies order goods and/or raw materials based on historical averages, best guesses, and vendor promotions rather than utilizing actual data. Are there documented processes for generating orders based on actual data?

- How well does your company collaborate with key customers, suppliers, or other partners with aligned, visible process controls? Many companies—including, perhaps, your customers—are secretive about their data and getting them to share with suppliers may be a chore. In other circumstances, even within a single company, the sharing is based on personal relationships, not documented processes. Are there documented processes for collaboration with business partners?

- How well does your company sense and respond to sudden or unanticipated shifts in demand, changes in supply execution, or unexpected events? Gather anecdotes where sudden shifts in demand occurred and describe what happened.

How well does your company comply with regulatory, safety, trace-ability, and national security requirements? Are these mechanisms in place and functioning smoothly, or are they a major, costly disruption?

What is the quality of your data resources? This is an important topic in its own right; see Chapter 13.

If your answers to these questions are not encouraging, collecting better demand data through RFID may not generate substantial return on investment. The mechanisms, systems, workflows, management structures, and habits should be established to utilize the data you have, before implementing new data collection systems.

Document your findings in the initial Business Infrastructure Document (BID), along with any changes you will recommend. Review these with the core team to determine how best to deal with any issues. Note that some or all of these considerations will be out of scope for the pilot. We discuss them here because these issues may take time to develop, and starting early allows you to move more quickly into final rollout. Any of them may affect the success of your rollout, depending on its severity.

10.2.5.2 Pilot Objectives Document (OD)

Review the pilot Objectives Document (OD), determine which project objectives are relevant to this pilot, and list them. Then add any objectives that are specific to this pilot but not part of the pilot OD. For example, a pilot objective might be to prove the effectiveness of the SKU testing protocol or the vendors selected. This pilot OD will be the scorecard for evaluating the pilot when it is completed.

Pilot objectives might be

- Select tags for 24-unit case of Product A.
- Select tags for 48-unit case of Product B.
- Select tags for pallet of 48-unit cases for Product B.
- Set up assembly line to tag all 24- and 48-unit cases shipped from Distribution Center D to Store E.
- Validate all tags as they are loaded onto trucks.
- Send Advanced Shipping Notice (ASN) to Store E electronically.
- Validate RF test protocols.
- Validate documentation methodology.
- Evaluate and document business infrastructure.
- Select vendors for project and test their ability to perform.

10.2.5.3 Pilot System Requirements List (SRL)

The SRL is a list of what it will take to deliver all the objectives. It lists what are called requirements, and they may fall into the following categories. Each of these categories will be described in detail in its own document. The SRL is a comprehensive list, along with as much detail as you have about the issue before engaging in extensive research. The documents described in the following sections are developed from the SRL.

- **Workflow requirements** List the workflows that will need to be utilized, required, created, changed, or removed by the pilot.

- **Facilities requirements list** List the construction activities, equipment moves, new equipment needed, power lines that will need to be added or moved, and RF shields that will need to be installed. You will have all of this information when you have completed the site survey.

- **Data requirements** List the data structures that will be utilized, impacted, created, changed, or removed by this pilot. Data structures are elements, records, and groups of elements that work together to accomplish some business purpose. Examples of a structure might be the group of data elements written to a tag at the same time or a line item on an ASN.

- **Software requirements** This part of the SRL will list all software the pilot will require to deliver the objectives. List the functions that each software package provides the pilot. Where necessary, allocate functions to known upgrades that are available for purchase, or describe where the function is on the vendor's product roadmap. Alternatively, you may recommend that new software be purchased.

- **Systems and storage list** This is a list of the computer systems that are required by the pilot. It should also include all storage resources that will be impacted. Your company may have an enterprise resource that is used across the company and is available for your project. Or, you may have to designate your own storage or utilize some other storage resource.

- **Products list** This is a list of the company's products that will be tagged for the pilot. It may include products manufactured by the company, products manufactured by others and distributed by the company, or products purchased for incorporation into the company's final products. If you have the information, include in the products list the source of the product and the contact information for the product manager. Note that only selected package variations of the products will be tagged for the pilot. Note also that in Phase 3 you will be selecting an application to integrate with based on its ability to create business value. Select a product and packages for the pilot that will support this activity.

- **Packages list**   This part of the SRL should list all the package variations for the products in the pilot by SKU. Note which ones will be tagged and which will not.
- **Network impacts list**   This will list all network components—routers, lines, bridges, wireless access ports—that will be required, utilized, or acquired for the pilot.
- **Human resources list**   This will list all the human resources that will be required to complete the pilot. Facilities, software, systems, and networks each could require human resources to do the work to accomplish the tasks of the pilot. Human resources are designated by skill required and estimated person hours that will be needed. For example, if you are going to upgrade the router in the facility, that may require 10 hours of a network specialist's time to install it and test it.
- **Tags list**   Once you have completed SKU testing on the packages listed, you will be able to say which tags will need to be ordered. Estimate the quantities you will need for the pilot.
- **Readers list**   List all the locations you expect to install readers. Categorize each location (portal, mobile, or handheld). Once you have completed the reader selection (Protocols 8 and 9 in Chapter 12), you will be able to list the readers you intend to purchase. Include in this list the vendor, contact information, and lead times.

Each element of each of these lists is a separate requirement. A single requirement may support more than one objective. It is useful to name and number each requirement so you can track its progress and its upstream and downstream dependencies and reference it unambiguously.

10.2.5.4 Pilot Workflows Description Document (WDD)

The WDD is a description of each workflow impacted by your pilot. Create a section of the WDD for each item in the list of impacted workflows in the SRL. Name each workflow and document who is the manager. You will need to decide whether the workflow will be necessary once the RFID system is installed and whether it will need to change. In all cases, be sure to include the manager of this area in your discussions. For all relevant workflows, document current performance levels and explore how they will change with an RFID system in place.

Document current performance of current workflows and how you expect the performance level to change. Each workflow description should include a detailed use case describing the steps actors (systems, workflows, or workers) must take to accomplish it.

Note that workflows are not the same thing as protocols. A protocol is the standard for executing the workflow, and you may find it advantageous to include the protocols as an appendix to the WDD. The same protocol may support many different workflows applied in various settings. A section in the WDD represents a specific workflow, one instance of the protocol. When you document workflows, you can add information specific to that instance such as network connectivity, which products are being processed, the names or titles of the actors, performance rates, and so on. If the workflow is supported by a standard protocol, you can incorporate it by reference. Recall that for each workflow, the list must include the input resources, relevant personnel, outputs, performance levels, and use-case description. The input resources may be data structures, as well as person minutes, equipment minutes, and consumable supplies. Input resources may also be IT resources such as network connectivity and performance or computers or storage. They may also be RFID resources such as tags, labels, or reader stations. For the last item, document the category of reader station that it is, along with any exceptions or special circumstances. List all these resources back in the SRL.

Each workflow should be documented with a use-case description showing exactly how it works. Note where the actual use case may differ from the protocol, if there is one.

If the instance of the workflow addresses more than one product, make careful note of which products and packages are included, as these will have to be tagged. Document how the workflow will change and set new performance expectations. For each workflow, list the resources you will need and the activities such as procurement, installation, integration, coordination, training, and documentation that will be required.

10.2.5.5 Pilot Facilities Description Document (FDD)

The FDD describes a facilities plan for the pilot that will serve as a template for other facilities once you begin planning the rollout. Start with a site survey to determine what it will take to install your system. The site survey will examine the facility for levels of RF interference. See Protocols 1, 2, 3, and 4 in Chapter 12.

The FDD consists of the following information:

- Name of facility
- Contact information for facility manager
- Roles facility plays in company's production system
- List of workflows operated at the facility
- List of products that are created or stored at the facility
- List of reader locations at the facility (may not be available until after site visit) by priorities
- Any known network connectivity issues
- Any known construction issues such as power, space available, equipment that needs to be moved, and so on
- Future plans for the site, including any planned construction and/or renovations

For the pilot, we create a facilities plan just for the one location of the pilot. We will use the same location for technical integration. For the rollout, we will need to use this protocol for several other facilities.

10.2.5.6 Pilot Data Description Document (DDD)

Working from the SRL, extract the list of the data structures, records, and elements that will be necessary to deliver the objectives of the pilot to create the DDD.

For each data structure or record, list

- Workflows that provide it or utilize it (sources and uses)
- IT products such as labels, tags, alerts, or reports that utilize it
- Data elements that comprise it

A single data element may be used in several structures. Describe each data element in terms of

- Name of data element
- Contact location for responsible person
- Its source, including method of acquisition
- Its destinations (workflows and applications that use it and all data structures in which it is used)
- Any transformations necessary to create it or render it usable in the pilot

- Contact information for person who "owns" the data element

- Security considerations, including value of the data element to terrorists, competitors, or hackers; recommend solutions

- Privacy considerations, including recommendations for ensuring the privacy of any consumers who may come into contact with the data element; recommend solutions

There are two types of data sources. First are the sensors or tags, such as those on items, cases, pallets, employee badges, and tags affixed to vehicles and those embedded in floors, walls, and ceilings. The second set of sources is the local, remote, enterprise, vendor, and partner applications, databases, and repositories where data exists that will need to be combined with the tag data to create useful records and structures. In some cases, your system may be able to query the application directly. In other cases, you may need to extract and create an intermediate data table that is updated on a regular basis.

For example, a record might be required that defines a carton stored on Shelf A-19 and stored in a database. The record could be created by this process:

The reader would read the tag and report the Electronic Product Code (EPC). The middleware would use the EPC to query the local EPCIS or some other local database to acquire the supplier identifier, part number, size (length, width, height, weight), and expiration date. It would query the ERP system to get the order number and store the complete record in the Warehouse Management System.

Destinations are various applications that will use the data records. Destinations may be local applications such as the Warehouse Management System; they may be remote applications, enterprise applications, public repositories such as ONS or GDSN; or they may be a partner's applications.

Destinations may also be reports or documents such as invoices, ASNs, customs documents, or other formal documents. Destinations may also be applications or applets that send alerts to individuals or systems when certain conditions are met.

Transformations include reformatting, table lookups, and calculations to create data elements needed for the application.

Combine data elements to create structures and records that the applications can work with. In the DDD, you will also determine the retention strategy for the records—where and how long they should be retained by systems within your control. Some records will need to be kept and others will be sent on and discarded.

For the next two phases, the DDD will be expanded to contain a map of the flows of data for the entire project. For the pilot, we will be concentrating on less ambitious goals.

Note that, for the pilot, many companies bypass the complexities of generating numbers and elect to use pre-encoded tags. This gives them unique numbers to work with, and they do not have to set up systems to manage allocation of their own EPCs.

10.2.5.7 Pilot Software Description Document (SDD)

For the pilot, you will need to install temporary software on the host computer so you can control the reader and also receive and store tag data.

If your pilot encompasses writing data to tags, you will need to provide software to manage this.

If your pilot encompasses completing smart labels, you will need to provide software to format the data for the smart labels and to create the bar codes that will go on them.

In the SDD, describe how you will meet these three software requirements.

The software used for production in your technical integration and roll-out phases is very different. An entirely different protocol for creating sections in the SDD is used in technical integration, and it is described in Section 10.4.2.2.

10.2.5.8 Pilot Systems and Storage Description Document (SSDD)

To create the SSDD, work with the list of computer systems from the SRL that will be touched by your pilot. Set up a section of the SSDD for each system. You need to evaluate whether it has the CPU speed, memory, interfaces, and peripherals needed to support your pilot or if it will need to be upgraded.

For each system listed, document

- Name of system
- Contact information for system manager
- Applications run on the system
- Workflows that utilize the system

For this document, you will also need to calculate the amount of storage your pilot will require and designate where it is available. Be certain to notify the document coordinator of any procurement that will be required.

Include your team's determination of viability of each impacted computer upon completion of your project. This includes your recommendations whether the system can function effectively as is, whether it can and should be upgraded to process RFID data, or whether it should be replaced.

10.2.5.9 Pilot Products Description Document (PDD)

To create the PDD, work from the list of products in the SRL. Depending on your scope, this may include products the company produces, products it ships, or incoming products that are inventoried, stored, and used in the production process. Be sure to consider whether the incoming products may be tagged by some other company and whether those tags can be leveraged by your project.

For the pilot, select a limited number of products to tag. In selecting products to work with, keep in mind that you will use it to support the Business Value Statements created in technical integration; see Section 10.4.2.6.

For each product, record the following information:

- Workflows involved in acquisition, manufacture, or distribution of the product.
- Facilities that create, acquire, store, or deliver the product.
- Contact information for the product manager.
- All packaging options for the product. Upon completion of the pilot, you will need to set up a mechanism to make sure the RFID team is notified whenever a new package is developed, so you can perform SKU testing and incorporate the tags.
- Any unique systems considerations for the product.
- Any legal or regulatory issues associated with the product that might impact your project.
- Any unique business partner considerations for the product.

For each product, capture the contact information for the person who will be responsible for managing the numbering schemes the RFID implementation will require. You will also need to know who will keep RFID management informed when packaging changes, so you can perform SKU testing on new packages before they are released.

When you have completed SKU testing, update the PDD to include a description of the product's RF characteristics.

10.2.5.10 Pilot SKU Description Document (SKUDD)

The SKUDD lists the SKUs for each of the products that your pilot will need to handle. Items may be packaged into cartons, cartons into cases, cases into pallets, and pallets into containers. Item-level tracking is not yet prevalent, but cartons, cases, and pallets are routinely tagged. Your pilot should include at least one example of packaging hierarchy so you can be sure your systems can manage container relationships. This list of information about packages will be critical. One objective of the Department of Defense (DoD) and the retailers is to track individual items reliably by processing their higher-level packages. This would enable them to do away with the time it takes to read all the individual items in a case or all the cases in a pallet.

Select a few SKUs to include in the pilot. In making your selection, keep in mind that during technical integration you will need to create Business Value Statements (BVSs) and deliver them; see Section 10.4.2.6.

Create a section in the SKUDD for each SKU that will be included in the pilot. In that section you will document

- Name of package.
- Name of product it contains.
- Facilities involved in creating, storing, or distributing the package.
- Workflows associated with creating, storing, or distributing the package.
- Who will be responsible for managing the number assignments for the package. Describe the numbering schemes and systems that will keep them consistent with standards.
- Any business-partner-specific-considerations.
- Any governmental or environmental considerations, requirements, or restrictions.
- Shelf-life, temperature restrictions, handling restrictions, hazards.
- Whether container is returnable.
- Whether package goes to end-consumer, which triggers privacy concerns.
- Security issues regarding competitors, hackers, insiders.
- Counterfeiting, disposal, or special theft issues that should be managed.

Use Protocol 10 in Chapter 12 to determine which tags will work with each package and what the optimal location and orientation is. As SKU testing is completed, update the SKUDD with tag selection, mounting and orientation information, and all RF issues detected. Try to give yourself a choice of tags that will work on each package.

10.2.5.11 Network Impacts Description Document (NIDD)

The SRL includes the list of network impacts for the pilot. Data volumes for the pilot may be negligible. Also, most issues of network connectivity are out of scope for the pilot. But the process of calculating data volumes and impacts will be important for the rollout, so you can go through it during the pilot to validate the process. During the pilot, document the data volumes in each direction and make sure the network infrastructure and storage are in place to support them.

10.2.5.12 Human Resources Description Document (HRDD)

The SRL lists the human resources required for each area of the pilot. In this HRDD, list and describe the tasks each individual will need to perform and a competency model for each title. For each task, estimate the hours, provide a succinct description of the task, and describe its rationale and expected outcome, unless they are self-evident.

10.2.5.13 Tag Selection and Configuration Document (TSD)

You need to determine whether a single tag will work for all of the pallets, cases, and items you plan to track in your pilot. If you can make this work, it will simplify the rest of your project, but it may be challenging. Use Protocols 6, 7, 8, and 9 in Chapter 12 to select tags yourself, or you may outsource the testing of tags.

Use the list of pallets, cases, containers, and any items to be tagged in the SRL. Create a section in the TSD for each unique tag you will use.

The following information should be included in the TSD:

- Test descriptions.
- Tag physical mounting issues such as adhesive, orientation, and survivability.
- Tag data requirements: the amount of storage required, what data elements will be stored on the tag, where they will be commissioned, which workflows will generate updates.
- Tag distance and orientation issues.
- Any incompatibilities of tag with any known readers.

• Tag certification status with regard to readers you plan to use.
• Data elements each will carry. (This will enable you to specify the data storage capabilities of the tag.)

NOTE *Security considerations for each data element on the tag are documented in the DDD.*

• How each tag's data will be provisioned. Provisioning profile for data at the element level is described in the DDD. For the pilot, an intermediate database may be created, just to prove that the processes all work.

10.2.5.14 Reader Selection and Deployment Document (RSDD)

Create a section in the RSDD for each location you plan to install a reader. For each location, include the information listed here. This list will be useful for the rollout, when you have numerous reader locations to test, design, populate, connect, and test again:

1. Check and describe power sources for each reader and encoder/ printer location.
2. Document the reader category for the location. This tells you the manufacturer, model number, reader configuration settings, and version of the firmware/software. If the reader allows it, the document should also list any software to be loaded on the reader.
3. Document each location's requirements: required read range, read reliability, and range of tags to be created and read.
4. Document recommended cables and connectors. Cable length is often an issue.
5. Specify antennas for each read point.
6. Design read zone or footprint for the location.
7. Be sure to take into account RF opaque materials like water and metals in the environment.

Before you actually go to the facility and install the unit, it is useful to mock up the installation in a clean environment and test and tune its performance. Secure a clean environment and perform the following tasks:

1. Install readers, antennas, reader networks, and interface to host computer.
2. Place, orient, and tune antennas.

3. Install accessories and other digital inputs and outputs.

4. Configure reader.

5. Test your readers' performance, record tag read rates under all expected working conditions.

6. Tune installation until desired performance levels are achieved.

10.2.5.15 Pilot Procurement Document (PPD)

The PPD has a section for each type of material, service, and equipment you will need to purchase for the pilot. For each item, it identifies the vendor, the cost, and the lead time. It lists any logistical issues such as delivery, storage, and staging and lists any professional services you will need in order to make the item operational, as well as sources and contacts for those services.

The PPD also lists all services you plan to purchase for the pilot. For each item, it identifies the vendor, cost, procurement data, and employment data.

The Controversy over Portal Readers

Should you put readers at the portals? Initial implementations favored placing portal readers at every single door, but with readers and antennas costing up to $10,000 per portal and some facilities having more than 20 doors to equip, some Wal-Mart suppliers have favored placing one or two readers at a central staging area instead. This reconfigured reader placement strategy obviously can result in sizable savings and seems adequate in many cases. International Paper has been using RFID readers on forklifts and hand trucks instead of the portal readers. The average Distribution Center (DC) has one lift truck for every five portals, so this approach can provide a much less expensive alternative to RFID deployments in DCs by eliminating the need for RFID portals at every dock door. However, it provides a slightly different view of the data, which must be taken into account when designing the applications. The solution is able to identify and track product on board the forklift from loading to unloading. With an automated shipping and receiving process, forklift operators can focus on driving the trucks and improve their productivity instead of manually scanning bar codes.

For each item to be procured, document

- Cost
- Contact information
- Lead time
- Miscellaneous charges including shipping, handling, and maintenance
- Warranties that are acceptable
- Vendor contracts negotiation status
- Deliveries scheduled

You should engage your supply chain management organization to assist you in all negotiations and contracts with suppliers for hardware, software, and services.

10.2.5.16 Pilot Training Document

You will need to provide training for the workers who are responsible for attaching tags to the packages. A training outline is provided in Section 11.6.

10.2.5.17 Onsite Documentation

You will need to provide onsite documentation for the reader site. The documentation's recommended contents are listed in Section 11.6.10.

10.2.5.18 Project Plan (PP)

The PP brings together all of the activities that are necessary for a successful pilot; these are laid out in the PP on a timeline so you can coordinate and manage them. We strongly recommend that you use a tool such as Microsoft Project because it enables you to more easily document and track the tasks, subtasks, dependencies, resources, and timelines. It makes managing the project much more manageable.

The Project Plan should be as detailed as possible. A sample high-level breakdown is presented in Section 10.3. The Project Plan schedules all the tasks that need to be completed, by whom, and by when.

One output of the Project Plan is a work plan for each person. It will show them what tasks they need to have completed by what date. Each person's work plan should be reviewed with that person and reviewed as important milestones approach.

10.2.6 Scope of the Pilot

You can see that we have designed the pilot to be a dress rehearsal for the full-scale rollout, except for technical integration. This means that it should include as many procedures as possible. They may appear to be over-engineered for the pilot, but they will generate the information you will need to execute the pilot, and they will give you a firm foundation for the rollout. The pilot is a rehearsal to ensure that your team is capable of executing the test protocols, documenting and managing the construction, and sourcing and installing and integrating the equipment in acceptable timelines and costs. As you work through the pilot, you may find yourself taking expedient shortcuts. Be certain to record them in the relevant document and notify the document coordinator, so you can revisit them in the planning for the rollout. Also, some elements of the full-scale rollout may be outside the scope of the pilot. For example,

- The rollout may contemplate purchasing and installation of middleware, but that is out of scope for the pilot. We have included this in Phase 3: Technical Integration.

- The rollout may include interfacing with the corporate network, but that may be out of scope for the pilot. We have included it in Phase 3: Technical Integration.

- The rollout may include many readers operating in close proximity to one another, but the pilot may only have one or two. We will deal with this in Phase 4: Rollout.

- The rollout may include interfaces with numerous applications such as WMS, ERP, APS, and various data warehouses, databases, data repositories, e-mail notifications, and so on. We deal with this in Phase 4: Rollout.

We recommend that the pilot be limited to just getting data to the local PC. This removes the issues of middleware and application integration from the pilot, as these may involve costly expenditures.

Your executives may require you to address some of these issues in the pilot, but even then, we still recommend that you keep them out of scope for the pilot. Get your readers talking to tags first and then to a single computer second. Spend your initial efforts getting the team working well together, the RFID physics tamed, the procurements and facilities, and tag-read rates high. Once that is done, you can turn your attention to middleware, application integration, software, and dense reader issues.

10.2.7 Resourcing the Pilot

The pilot as designed is a wealth of planning, preparation, and documentation so that the actual installation will go smoothly and succeed. It relies heavily on reliable instruments like analysis, documents, checklists, walkthroughs, and preparation. The team that completes the preparation as described here will be well equipped to succeed in the pilot and lead the project through the rollout.

A complete RFID project documentation kit is available at www.rfidrunner.com. This kit contains templates for each of the documents listed in this chapter, as well as a Microsoft Project template for the project. The website also allows you to ask questions and leave comments for others about your experiences with this methodology, recommendations, tips, and lessons learned.

10.3 Phase 2: Step 2, Execute the Pilot

Phase 2 of the pilot is where all that planning and documentation pays off. It gets the readers installed and working with the local host computer. It enables you to demonstrate and prove your team's planning, documentation, testing, construction, and sourcing skills. Section 10.3.1 provides a final checklist to be completed before the execution can begin. The checklist is divided into three sections: mandate, documents, and activities. The activities rely on feedback from the documents, which should be completed before you can begin the parts of the installation dependent on them. This is very important. Do not begin pilot deployment until the checklist is completed!

We have organized the actual execution in terms of preparation, facilities, procurement, installation, data management, systems, and software.

10.3.1 Preparation Checklist

This checklist summarizes the preceding material in a convenient checklist form. If you can check off all of the following items, your pilot is ready to begin. When the following tasks are completed, you can execute the pilot with confidence. We have divided the checklist into three sections.

- Your mandate
- Documents
- Activities

10.3.1.1 Your Mandate

- Pilot scope is documented and approved.
- Pilot objectives are clearly documented and approved.
- Core team composition, meeting, and reporting structure is in place.
- Executive sponsorship is secured.
- Stakeholders have been identified, contacted, briefed, and interviewed.

10.3.1.2 Pilot Documents

Review the list in Section 10.2.3 and be sure each document is complete.

10.3.1.3 Activities

You can begin to execute the pilot without actually having completed all of the following tasks. If they are scheduled, and activities that depend on them will not begin until they are complete, you can accelerate the pilot. However, do not attempt any task that utilizes data these tasks are designed to develop.

- RF site survey(s) completed for pilot Site
- All stakeholders briefed; issues logged and dealt with
- Vendor partners selected, briefed, and scheduled; issues logged and dealt with
- Reader site mocked up and tested
- Reader configuration settings documented
- Reader testing completed
- Antenna footprints documented
- Antenna output tested for regulatory compliance
- Tag(s) selected and procured
- Static testing completed on all packages in the pilot
- Dynamic testing completed on all packages in the pilot
- User training completed

10.3.2 Execution of the Pilot

The pilot project plan describes the steps necessary to execute the pilot. The activities will be scheduled in the pilot project plan. They fall into the following categories:

- Facilities
- Procurement
- Installation
- Data management
- Systems
- Software

Each of the following sections lists the steps that must be checked off in these categories.

10.3.2.1 Facilities

- Construction, power installation, and network installation completed
- Conveyors or shelves moved as necessary
- RF shielding installed as necessary
- Network cables and wireless access ports installed

10.3.2.2 Procurement

- Reader(s) selected, ordered, delivered, configured, and tested
- Tags ordered, delivered
- Antennas ordered, delivered, and tested
- Cables, connectors, racks, chokes, baluns, and clamps ordered
- Personal (host) computer ordered, then delivered, configured, installed, and tested

10.3.2.3 Installation

- RF site analysis for facility completed
- RF site analysis for each reader location completed
- Any sensors installed and tested
- Reader(s) tested, installed, and tested in place

- Antenna racks installed
- Antennas mounted on racks
- Antennas connected to reader(s) and tested
- Reader connected to PC
- Static read tests performed, antennas adjusted as necessary, readers configured
- Dynamic read tests performed, antennas adjusted as necessary, readers configured
- Write tests passed
- Tag kills tested
- Workers trained to handle tags
- Workers have begun attaching tags
- Read rates satisfactory

10.3.2.4 Data Management

- Tag data elements sources accessed
- Confirmed that tag data elements written to tags for each SKU
- Confirmed that all data elements printed correctly on smart labels
- Reports produced correctly
- Queries tested
- Tag security tested

10.3.2.5 Systems and Storage

- Any required system upgrades completed, tested
- Any required storage upgrades completed, tested
- Any required network router installations and upgrades completed, tested

10.3.2.6 Software

In order to demonstrate and test the reader, you will need to install software on the host computer, but any consideration of applications software is out of scope for the pilot.

- Reader control software installed and running
- Software to receive and store tag data installed and running

10.3.3 End of Pilot

The pilot is completed when tag contents can be read reliably by the reader and recognized by the host computer. The end of the pilot is marked by five activities, and it is followed by approval for the next phase:

- Evaluate the pilot
- Capture and document lessons learned
- Preliminary assessment for Phase 3
- Brief Executive Steering Committee (ESC)

10.3.3.1 Evaluate Pilot

With your core team, evaluate the pilot against the criteria listed in the pilot Objectives Document. Each objective should be rated exceptional, satisfactory, or unsatisfactory, with any appropriate discussion. Review evaluations with the project team and solicit input.

10.3.3.2 Capture and Document Lessons Learned

Immediately upon completion of the pilot, you should canvas the stakeholders to review all opportunities for improving your team's processes. In your discussions, review and document any insights they may have regarding ways RFID may support their workflows. Then schedule a team meeting to discuss these inputs and gather contributions of your team members. You should lead the team through a structured review of the pilot and gather all comments in terms of what went right, what went wrong, what worked well, what can be improved, and how to go forward.

10.3.3.3 Preliminary Assessment for Technical Integration

The purpose of the preliminary assessment for Phase 3 is to gather enough information to make recommendations to the executives during their next briefing. To make recommendations, you will need the following information:

- Set objectives for Phase 3; see Section 10.4.3.
- Select application program to integrate.
- Select a list of possible candidates for middleware purchase.
- Estimate the costs for application upgrades, integration, middleware purchase, and services you will need to complete the technical integration.

Do not attempt to develop this information on your own. It should be the product of the core team, with all members having a chance to contribute. It should also be reviewed by relevant project team members (stakeholders) before being presented to the ESC.

10.3.3.4 Brief the Executive Steering Committee (ESC)

The briefing of the ESC has two elements: a formal presentation by your executive sponsor and appropriate members of your team and a printed document that you can hand out. The purpose of the briefing is to inform them of the success of your project and present your recommendations for moving to the next phase.

The success of the project is the extent to which it met or exceeded all of its stated objectives. Review the objectives and your assessment of their level of achievement. Also review any important lessons learned.

To get approval for the next phase, you will need to set objectives, time lines, and estimate costs. Your vendor partners will be able to furnish cost estimates for most of the items you will need.

10.4 Phase 3: Step 1, Plan Your Technical Integration

Once the tags are communicating with the reader reliably and the reader is communicating with the host computer, you can begin to integrate the tags, readers, and a single application. Upon completion of Phase 3, your network of readers, represented in the pilot by the reader(s) you installed in the pilot, will communicate with one or more application programs. EPCglobal recommends that this be accomplished using a middleware package, and there are several available on the market. As discussed in Chapter 3, RFID middleware gives you a scalable mechanism for communicating tag data to any number of application programs, managing the individual readers, and communicating with external data sources and destinations such as EPCIS.

Surprisingly, technical integration is the phase of the project in which we introduce the possibility of business value for the project. Upon reflection, however, this makes sense, for the business value of most technical projects arises out of their integration with other systems and with workflows. This is especially true for RFID, so during Phase 3, we will identify and deliver business value for the first time, beyond just the functioning of new technology.

10.4.1 Teams for Technical Integration

Begin Phase 3 by reviewing your core team membership for the project. Make any necessary changes to the team going forward. Make sure you have knowledge of the applications and middleware and of the integration skills on your team.

Once your team composition is determined, schedule a kickoff meeting for Phase 3. At the kickoff meeting, introduce any new members and review the roles of each member. Review the project objectives and the objectives for Phase 3 as you presented them to the Executive Steering Committee. Assign responsibility for each document and discuss any issues. Responsibility for updating the System Resource List may be divided so different sections are different people's responsibility. Lay out activities for the project as a whole and for the coming month, including deliverables and specific responsibilities. You should also schedule a meeting of the project team (stakeholders) to cover the same issues at a higher level of generality.

10.4.2 Plan the Technical Integration Phase

As with the pilot, the planning of the technical integration phase consists of investigation and preparation of documents. With a few exceptions, the documents created during the pilot will be augmented with new information to reflect the new objectives and ongoing discoveries.

To complete the planning, you will need to review all the project documents and update as necessary. We will list here only those where there are special issues for this phase of the project:

- Business Infrastructure Document (BID)
- Software Description Document (SDD)
- Systems and Storage Description Document (SSDD)
- Network Impacts Description Document (NIDD)
- Procurement Document (PD)
- Business Value Descriptions Document (BVDD) (new)

10.4.2.1 Business Infrastructure Document (BID)

Now is the time to finalize the BID. Review each element in light of its impact on your project (see Section 10.2.5.1). Pay particular attention to the methods of item, case, and pallet identification your company uses. You will need to conform the methods of identification to the standards you expect to deploy. Document discrepancies and make recommendations. The readers must interface with the enterprise network in a way that is scalable for the populations you plan to deploy. Document how you plan to do this. The data volumes must be considered and accommodated. Investigate and document any issues of data quality and make recommendations to get them resolved.

10.4.2.2 Software Description Document (SDD)

For the pilot, we worked with temporary software to prove the connectivity of the readers and host. For this technical integration phase, we will embrace an enterprise software description methodology. The Software Description Document (SDD) will need to be extensively revised for the technical integration phase.

It is likely that you will integrate with a software package already running in your company. Select an application program with the understanding that in the next step of the project, you will need to create one or more Business Value Statements (BVSs) and deliver them. The application program you select will need to support your BVSs; see Section 10.4.2.6. Also, be sure the application program you select has a test instance and test database available so you can test your installation without impacting the live database. Add the name of the application program you have chosen to work with to the software section of the SRL.

Create a section of the SDD for your application program. Designate the system as a source of data for tags or a destination of data from tags or both. Identify the data structures that will be received and sent from and to the RFID reader. Also list and describe each data read or write scenario so you can test the integration to confirm that the data is being written correctly to tags, reports, software packages, and messages.

Provide the following information about the software package:

- Name of system
- The administrative contact (the person responsible for administering the system)
- Contact information of person responsible for users of the software

NOTE *A classic error of project management is failure to involve the users in the design and implementation of new systems. The users' attitudes can determine the success of a system, and their input can help you make it more productive. Do not fail to inform and consult these important stakeholders.*

- Status of the software (whether it is on maintenance, whether you get automatic upgrades and support, and so on)
- Ability of application to interface with and provide data to your middleware (hint: list the middleware packages it has been tested or certified with)
- Ability of application to create data-level interfaces and event-level interfaces to your middleware
- Ability of application to work with real time, as opposed to batch data inputs
- If relevant, ability of application to work with serialized item information
- Roadmap for the software describing when the required functions will be available

Work with the administrative contact to plan the installation of any upgrades that your project requires. List the tasks and responsibilities and create a timeline. Be certain to include user acceptance testing in the installation protocol.

Create a section of the SDD for the middleware package. List each function you expect the middleware to perform and create a test case so you can confirm that it performs correctly. Middleware functions are listed in Section 3.9.1.

Notify the document coordinator of any procurement that will be required.

10.4.2.3 Systems and Storage Description Document (SSDD)

The SSDD may need to be augmented for the technical integration phase. You may need additional computers, memory, storage, routers, peripherals, and so on. You should be working with only a single application program, but you will need to select a middleware package, and that is very important. The middleware may be distributed and therefore run on more than one computer. Some components may run at the facility and others at a data center that may be at a location remote to the site.

10.4.2.4 Network Impacts
Description Document (NIDD)

Network connectivity was out of scope for the pilot, so this document was not fully researched. Technical integration, however, is the foundation for the entire project's connectivity, so you need to carefully describe how readers will connect to the enterprise network.

The System Resource List (SRL) must contain a list of network resources your project will need. Other projects have found this to be a very challenging effort, as the quantities of data produced by the RFID project can be very large.

You should calculate the data flow rate your technical integration phase will generate across the network and confirm that the connectivity and capacity is there to handle it. In the event it is not, work with your IT department to secure the necessary upgrades. Update the NIDD to reflect your requirements, recommended solutions, and any costs and timeline issues.

10.4.2.5 Procurement Document (PD)

The PD for technical integration must deal with any needed upgrades to the application software, any controllers you may have decided to incorporate, the middleware, and any services you will need to get them installed and integrated.

- If outside vendor support is required, describe the support you need, and document the costs, lead times, and availability.
- If any software upgrades or new software packages are required, describe them, their costs, lead time, and availability.
- Research your middleware package and the services to get it installed; document the selection and research costs and availability.

Provide these documents to your supply chain management function as soon as possible so that they can perform their tasks in contracting with reliable and scalable suppliers.

10.4.2.6 Business Value
Descriptions Document (BVDD)

We are going to introduce a new document to our library, the BVDD. Objectives, as articulated in the OD, are important and, indeed, the foundation of the entire project. However, they are too general to be measured and therefore cannot serve as ready measures of business value achieved. The BVDD will turn objectives from the OD into specific, measurable value statements that can be used to calculate the actual value of the project as we deliver it. If one of the objectives is to reduce shrinkage within the company, we would need to determine the level of shrink in a given facility before the project was initiated and then create a Business Value Statement (BVS) such as, *"Reduce shrink in Texas DC from 8 percent to 4 percent."*

If one of the objectives is to increase product availability, we would need to determine the existing incidence of stock-outs of a particular product in a particular location. We could then create a BVS such as, *"Reduce stock-outs of power saws in Abilene DC from 10 percent to 3 percent."*

As you can see, there can be any number of BVSs for any single objective. The advantage of working with BVS is twofold: first, it forces us to document the level of performance prior to the beginning of the project; and second, it gives us a way to measure our project's contribution to the company as the project proceeds.

10.4.2.7 Workflow Description Document (WDD)

Business value will be created by enabling changes in workflows. Create a section of the WDD for each workflow that will be impacted by the technical integration. Describe the changes, including new business flow diagrams and use-case documents, to show how you will generate the business value you committed to.

10.4.3 Scope of the Technical Integration

As you can see, we have designed the technical integration to accomplish a few very well-defined objectives and to continue to develop our enterprise-class methodologies and documentation base. The procedures are designed to perform the tasks in a very orderly fashion, which will increase your chances for success and ultimately reduce the costs of the project. We have urged you to accomplish the technical integration's objectives in the most economical and straightforward way possible, without taking shortcuts that will impair your ability to deliver the rollout successfully.

10.5 Phase 3: Step 2, Execute Technical Integration

The work of technical integration can begin once the planning and documentation is completed. It gets the tags and readers communicating with the middleware and the middleware communicating with an application. It enables you to re-engineer a few workflows and measure your impact. It enables you to begin generating business value for the company.

As in the pilot, we have organized the actual execution in terms of preparation, facilities, procurement, installation, data management, systems, and software. First, we provide a checklist to complete before beginning execution.

10.5.1 Preparation Checklist for Technical Integration

This checklist summarizes the preceding material in a convenient checklist form. If you can check off all of the following items, your technical integration is ready to begin. You should begin only when all of the following activities have been completed:

- Scope documented and approved
- Objectives documented and approved
- Core team in place and meeting regularly
- Executive sponsorship secured
- Plan documents reviewed
- The following documents in particular have been updated or created:
  - Business Infrastructure Document (BID)
  - Data Description Document (DDD)
  - Software Description Document (SDD)
  - Systems and Storage Description Document (SSDD)
  - Network Impacts Description Document (NIDD)
  - Procurement Document (PD)
  - Business Value Descriptions Document (BVDD)
- Stakeholders briefed
- Vendor partners selected, briefed, and scheduled; issues logged and dealt with

- Pilot working satisfactorily
- User retraining completed if necessary

10.5.2 Execution of Technical Integration

The Project Plan (PP) describes the steps necessary to execute the technical integration. The activities will be scheduled in the PP and fall into the following categories:

- Procurement
- Installation
- Integration
- Data management
- Business value creation

Upon completion of the Project Plan, transition to operation.

10.5.2.1 Procurement

- Purchase any required upgrades to application software.
- Purchase middleware product.
- Purchase any services needed.

10.5.2.2 Installation

- Install upgrades to application software; test according to protocol in SDD.
- Install middleware product; test according to protocol in SDD.

10.5.2.3 Integration

Perform the following tests on the test database, to make sure the system can process single transactions correctly without corrupting the live database:

- Subscribe application to a tag event; initiate event and observe application processing correctly.
- Flow data to encoder/printer and print smart label.
- Flow data to tag writer and commission tag.
- Test each write event listed in SDD.
- Test each read event listed in SDD.
- Test all middleware functionality; use function list in SDD.

10.5.2.4 Data Management

The integration activity in Section 10.5.2.3 tests to make sure that the connectivity is working correctly on single transactions. Data management extends these tests to make sure that the data is flowing correctly, updating databases, triggering events, producing reports, or sending alerts. At the completion of data management tests, connect the system to the live database.

10.5.2.5 Business Value Creation

Supervise the transition to new workflows as the RFID data becomes available.

10.5.3 End of Technical Integration

Technical integration is completed when you can operate a workflow involving workers, packages, readers, antennas, tags, and application software all working together to produce a business outcome that has value to the organization. The methods for engineering each element of this workflow have been created, documented, and tested so they are reliable and repeatable. When this has been done, you are ready for the rollout. As with the pilot, the end of technical integration is marked by four activities and is followed by approval for the next phase:

- Evaluate technical integration.
- Capture and document lessons learned.
- Preliminary assessment for Phase 4.
- Brief Executive Steering Committee (ESC).
- Get approval for Phase 4; see Section 10.5.3.3.

10.5.3.1 Evaluate Technical Integration

With your core team, evaluate the pilot against the criteria listed in the OD. Each objective should be rated exceptional, satisfactory, or unsatisfactory. Also evaluate the Business Value Statements (BVSs) and rate their accomplishment. Solicit and document any unexpected values and costs the project may have generated.

10.5.3.2 Capture and Document Lessons Learned

As soon as possible upon completion of technical integration, you should canvas the stakeholders to review all opportunities for improving your team's processes. Solicit insights they may have regarding ways RFID may support their workflows. Discuss these at the next team meeting and gather contributions of team members. Generate a list of changes to initiate, going forward.

10.5.3.3 Preliminary Assessment for Phase 4

The purpose of the preliminary assessment is to gather enough information to make sensible recommendations to the executives during their next briefings. To make recommendations, you will need to develop the following information:

- Set objectives for Phase 4; see Section 10.6.2.2.
- Create roadmap for adding readers in the pilot facility.
- Create roadmap for adding facilities.
- Create roadmap for adding applications.
- Create roadmap for adding packages and products.
- Create roadmap for adding workflows.
- Report on progress of any dependencies from the BID and discuss any issues.
- Estimate a timeline for these roadmaps.
- Estimate costs for RF equipment, tags, facilities changes, network upgrades, computer equipment, integration, training, and project management.

These roadmaps should be the product of your core team, with input from the project team members before being presented to the ESC.

10.5.3.4 Brief the ESC

The actual briefing is conducted verbally and supported by a printed handout. The purpose is to report the success of the project thus far and to get approval for the next phase.

Review the project objectives for technical integration and your assessment of their level of achievement. Also review any important lessons learned. Then present objectives, timelines, and estimated costs and get approval for Phase 4.

10.6 Phase 4: Step 1, Plan Your Rollout

Planning your rollout should be very similar to planning the pilot and technical integration, except there are more facilities, more readers, and more applications. At the risk of some repetition, we will present here a complete project rollout plan. These documents utilize the habits, processes, and disciplines you learned and practiced during the previous two phases.

We will present Phase 4 as a monolithic rollout, but it is more likely that you will want to break it into individual blocks of activity in logical segments. Each separate block should be approached with the same dynamic: first plan and document, then execute. It has served us well so far and will keep you out of trouble. Remember that the document preparation process is built to include stakeholders to make sure their needs are met, and you can benefit from their insights and contributions.

10.6.1 Project Team for Rollout

Begin rollout by reviewing your core team membership. Make any necessary changes to the team going forward. You may have to add headcount for the increased levels of the workload. (Given the larger number of individuals involved in this rollout phase, constant two-way communications will be critical to assure success.)

Once your team composition is settled, schedule a kickoff meeting for rollout. At the kickoff meeting, introduce any new members. Review the objectives, roadmaps, and timelines as you presented them to the ESC. Discuss any issues. Assign responsibility for each document and agree on delivery dates and milestones. Review the Business Infrastructure Document and discuss issues. Lay out activities for the coming month and for the project as a whole. (Communicate this plan in simple to understand terms as soon as possible to all members of the rollout team.)

Schedule a second meeting with the project team to cover the same information in less detail. Solicit inputs, comments, and issues.

Schedule weekly meetings to review progress, issues, risks update, achievements, and milestones. Also present and distribute all project documents in final draft that week for review, coordination, and comment. (Make sure the project website is always maintained and contains a two-way communications capability.)

10.6.2 Plan Rollout

The project methodology is now well established. The planning step consists of investigation and preparation of a set of documents. For the most part, the documents created during the previous two phases will be reviewed and updated. We will discuss the documents where the issues are significantly different for rollout than for the prior phases.

10.6.2.1 Business Infrastructure Document (BID)

Review and schedule appropriate action on the issues revealed in the BID. Recall that many of these issues were out of scope or moot until now. At the rollout these issues must be addressed in detail to assure the rollout will succeed in producing the desired return on investment (ROI).

1. Verify that methods of numbering, tracking, and cross-referencing comply with relevant standards. This issue is resolved when there are documented systems and procedures in place to issue and track ID numbers, ensure their uniqueness, and communicate them to the correct reader.

2. Verify that reader network interface with enterprise network is documented.

3. Verify that data volumes are estimated, their impact on enterprise network is stated, and any necessary upgrades are scheduled.

4. Verify the existing business systems' ability to work with real-time data has been evaluated and the necessary upgrades have been scheduled.

5. Verify that data quality analysis has been completed and the necessary activities have been scheduled.

10.6.2.2 Objectives Document

Return to the OD for the project as a whole. Review these objectives and make any additions or changes necessary. Recall that these are high-level statements of purpose and will need to be expressed as BVSs in order to be actionable.

10.6.2.3 Business Value Description Document (BVDD)

Review the OD and support each objective with specific BVSs. Solicit these from your stakeholders and involve them in discussions of what will be required to achieve them. Two examples of BVSs are listed in Section 10.4.2.6.

10.6.2.4 System Requirements List (SRL)

Review the BVDD and support each BVS with specific resources. These will include data structures, workflow additions and changes, human resources, network resources, and systems.

Add to the SRL a list of all facilities where you plan to install equipment. Characterize each resource by its facility name and location.

Add to the SRL a list of all products that will be tagged. Depending on your scope, it may include products your company manufactures, those made by others that you distribute, or those incorporated into your own finished products. It could also include materials that your company consumes, such as cleaning supplies.

Add to the SRL a list of all package variations that will be tagged.

10.6.2.5 Workflows Description Document (WDD)

Use the list of workflows in the SRL and create a section in the WDD for each workflow. Use the protocol described in Section 10.2.5.4.

10.6.2.6 Data Description Document (DDD)

Review the SRL and set up a section in the DDD for each data structure that will be required to support the workflows. Use the data description methodology in Section 10.2.5.6.

For the rollout, you will need to augment the DDD with a data map showing flows of data from tags to applications. Besides listing the data structures, the DDD contains a map of the flow of data for the entire pilot. To build this map, follow this protocol:

1. Determine each software package and work process involved.

2. Name and describe the data records that must be sent to that software.

3. Document the source(s) of each data element within the record.

4. Name and list the data elements of each smart label that must be printed.

5. Name and list the data elements of each tag that must be encoded or smart label that must be printed; map the sources of the data that will be printed on the label and written to its embedded tag.

6. Evaluate issues of synchronization, making sure that each record consists of data collected in a consistent time frame and under consistent definitions. This is particularly important if the record combines data from multiple sources such as tags, sensors, and applications.

7. Document the required level of security for each data record or structure. Note that the Gen-2 tags have capabilities to protect individual data elements within a tag with passwords and, indeed, to control access to the entire tag by use of a password. You will need a security architecture to make sure the reader/encoder has the right passwords to be able to write the data to the tags. The security architecture will also need to provide each reader with correct passwords in order to be able to read the desired data from the tags.

   To document the security profile for each data element, you must decide

   - Whether the data element is important data in its own right, or just a pointer to a secure database
   - Whether the data element must be encrypted, and if so, to what level
   - Whether the data element must be password protected
   - How tag passwords will be managed
   - Whether the data element is governed by any national or local laws
   - Whether the data element would be valuable to terrorists or criminals, either outside or inside the company

8. Specify any format transformations that must occur. Some readers create fields that are the same size, with the actual contents padded on the left with zeros or with spaces. Specify how to reformat the fields and which software will do this.

9. Specify any other transformations, including table lookups, calculations, or adjustments.

10. Specify any summaries, averages, and so on that must take place.

11. Specify any thresholds that should trigger an alarm, alert, or other immediate action. Map the flow of data to the alarm mechanism.

12. Specify any thresholds that should trigger a business action such as placing an order, sending an invoice, or sending a notification. Map the flow of data to the correct software.

13. Specify thresholds that should trigger a system administration notification, such as bad reader, faulty communication, damaged antenna, or extraneous radio frequency interference.

14. Build a test set that is readily available to help you pinpoint the exact nature and location of any failures.

15. Specify how your system will deal with duplicate reads. Consider tags that may be read by adjacent readers. Consider tags that go through a reader inbound and then go out again. Consider tags that may appear to go through the same portal twice because the item was moved out through an unmonitored exit.

16. Consider and document any read and write operations that take place after commissioning. Note which work process produces or consumes the data elements.

In the DDD, your test strategy calls for onsite installers to verify each data flow before signing off on the installation. Include the source for test data in this process.

10.6.2.7 Software Description Document (SDD)

Review the SRL and set up a section in the SDD for each application program, software agent, or report writer module that will be providing or consuming data from your project. Use the software description protocol in Section 10.4.2.2.

10.6.2.8 Systems and Storage Description Document (SSDD)

For the technical integration, we selected a single computer system to prove that we could provide an enterprise-class interface. Now, review the SRL and create a section in the SSDD for each computer system impacted by the RFID project. Follow the protocol described in Section 10.2.5.8.

Calculate your storage requirements and be sure the capacity you need is available and/or can be acquired.

10.6.2.9 Facilities Description Document (FDD)

For the technical integration phase, we only had to deal with a single facility. For this phase, we will have to deal with several. Working from the list of facilities in the SRL, create a section in the FDD for each facility where you plan to install readers. Follow the protocol described in Section 10.2.5.5.

10.6.2.10 Products Description Document (PDD)

Create a section in the PDD for each product that will be tagged. You can get this list from the SRL. Use the protocol described in Section 10.2.5.9.

10.6.2.11 SKU Description Document (SKUDD)

Create a section in the SKUDD for each separate package variant in which the product is available. These are normally referred to as SKUs. Document each SKU per the protocol described in Section 10.2.5.10.

For each SKU, you will have to select a tag and determine the best mounting spot and orientation to insure optimal read rates. Use Protocol 10 in Chapter 12. In the SKUDD, document which tags work best and diagram the location and orientation.

For each SKU, note any special conditions the package may present. For example, if it is moist, there will be adhesion issues. Similarly, cold or heat will affect tag adhesion, survivability, and performance.

Ultimately, you want to generate near-perfect read rates using as few different tags as possible across the project, so try to give yourself several choices for each SKU in the SKUDD.

10.6.2.12 Tag Selection Document (TSD)

Create a section in the TSD for each tag you wish to use. For each tag, use the protocol in Section 10.2.5.13.

Make every effort to make a single tag work for all of the packages.

Ultimately, you need to have a tag strategy, and over time, the tags are the most costly portion of the RFID installation. The more you can leverage a few tags, the lower your overall cost will be. In the TSD, project forward the quantities of tags on each package and see what your volumes will be. This data will be useful when you negotiate prices for the tags.

10.6.2.13 Reader Category Document (RCD)

You will need to develop standard configurations of readers for the various categories you will need to install. For example, you might have categories of readers such as large portal, small outdoor portal, forklift, indoor handheld, and so on. The RCD was not included in the pilot because the number of readers was limited. But as the number of readers increases, having standard product selections and configurations will simplify your project. Using a standard nomenclature for each reader category will also be surprisingly helpful. Be sure to include testing time to develop these standards. For each category in the RCD, you will need to create standard onsite documentation.

As you proceed with deployment, be alert for exceptions—occurrences when a standard reader does not fit in its category. For instance, if you are unable to make your standard portal reader work in a portal situation, document the event and determine the circumstances. Notify the document coordinator and update the RCD.

10.6.2.14 Reader Selection and Deployment Document (RSDD)

Expand your RSDD using the protocol in Section 10.2.5.14. In addition, for the rollout, you need to augment the RSDD with a section on managing readers. The middleware gives you reader management capabilities:

- It should track read rates and issue notifications when there are problems with a particular reader.
- It should enable you to reconfigure readers from a central location.
- It should enable you to update the reader's firmware or software from a central location.

Your RSDD should document how these events are to take place in your project, who is responsible, who is notified, and what actions are to be taken.

10.6.3 Procurement Document (PD)

Create a section in the PD for each equipment type, supply type, tag model, and services type you will need to purchase. Use the protocol in Section 10.2.5.15. Record the quantities of each item you will be purchasing. As you work out your project plan, document the dates, quantities, models, and locations to which you will need the items delivered. Include in the PD the availability and delivery lead times.

10.6.4 Project Training Documents

For the pilot, we only provided training to the workers who would be involved in creating or working with the tags. As the project rolls out, however, new training issues will arise and you will need the following:

- Additional training modules for the additional reader categories
- System administration training for the RFID network
- System administration and configuration for the middleware
- Configuration of application-level events (ALEs) for each application that will use them

10.6.5 Onsite Documentation

For each reader site, you will need to provide a binder, CD, or access to a web page that includes relevant documentation. We recommend that the documentation be available on the project website as a preferred method, since binders and CDs may become misplaced or otherwise inaccessible. The required contents are listed in Section 11.6.10.

10.6.6 Business Integration

As readers are being installed, you can begin business integration. This is the process of integrating RFID data into the various workflows and adding new workflows to leverage the data. One of the unexpected early lessons of the Wal-Mart initiative is that business advantage can be achieved even without complete data, perfect read rates, or substantial numbers of facilities enabled with RFID. With just a few stores on their RFID network, Wal-Mart was able to document a significant decrease in stock-outs and lower inventory levels.

Full business integration is out of scope for this book, but you can probably gain early business advantages as you install your systems.

10.6.7 Project Plan

Let's look at how these issues play out in the project plan. The plan is divided into these sections, as described in Section 10.7.

- Testing
- Facilities
- Installation
- Data management
- Reader management

10.7 Phase 4: Step 2, Execute the Rollout

The rollout is the final demonstration of your capabilities, the final delivery of your system. As with the previous two phases, each area must be planned to the last detail, documented and tested before delivery is begun. A task is completed only when the following criteria have been met:

- Equipment performance is consistent.
- Performance levels are satisfactory.
- Personnel are trained in all aspects of normal operation, maintenance, and simple remedies for common problems.
- Data is being delivered as described in the data map.
- Workflows are in place to make use of the new data productively.

You need to produce a complete, detailed project plan for the rollout. This plan will list every activity that needs to be performed, the date it begins and the date it ends, along with the person responsible and the resources they will require. The plan should be developed using a tool like Microsoft Project.

The plan should be broken into individual phases according to a logical order. For example, you might break it down as follows:

Phases	Activities
Implementation	
Testing	Static reader testing
	Dynamic reader testing
	Tag selection
	SKU testing
Facilities	Facility installation schedule
	RF site analysis
	Power line upgrades
	Construction
	Facility equipment moves, changes

Phases	Activities
Installation	RFID equipment installation: readers, antennas, encoder/printers RFID supplies: cables, connectors, racks, tags, labels Network upgrades Test readers Train workers Read rates verification Tag validity verification Correct tag writing verification Labels completed correctly verification
Procurement	
RFID Equipment	Readers Antennas Tags
RFID Supplies	Labels Other
Software	Purchased Developed
Services	As needed
Management	
Data Management	Correct tag numbering and maintenance of databases Correct data elements delivered to labels Application logs audited to make sure data is flowing in correctly Persons responsible for various applications know how to subscribe their applications to the RFID middleware Reports produced correctly Alerts and automatic activities functioning correctly
Reader Management	Read rates being monitored Alerts (or other appropriate actions) being taken when there are problems Updating of firmware tested Dense reader issues identified and resolved

10.8 Review and Evaluate the Project

When the installation of readers in each facility is completed, you will need to evaluate the project against its original objectives and determine whether they were met.

In your review and evaluation, document any unforecasted benefits the project created for the company and any unpredicted costs.

10.9 Implementation Notes

This section makes several suggestions for you to be aware of in your implementation. It draws on dozens of actual installations and offers sound, practical advice.

10.9.1 Test

- Test your RFID system as rigorously as possible.
- Test the components in a laboratory and document the limits.
- Test them again at their locations under as many adverse conditions as you can.
- Test the interfaces.
- Test the tags.
- Test various orientations.
- Make sure you know how your systems work and how they fail.

10.9.2 Do Your Analysis up Front

Document your process flows, data flows, and use cases before you make your decisions. It is surprising how quickly implementers rush to installations without having done their homework.

10.9.3 Your First Procurement

The very first thing you should buy is a spectrum analyzer. This device, as described in Chapter 12, will keep you from guessing what is happening in and around your read zones.

10.9.4 Pick a Problem and Solve It Well

RFID projects often fail because the scope of the project is too wide. Project managers try to solve too many problems at once, leading to overambitious project expectations that cannot be met. A good way to ensure success is to pick a small problem to solve first. This lets you learn the fundamentals, get experience, and have a quick win. (Make sure you gain clear alignment at the senior management level regarding realistic expectations and time lines for the project.) One example might be to just track full pallets entering and leaving a distribution center. Sometimes these pallets go across the dock without being broken down into cases, so they are not counted accurately. The bar code process works better when pallets are broken down into cases. Set up RFID portals for these pallets and work out the systems and kinks in the process before moving to more challenging applications.

10.9.5 Add Extra Checkpoints

Tags may be rugged, but they can still be damaged as the items are handled. Put readers at convenient points to check the tags and make sure they continue functioning. Even slap-and-ship operations need a checkpoint as the item leaves the facility to verify that tags are working properly.

10.9.6 Mark Your Read Zones

Make sure you know the boundaries of your read zones. You might consider marking them on the floor, so people always know when they are in one. The process of documenting the read zone will enable you to measure and set the power levels correctly. Doing this will help prevent ghost reads from other areas.

10.9.7 Angle Your Antennas

Do not point antennas at one another. Create the read zone in front of, or behind, the portal, and set the antennas at 30 to 45 degrees off, as shown in Figure 10-2. That will help prevent them from interfering with one another. It also lets you detect the order in which tags are read, which could be important in some applications.

Figure 10-2
Canted read zones

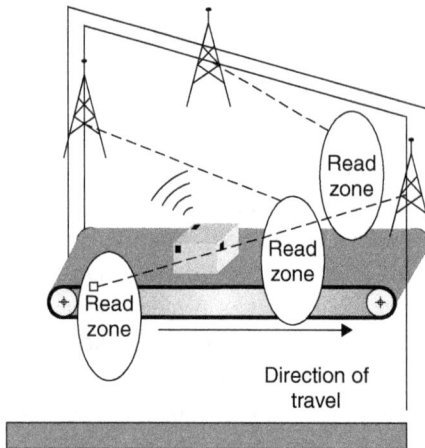

Read zone

Read zone

Read zone

Read zone

Direction of travel

10.9.8 Create Involvement Opportunities

Include as many people as possible in your project. These people will gain insight about what the project is trying to achieve and how you are trying to achieve it. Once these people are informed, they will become some of the best advocates for the project. One very successful project held open meetings once a week and served cookies. People who were interested came, learned about progress and problems, and were able to speak with one another about possible applications and lessons learned. The meetings became a major force for bringing together the whole company, an institution in its own right, and created a forum where people could explore whether the technology had anything to offer them.

Another project held a contest within the company to come up with a name and logo for the system. Over 300 people entered, and the winner was awarded a portable DVD player and her name and picture were in the company newspaper.

10.9.9 Set Up a Website

Create a project website where you show the project plan, issues list, risks list, meeting schedules, contact information for the core team members, pictures of installations, and performance metrics. A critical requirement of such a website is to allow two-way communications between both team members and other members of the company.

10.9.10 Set Up an RFID Dashboard

This can be a web interface that gives information about the RFID network. It could show performance statistics for each reader, aggregated by facility. The dashboard should show very high-level statistics, which can give a good picture of the functioning of the network. One problem with RFID is that it generates huge volumes of data. The dashboard should summarize that data as succinctly as possible. The ability to drill down into the data is helpful.

10.9.11 Validate Tags

Tags functioning is one part of your project's success, but it is not enough. Tag contents should be checked for accuracy and validity, not just readability. You need to ensure that the right data is on the tag. This is done using business logic at some point early in the process. Your process needs to trap readable tags with the wrong data with a high degree of accuracy.

10.9.12 Get the Right Tag

Choose the right tag for your package. Make the effort up front to look at all possibilities and ensure that the tag you pick will perform in all the circumstances and environments in which it will be read.

10.9.13 Antenna Coverage

Don't be afraid to cover a read zone with more than two antennas. The closer you can get the antennas to the tags, the more likely you are to have good read rates. If putting four or even more antennas on a read zone is what it takes to get you the performance you need, it is well worth it!

10.9.14 Metal Products and Packaging

We have discussed throughout this book the problem that metal creates for passive UHF tags. A package may contain a foil lining, so it may not be apparent that metal is present in some packages. When you are working with metal, place a quarter inch of cardboard or Styrofoam between the metal object and the tag.

10.9.15 Manage Expectations

Implementers need to make sure stakeholders understand how the system will save them time and money. But they also need to understand the risks, costs, and timetables. Do not implement business process changes until you are certain that the system will perform satisfactorily.

10.9.16 Remote Monitoring and Management of Readers

The ability to remotely monitor and manage the readers is essential. As these RFID systems become indispensable, you cannot afford the downtime associated with hands-on diagnosis, restarting, troubleshooting, and repair. Similarly, if there are 250 readers in your company, manually upgrading their firmware will take substantial amounts of personnel and business downtime. Having a system to do these upgrades remotely will pay off significantly.

10.9.17 Check the Regulations

There are numerous regulations around the RFID project, and you should be very careful to observe them. Radio frequency transmissions are regulated, as described in Chapter 1. Various hazardous substances and pharmaceuticals have strict federal and state regulations that must be complied with. Even the data itself may be regulated. For example, medical records are subject to government privacy regulations. Substantial fines can be imposed for violations even if they are unintentional.

10.9.18 Unexpected Benefits

There have been numerous cases where the sheer visibility of new information or new scrutiny of processes long taken for granted has revealed unexpected benefits. Be alert for these and be certain to document and leverage them. Note them for your project evaluation at the end.

10.9.19 Identify, Document, and Manage the Risks

Technical projects generally experience four sets of risks: the project management risk, the technical risk, the economic risk, and the schedule risk. In the case of your RFID project, all four are noteworthy. The project management risk is the risk that some key factor or resource has not been considered. This can impact the timeline and costs. The technical risk is the risk that the technology may not perform as expected, equipment originally selected and budgeted may not be adequate, or assumptions about the amount of equipment may prove too low. The economic risk is the risk that the cost of some services or equipment will be higher than budgeted. The schedule risk is the risk that any of the tasks take longer than planned. Maintain your risk log, be as specific as you can in identifying the risks, and constantly work to eliminate them before they become a problem.

10.9.20 Project Management

If possible, find someone who has had extensive experience managing a project of equal or greater complexity and scope. You should look for a clear, well-documented project methodology. Your potential vendors for this service should be able to show you project documentation from other engagements, and you can evaluate their quality. The project management process should emphasize clear consistent communication including:

- Progress made on each activity
- Milestones scheduled for the reporting period and milestones achieved
- New activities begun or activities completed
- Cumulative cost against budget
- An updated list of risks and the status of remediation efforts
- Any new issues

This author prefers to see weekly project meetings supported by updated documentation. Each item should be color-coded green (on time, on budget, no identified risks), yellow (on time, on budget, there are risks), or red (late, over budget, or has risks for which no remedies have been identified).

10.10 Chapter Summary

This chapter presents a detailed protocol for planning and documenting your project. It begins after the justification has been prepared, presented, and approved. It ends when the RFID system is rolled out throughout the entire organization. The four phases of the project are preliminary planning, pilot, technical integration, and rollout. The chapter presents protocols and document descriptions for each phase, and shows how to gather and organize the team, the documents, and the resources you will need.

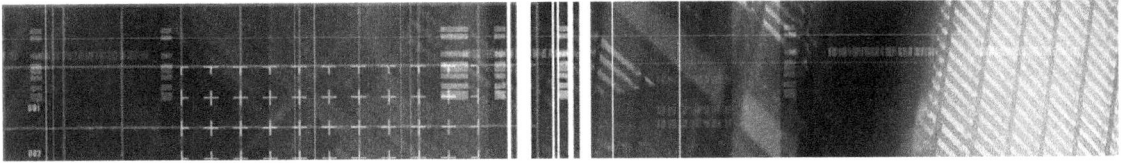

CHAPTER **11**

Tag-and-Ship

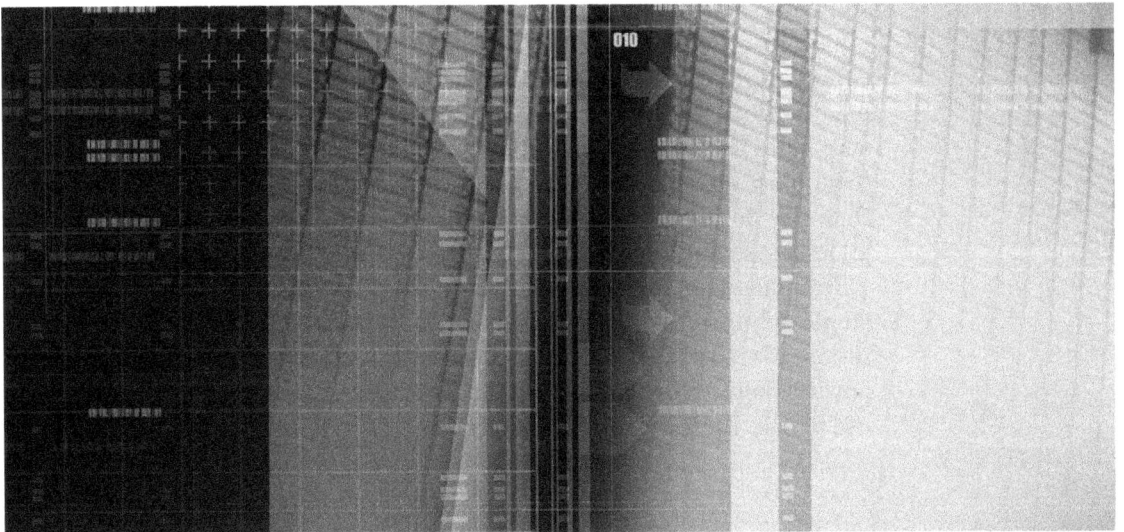

Companies confront the inevitability of RFID in the supply chain with a range of responses. Wal-Mart put the requirement on its 100 largest suppliers; 137 of their suppliers actually began compliant shipments—the additional 37 volunteered. The volunteers did so for one of two reasons: many wanted to be on-board with a large and influential customer, and many others used the mandate to justify and accelerate their adoption of a promising new technology. Of the 100 initial inductees, several embraced the initiative as a way to drive time, error, theft, and cost out of their business processes. At the other end of the spectrum, many of the original 100 companies opted for solutions that would have the lowest out-of-pocket cost and least disruption to their existing processes. These latter companies adopted strategies that are referred to as *tag-and-ship*.

11.1 Tag-and-Ship Is Controversial

Tag-and-ship has numerous detractors in the professional ranks. Experts point out that tag-and-ship only represents an additional cost, with no payoff for the investing companies (other than meeting the requirements of their largest customer!). They point to today's high ongoing costs of time, errors, and shrinkage, to say nothing of losses from counterfeiting, stockouts, misplacement of tools and equipment, misrouting of orders and management of recalls as opportunities not seized by a tag-and-ship strategy. They also cite improved handling and increasingly detailed requirements for record keeping for hazardous and controlled materials. RFID, they say, is the beginning of a technology that can address all of these issues.

Critics of RFID say that its ability to reduce these costs and actually solve these problems remains unproven. They say that, to date, there is no proven payoff for RFID investments. They also say that tag read rates are too low and too unreliable to provide a sound basis for management actions and that the standards are incomplete and not universally accepted. They state that the set of products being tagged is too small to support promises of large improvements in productivity. It's too soon, they say, to invest heavily in the technology. They point out that most pilots are unsuccessful.

Proponents of RFID say that tag read rates are low because companies employed unscientific tag and antenna selection methods and tag orientation, placement, and tuning processes. They also state that there is no payoff when companies ignore the RFID-generated data and fail to make use of it in their business processes.

Critics of RFID recognize that the retailer and DoD mandates have made investment in the technology part of the cost of doing business. And, as prudent managers, they seek to keep the cost and impact as low as possible. These managers have adopted tag-and-ship.

Proponents say that the current processes are remarkably inefficient and drive costs of inventory, errors, shrinkage, stock-outs, misplacement, misrouting, and fraud higher than necessary. They say that no other solution is on the horizon, and companies that make the effort to leverage RFID and build systems around it will develop a sustainable competitive advantage, while those that do not will see cost structures that are ultimately fatal in the marketplace.

Critics say that there have been numerous technology efforts to "solve" the supply chain problems in the past, including Material Resource Planning (MRP), Electronic Data Interchange (EDI), Enterprise Resource Planning (ERP), and Business-to-Business (B2B) portals. RFID, they say, is only the latest technology hype, marketed to solve problems they have lived with for years.

Proponents say that this time is different because of the pressure by the DoD, Wal-Mart, and other retailers. The companies that fail to leverage this technology to the fullest will ultimately fail in the marketplace. They point to the success of Dell, Wal-Mart, and Tesco as the first wave of savvy organizations leveraging their ability to employ technology to drive competitors out of business.

11.2 Methods of Tag-and-Ship

Three methods of tag-and-ship can be considered. Companies can replace a few bar code printers with smart label encoder/printers. This method has been called "slap-and-ship," as it highlights the manual process of applying the newly printed smart label to the packages bound for particular retailers. The term is faintly derogatory.

The second option is an alternative to slap-and-ship. It consists of applying an RFID tag, in addition to and separate from the bar code label. The company uses a separate RFID reader (recall that a reader can also commission a tag) to encode the new tag with the data required by the customer. For both methods, the bar code is printed and applied. The bar code thus serves as a backup for the RFID tag and enables the rest of the systems to work with the package without having to change them. This chapter will describe both methods.

Both of these methods contemplate leaving the entire manufacturing and distribution process untouched and accommodating the mandate by just adding a step at the end, where the tag is applied. The new process is applied only to pallets and packages bound for the demanding customer. The rest of the packages are processed as before.

A third option is not available today, but it is on the horizon. Its detailed characteristics are not yet distinct, but it seems clear that in the near future, companies that print your disposable packaging will be able to print antennas—and a few years after that, the integrated circuits themselves and even batteries!—directly onto your packages.

11.3 Steps to Tag-and-Ship

Companies that adopt tag-and-ship will need to take the following steps. We are assuming that you are working with a customer mandate, and the mandate specifies the frequency and characteristics of the tag.

1. Determine level of integration with your existing systems you wish to achieve.
2. Get EPCmanager number (Company Code).
3. Conform your identification numbering system to EPCglobal standards.
4. Select equipment.
5. Select tag for each package. Perform SKU testing to find where on the package to locate the tag.
6. Set up new workflow.
7. Train workers.
8. Launch production.

Each of these steps is discussed in a following section.

11.3.1 Level of Integration

Decide to what extent this implementation will integrate with existing systems. Points of integration to consider are

- Management of unique identification numbers
- Update of Warehouse Management System

- Production of Advanced Shipping Notices (ASNs) or other required documentation
- Update of ERP database
- Notifications and alerts of exceptional occurrences

These are thorny issues in their own right. The assignment of unique identification numbers, for example, should follow EPCglobal guidelines and be undertaken within their structure. At a minimum, you will need to apply for an EPCglobal manager number and set up identification systems to ensure that each item tagged has a number that is unique and trackable.

Even if you retain all your "real" processing in the bar code system, you will need to draw on your warehouse management system and your ERP system to create documents your trading partner may require. If the identification numbers are different for the tags than for the bar code, you will want to track the tag numbers in either or both of these systems.

11.3.2 EPCglobal Manager Number

If your company does not already have one, subscribe to EPCglobal and register your EPCglobal manager number. Annual subscription costs are calibrated to the revenues of your organization, starting at $5,000 for the very smallest companies. If you already have a UCC number or an EAN number, you can continue using that same number, but you will have to notify EPCglobal when you subscribe. Additional fees will be due.

Having an EPCglobal manager number ensures that your numbers are unique on the planet and that when the time comes you will be able to exchange identification data with your trading partners.

11.3.3 Conform Identification Numbering System

Establish or review your mechanism for assigning and maintaining SKU numbers. EPCglobal calls these *object classes*. Most companies have a "product master" database. Many companies have more than one, and often they are inconsistent. Different business units typically put up databases for their own requirements, and often there is no one clearly responsible for the completeness, accuracy, availability, consistency, or maintenance of this data. Many companies find this is a good time to establish one single product master database with clear ownership, availability guidelines, accuracy requirements, update turnaround requirements, data integrity responsibility, and cost responsibility. The product master is a company-wide resource and will become ever more important in the years ahead.

You should review the processes by which identifiers are assigned and managed. Verify that they conform to EPCglobal guidelines. In lieu of replacing existing identification systems, some companies merely set up a translation table to convert from their existing scheme to the EPC.

11.3.4 Select Equipment

Select an encoder/printer (if you plan to utilize slap-and-ship) or a reader (if you plan to attach just the tags). If you are adding RFID functionality to an existing bar code station, contact the equipment manufacturer and find out whether there is an upgrade for the equipment that provides RFID capability. Some manufacturers will add an RFID module to their bar code printers. If you need to purchase new equipment, see Chapter 3 for selection criteria.

In addition to at least one encoder/printer, you will need at least one reader to verify the tag performance and tag contents. Select a reader that can work with the tags you plan to use. Use the reader as close to the point where goods are loaded onto trucks as possible. This will ensure that the tag leaves your premises in good working order and contains the correct data. It will also reduce a common source of error: goods going through the portal but unable to be loaded on the truck for one reason or another. If the reading of the tag is interpreted to mean "on the truck," you should make every effort to make that true.

11.3.5 Select Tag and Tag Location for Each Package

Select the tag or label for each package type you will be tagging. Determine where on the package you will locate the tag. See Protocol 10, Chapter 12 for the details of this process. Note that if you are affixing a tag, you have greater flexibility in orienting it on the package than you will have with a label. This may make a difference, especially in weaker portions of your read zone. Determine the level of radio frequency interference around your reader. Take steps to reduce or eliminate it. See Chapter 12 for this protocol.

Even if you plan a pilot implementation involving just a few SKUs, you still must do formal SKU testing to select and position tags for each different package. You may want to engage a qualified consultant to do it for you. But as your scale increases and the variety of packages you need to tag grows, you will want to establish either your own RFID test range (see Chapter 12) or a consultant relationship with someone who can regularly provide those services.

Forecast your usage and order your stock of tags or labels.

11.3.6 Set Up Workflow

You will need to set up a new workflow to commission the tag or label and attach it to the package. It will need to be designed and documented. Carefully define the conditions under which it should be performed. Usually, this amounts to naming the destination (vendor and location) and the product. Some companies have their computer system mark packages that must be tagged with red signboards, upstream in the process, when the determination is made. The condition should be as clearly and unambiguously marked as possible, especially when it is part of a high-volume operation.

Carefully document the steps of the workflow, and name the actor for each step.

11.4 Slap-and-Ship with Smart Labels

Many bar code printer manufacturers provide smart label machines to replace their own thermal bar code printers, or they offer upgrades to their existing machines. These machines print smart labels that range in size from 4×2 or 3×3 to 4×8 inches and contain an embedded RFID tag. The bar code label printer is replaced with an encoder/printer, and all upstream processes remain unchanged. The only additional step is the changing of rolls of print stock in order to accommodate two different printing protocols—those without RFID tags embedded, and those with tags embedded. Software migration tools (for example, from Paragon Data Systems) allow users to convert from the printing of bar code labels to the encoding of smart labels without changing legacy systems.

The advantages of slap-and-ship with smart labels are

- The labels are convenient and inexpensive, although they are about ten cents more expensive than just tags by themselves.
- They provide machine-readable identification (both RFID and bar code label) and human-readable identification tags on the same package, at the same time, and with the same labor step.
- They perform their own error detection.
- The encoding step requires no separate engineering; its process is built into the structure and usage of the encoder/printer. (Note that the analytical step of determining where to locate the label still needs to be done.)

The disadvantages are

- Workers will have to change roller stock each time goods are being shipped to your RFID-expecting customers.
- Some packaging solutions can be solved better with tags than with labels, specifically in challenging environments or on very small packages.

11.5 Tag-and-Ship with RFID Tags

Some companies leave the bar code system in place and add a separate RFID tag applicator system. The processes all continue to work as they did before, and bar code labels are applied (as before) to all the cases or pallets. These companies then apply stand-alone RFID tags only to the items that will be shipped to customers who demand them. The advantages of this approach are

- RFID tags are applied only when needed. The task of switching rolls of print stock is removed.
- The tags are five to ten cents cheaper than comparable smart labels (but you still have the cost of the labels).
- All upstream processes are undisturbed. The system can be placed downstream from the bar code labeling station.
- RFID tags are available in a wider variety of types and sizes than smart labels. This offers more flexibility in locating and orienting the tags on the package.

There are some disadvantages to this approach:

- It adds a step and a cost to the production process. It also adds an opportunity for error.
- It does not necessarily comply with industry and consumer guidelines for having a visible indication that a package has an RFID tag.
- Engineering the tag write process must be done with care; it must be clear which tag is being commissioned at any point in time, so you must not ever have more than one tag in the write zone at a time.
- If tags will go on items to be delivered to the general public, you will need to provide a mechanism to remove or kill the tags.

11.6 Training and Onsite Documentation

Workers at a tag-and-ship operation will need basic RFID training. This will include training in

- Tag placement
- Removing the adhesive layer
- Monitoring stocks of tags
- Arrangement of cartons on pallet
- Correct use of encoder/printer
- Control of ambient electronic noise
- Data integrity
- Orientation on onsite documentation

11.6.1 Tag Placement

Every different package type will have its own optimal location and orientation for the tag and may have its optimal tag as well. These are based on the package configuration and the contents. You can determine these using Protocols 6, 9, and 10 in Chapter 12. You will need to provide detailed directions to your workers and training in how to make sure the right tag is in the right position on each package type they will be tagging. Recall that in some cases, placement and orientation must be very precise.

11.6.2 Adhesive Layer

Smart labels have a protective sheet that must be removed before they can be attached to the package. This is similar to other labels, but smart labels are more delicate than the bar code labels that workers may be used to. The antennas are delicate and may become damaged if handled roughly. Training your workers to remove the protective sheet will ensure they know how to work without damaging the antennas.

11.6.3 Monitoring Stocks

If you are running several different kinds of packages through your tag-and-ship operation, you may stock several different kinds of tags. Even if you only have one kind of smart label or RFID tag, you will need to monitor stocks to make sure your ability to fulfill customer orders is not stopped by running out of tags or labels. Your workers are the first line of defense against this, and they should know lead times and ordering procedures.

11.6.4 Arrangement on Pallets

The arrangement of packages on pallets can influence the ability of the readers to read the tags. For each type of package, there will be an optimal pallet packing strategy, and you should have determined it during your testing. If a pallet of mixed goods is being sent, your workers will have to know the principles of RFID and the degree of interference your various products produce. In that way, they can correctly pack standard pallets, and they will be able to make informed judgments about nonstandard ones.

If you are receiving tagged pallets, some experimentation may be required to maximize read rates. If you are using a handheld reader, specific areas may be more successful than others. If you are using fixed readers, orientation, height, placement, and speed can all influence your success rates. You will want to experiment with the variety of packages and pallets you will be receiving and train your workers in how to maximize read rates for each. Some package combinations will require that you disassemble the pallet, read the package, and then repalletize the contents.

11.6.5 Correct Use of Encoder/Printer

Encoder/printers are very similar to bar code printers, but there are some differences. One issue is that they normally perform a verification test on the smart labels after they've written them. They give a visual indication of the results of that test, and your workers will have to be trained to recognize it. They should also learn your procedures, as you will want to monitor the quality of your tags, your equipment, and your vendor continually. Any change in the ratio of failed to successful write operations should be investigated. Another issue is that you will probably be changing label stocks more frequently as your different packages are run through the station. Training to minimize waste and wasted time will be helpful.

Firmware or reader software upgrades may be accomplished over the network, with little or no worker involvement if you have those facilities. But if your installation is not yet that sophisticated, you will need to train workers who know how to upgrade the readers.

11.6.6 RF Interference

Your installation process should have identified, removed, and/or accounted for all fixed sources of RF interference noise and radio interference. Mobile sources may still show up and cause problems. Cell phones, vehicles, carelessly placed pallets of goods, coins in pockets, and FedEx delivery persons can all create problems for your tag-and-ship operation. Train your workers to recognize the effects of ambient RF interference and to locate and remove its sources as they occur.

11.6.7 Data Integrity

Workers must be trained to recognize and understand threats to data integrity. Consider scenarios like this one: your workers drive several forklifts-full of pallets to the truck to be loaded. They lay the pallets in a staging area next to the truck so they can see how they will fit. They load the truck so the driver at each stop has access to the cases for that stop. Sometimes there is not enough room on the truck for all the pallets, so the pallets await the next truck. Without training, the workers might not recognize that the ASN was created based on the forklift driving the pallets through the portal and that one pallet not loaded means the ASN will be incorrect. This will cause that receiver to report a missing pallet. Train your workers to understand this flow of data driven by their actions so they can ensure its integrity.

11.6.8 Pallet Packing

When workers position packages onto pallets, there are RFID considerations. Tags buried in the center of a pallet of foil-wrapped cat food stand little chance of being read reliably. If your product set contains opaque packages, you need to develop workflows to ensure tag readability, and you need to train the workers on how to employ them and how to check a package. See Protocol 8 in Chapter 12 for the pallet-checking protocol.

11.6.9 Lessons Learned

Errors, problems, and lessons learned should be collected and documented regularly by your workers and shared with management, suppliers, and customers.

11.6.10 Onsite Documentation

You will need to provide documentation for the reader site. The documentation should include:

- Reader documentation provided by the manufacturer
- System support contact information
- Warranty and service contact information and account numbers
- Equipment serial and model numbers
- Supplies and tags contact information and ordering procedures and contacts (internal or external)
- Instructions for restarting the reader in the event of power outages
- Meanings of any indicator lights or other alerts
- A preventive maintenance schedule
- A preventive maintenance log

⬚ A troubleshooting guide

⬚ Installation notes and cautions

Documentation for each site should be available on a website in addition to a traditional binder or CD located at the site.

11.7 Moving Beyond Tag-and-Ship

Nearly every trade show presentation, seminar, and white paper directs readers to look beyond tag-and-ship to get a return on your RFID investment. Very few, however, provide any realistic guidance as to how to accomplish that.

One extension of tag-and-ship might be to attach the same inexpensive tags to assets such as tools, laptop computers, forklifts, or vehicles. Create a database linking the tag to a record of information about the assets. Management of these assets with a manual system has obvious costs and limitations. If necessary, install a reader at the portal behind which the asset is stored and capture asset checkout and check-in. The resultant data set can be very useful. For example, checkout implies asset utilization. You can compare utilization in various locations and move units around to more accurately support the actual needs. You can track whether maintenance is done on schedule by putting a reader at the maintenance facility. You can begin to decide on fleet reduction or expansion based on actual usage data. Within a warehouse, a tag on the forklift tells you each time the equipment itself passes a reader. This data can be mined to determine the utilization of the equipment, the return on your forklift investment, and the accuracy of your other record keeping. It will also give you a point to begin searching when the equipment is missing.

11.8 Chapter Summary

Tag-and-ship is a business strategy that enables a company to comply with customer mandates. The controversy arises from the fact that, while it is the lowest-cost and least-disruptive way to comply with customer mandates, it generates no business value for the investment.

Any effort to obtain business value (better inventory control, reduced cycle times, reduced shrinkage) will entail higher implementation costs and greater disruption of existing processes.

There are three methods of tag-and-ship. The first, called slap-and-ship, involves manually attaching a compliant smart label for demanding customers at the end of the existing process. The second consists of attaching an RFID tag, in addition to the printed label. The third is not available as of this writing but is clearly coming. It involves printing the antennas and circuits directly onto the packaging along with other printed materials.

Even the simplest slap-and-ship operation will require training and documentation. You will also require an EPCglobal manager number, and you will still need to do SKU testing to ensure that your customer can read the tags.

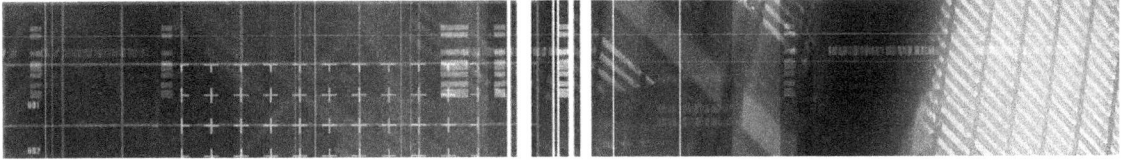

CHAPTER 12

Engineering Tag Performance

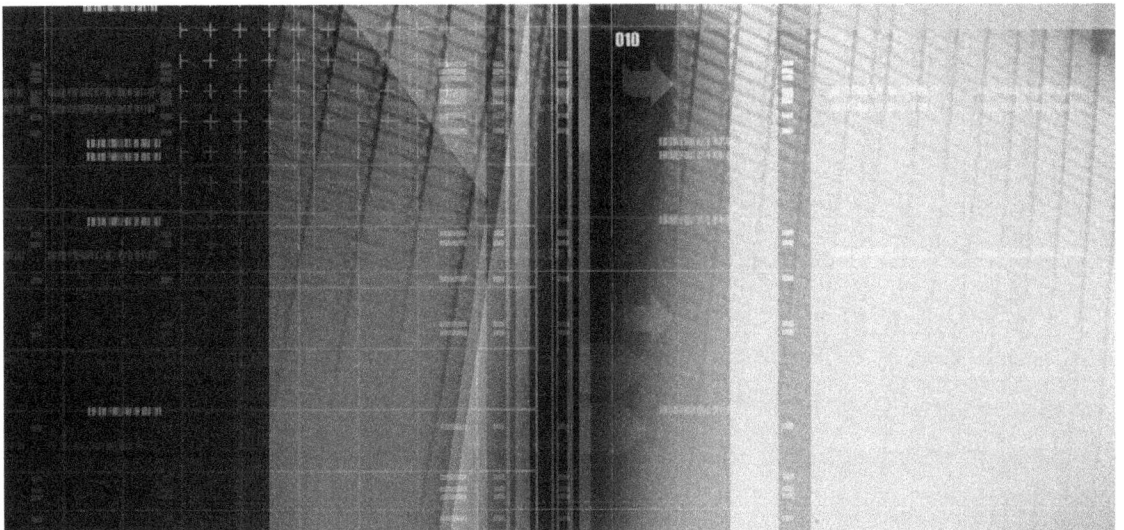

Tag performance rarely happens by accident. It is the product of careful processes including product selection, site analysis, and testing. It requires careful calibration of readers and antennas. Above all, it involves testing in a rigorous, scientific manner so that your systems will read tags reliably and consistently. This chapter describes how to do the analysis, calibration, and testing to engineer the performance of your tags.

The testing must be done at the correct frequency ranges in the location(s) of your application. Determine in which region(s) your usage will occur, and then use Table 12-1 to determine the correct frequency settings (UHF only).

Throughout this chapter, we will refer to this as the *local frequency*.

12.1 Your Test Department

You will need a way to perform a variety of tests throughout your project. You will need to test the following items:

- Readers and antennas, in order to decide which to purchase, and which to use in particular circumstances. These tests will also enable you to design read zones, tune your antennas and readers, and discover the performance characteristics of your RFID systems.

- Tags, in order to find out which will work well on your various packages, cartons, and products. These tests will also enable you to document performance and discover the small differences in placement and orientation that enable tags to communicate with readers at satisfactory levels under true operating conditions.

- Sites, to locate and control sources of RF interference.

- Software, to engineer the end-to-end transmission of tag data to applications and also to test the ability of the software to monitor and manage the reader network.

Table 12-1

Local UHF Frequency Test Settings

Region	Frequency
North America	902, 915, 928
Europe	865, 866.5, 868
Asia	950, 953, 956

If you decide to outsource some or all of the testing functions, the tests will still have to be made, and the material in this chapter will help you select and audit your vendor. If you decide to set up your own test range, this chapter will tell you how to select a site, build a shielded chamber, and equip it for all the tests you will need to perform. Section 12.1.1 describes the trade-offs to consider when you decide whether to do your testing in-house or to outsource it.

Section 12.2 describes the site analysis tests. Use these tests first to select and plan your test range. Use them also to examine various locations before attempting RF operations in them. The site analysis tests are also used to determine the levels of interference in any location where you plan to conduct RF operations. Site analysis tests can be run without having your own test range. Section 12.3 tells you how to set up your test range. Section 12.4 gives you protocols for testing readers, tags, and packages. The most common way to test readers is by measuring their read range, but there is much more than read range to the performance of an RFID reader. Protocols 12 through 17 provide a number of ways to assess reader performance. Note that Protocols 13 and 15 through 17 are derived from the ISO Standard 18046.

12.1.1 Whether to Build
Your Own Test Range

You learned in Section 10.1.7 that your project would need a disciplined way to test locations, readers, and tags. There are two ways to meet these requirements. You can engage a commercial testing organization, or you can build your own test range and staff it yourself. If you decide to build your own, you will need to locate a site and build out the facility. The protocols for this are provided in Section 12.3. If your own test range is outside your scope or budget, you will need to set up a relationship with a commercial or academic testing service that can help you test antennas, tags, and readers and perform SKU tests on all the packages you plan to tag.

Your company may have only a few packages to tag, and they may not change very often. In this case, a service is the most economical option. It may be the case, however, that your company is particularly sensitive about copies of new packages being seen or held by outsiders. If this is a serious concern at your company, be aware that an outside testing service will need advance copies of new packages in order to perform SKU testing as part of your preparation to manufacture them. In this case, you may need your own test range in order to preserve security around new package introductions.

If your company is constantly releasing new packages and new products that will need to be tagged, it may make more sense to establish your own test range. Section 12.1.2 lists and describes the equipment you will need in order to run your own test site, and the balance of the chapter provides protocols for the various tests you will need to run.

The scope of the test department must be clearly established. One option is to limit its scope to site analysis, equipment evaluation, and tag placement studies. The protocols for these tests are included in this chapter. But your project will also need to test middleware functions and software configurations and interfaces. If you plan to include the software configuration in your test department's scope, you will need to provide external network connectivity for the laboratory or install test copies of the middleware and all relevant application packages. Protocols for software tests are dependent on the products you work with and thus are not included in this book.

12.1.2 Equipment for Your Test Range

This section lists and describes the various types of equipment you will need to equip your own test range. Some of the equipment is specialized. Some is generally available. And you might have to build a few of the items yourself. Bear in mind that RF operations can be affected by even small amounts of moisture or metal, so items like carts, tables, and clamps should be made of kiln-dried wood or plastic. An abbreviated equipment list, for site analysis only, is included in Section 12.2.1. The following equipment list will equip your test range to test readers, antennas, and encoder/printers and also various packages that may need to have tags attached:

- Spectrum analyzer
- RFID readers
- Antennas and rack
- Laptop computer
- Hubs and connectors
- Signal generator
- Electronically variable attenuators

- Tripods and rolling cart
- Rotating table
- Motion platform
- Forklift
- Software
- Network connections
- Uninterruptible power supply
- Package samples

12.1.2.1 Spectrum Analyzer

If you are going to do any serious testing of your RFID performance, the first thing you will need is a good spectrum analyzer (SA). The SA is an instrument that displays the strength of an RF signal. When used in your site analysis, it will show you the strength of the transmissions that exist in the measurement area around the frequencies you are interested in. When used in various reader tests, it will show you the strength of the readers' signal at various frequencies and distances. Your SA will be central to your ability to diagnose problems and measure performance. Be sure to get an SA that will cover the local frequency spectrum(s) you will be using. For your local frequency range, refer to Table 12-1. Your SA should have an RS-232 or Ethernet plug so you can connect it to your laptop computer. You can purchase your SA new from Tektronics, Agilent, HP, or Advantest. Tektronix offers an SA with a set of software designed specifically to help in making RFID measurements. You can purchase SAs used from www.testequipmentconnection.com, or you might look on eBay.

12.1.2.2 RFID Readers

You will need at least one utility reader in your test range that you can use to support various testing protocols. For reader tests, you will need a sample of each reader you will be testing. The utility reader should be agile—that is, it should be able to read as many protocols and frequencies as possible. It should interface well with a normal laptop computer, over Ethernet, or an RS-232 connection. USB connections are also an option, but they are not common. The reader should have settings that allow you to set the power output, inhibit frequency hopping, and enable you to set particular frequencies, with the range to cover the local frequencies used in your expected area(s) of operations. Get a reader that has a built-in multiplexer if you can. Otherwise, you may need to purchase a multiplexer separately in order to attach and control several antennas.

12.1.2.3 An Assortment of Antennas and Racks

You will need two dipole antennas and at least two panel antennas. For the dipoles, you can use either a ½-wave antenna, or a ¼-wave dipole antenna with a ground plane, as shown in Figure 12-1. The ground plane is a metal plate at least 33 cm square (for UHF). The antenna is mounted at the center of the ground plane. You will also need a circular polarized antenna.

Construct a rack out of PVC pipe or some other RF-translucent material to hold your patch antennas. The 2-inch pipe is strong enough to support most antennas, and the pipe can be cut with a simple handsaw. Various connectors allow you to use Tinker Toy–like construction to build racks to hold your patch antennas at various heights. Secure a variety of clamps to hold the antennas in place. The clamps should not be made of metal. Use plastic clamps or tie-downs.

12.1.2.4 Laptop Computer and Printer

You will find it most convenient to execute these protocols using a laptop computer. Modern spectrum analyzers come with software to interface with a Windows computer that can capture and store the data. Make sure the software works with the reader and that it can store the data on your hard drive and export the data to Excel. You will also want an inexpensive color printer.

Figure 12-1
Dipole antenna
with ground plane

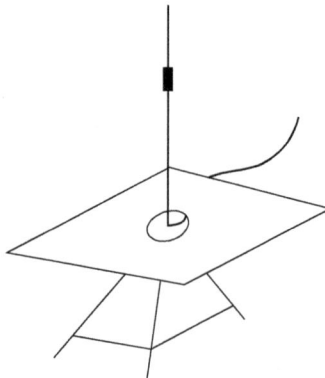

12.1.2.5 Hubs and Connectors

You will need RS232, RS485, Ethernet, and USB hubs and cables. You will also need several serial-to-Ethernet converters. In the United States, the Federal Communications Commission requires manufacturers to use nonstandard antenna connectors, so it may take some ingenuity on your part to get your hands on the all connectors you will need.

12.1.2.6 Signal Generator

You need a signal generator that can output up to 5 watts of power at the RF frequencies you are interested in. It will be connected to the ¼-wave dipole antenna to generate precise levels of power. A suitable signal generator is available from Rohde & Schwarz; see www.rohde-schwarz.com.

12.1.2.7 Electronically Variable Attenuator

You will need two electronically variable attenuators. These attenuators enable you to control the reader's transmission output power in a way that is precise, measurable, and repeatable. By recording the attenuator settings at which readers become unable to communicate with tags, you can measure and record the power output of a unit for any situation.

Various tests require precise control of reader output. The alternative is to increase the distance between the reader and tag, which is less precise and more time consuming. Unless you are in an anechoic chamber (see Section 12.3.3), physically moving the tag farther from the reader introduces additional likelihood of interference, humidity changes, and multipath effects that can affect the readings. Physically moving the tag also takes more time and is not subject to software control for automated testing.

12.1.2.8 Tripods and Rolling Cart

You will need at least two tripod stands. A strong plastic camera stand is perfect for this. Make sure it can go as low as one foot and as high as six feet off the ground and can hold your antenna and ground plane.

The rolling cart enables you to move your equipment around conveniently. It should be made of kiln-dried wood or plastic. The cart should have an upper shelf 3 feet above the ground and a lower shelf 12 inches above the ground. If you are doing distance tests, the cart is a convenient mechanism to hold the package as you move it farther and farther from the antenna.

12.1.2.9 Rotating Table

You will need a table that can hold the various packages you will be testing and rotate them in precise, measurable fashion. The solution is a rotating table, with calibration markers so you can record the angles at which the tag read rate is improved or degraded. When you do your SKU testing, being able to record these angles is particularly valuable.

12.1.2.10 Motion Platform

You will need to test packages moving through the read zone, and you will need a platform that can carry all the packages you will be testing. Determine your own use cases, but packages could range from very small, up to the size of full pallets. Some companies have installed loop conveyor systems, which move the same package repeatedly through the same read zone without manual intervention. Odin has designed what they call a "Cyclotron" that has several advantages. The Cyclotron, shown in Figure 12-2, takes up less space and enables you to test at up to twice the 600 feet per minute (fpm) test velocity. Odin makes the design available to any company that wishes to make one.

12.1.2.11 Forklift

You will need access to a forklift in order to test pallets being carried across your portal.

12.1.2.12 Software

For general utility, you will need a copy of Microsoft Office. This gives you the Word processor and the Excel spreadsheet program. Excel's ability to average data columns and to create graphs will be useful, and its radar

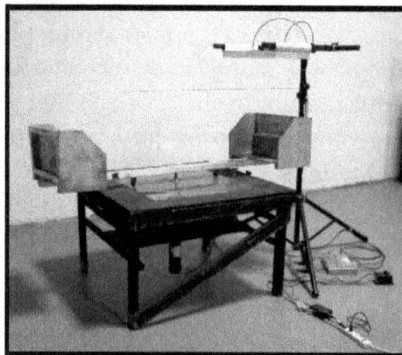

Figure 12-2
Cyclotron
(Photograph
courtesy of Odin
Technologies.)

plot is good for diagramming antenna characteristics. You will also find PowerPoint useful.

Copies of the middleware that your company will be using will extend the utility of your test range into two new areas: software monitoring/control of readers and software interface to applications. This is a project scope decision and not to be taken lightly, but as time goes on you will want to establish your capabilities in these areas someplace within your organization. There are two reasons:

- First, the true value of the RFID implementation lies in the data capture, filtering, culling, and reduction of data. Being able to exercise, model, and test this in your own test range is useful.

- Second, the middleware offers capabilities to control readers, and your test range might be the perfect facility for working out those interfaces and processes, as well.

12.1.2.13 Networks and Server

Plan to have a server in your test range. Include Ethernet connectivity at convenient points along the wall and to the various testing stations. This will enable you to easily share data from reader to server and to and from your laptop computer. The server enables you to centralize and manage (and back up) the various sets of data you will be creating.

Your IT department would probably prefer that you keep your test range network separate from the enterprise network for security reasons. But if your test range is assigned the mission of testing software interfaces, you may find it advantageous to connect to the corporate network so you can access software located outside your test range.

12.1.2.14 Uninterruptible Power Supply (UPS)

Your UPS is needed to power your SA and reader when you do mobile tests. It should be able to power them both for about an hour.

12.1.2.15 Package Samples

You should have in your test range sample packages to support testing. Samples will include RF neutral empty cardboard boxes with the same configuration as your actual packages. You will want RF-challenging packages such as metal, foil, liquids, graphite, and sodium. You will also want actual examples of products you will be tagging and shipping packed exactly as they are when they leave the factory.

12.1.2.16 Miscellaneous

You will need colored tape, 2-inch PVC pipe, joints, and so on. You will need various hand tools, a file cabinet, file folders, an assortment of three-ring binders, and a set of bookshelves. You should have a thermometer so you can record and monitor the room temperature and a humidity meter so you can manage the humidity conditions of your test.

12.1.3 Reader Categories

It is useful to develop categories for the various reader installation situations you will encounter and standard reader solutions for them. For example, you may categorize readers as portal readers, forklift readers, access control readers, shrink-wrap readers, handheld readers, and conveyor readers. Either using your own test range or working with an outside vendor, you can test various manufacturers' readers in these configurations and select readers for each category. Your solution description for each category will give reader manufacturer and model number, antenna multiplexer, cable set, antennas, settings, and so on. As you develop this reader category document, you can test your standard readers with particular products, cases, and packages and include usage notes. For example, you may find that a case of detergent can be read better if the antennas are within 18 inches of the case. Having this kind of information accessible means you will not have to start from scratch and test anew as you begin to equip each facility.

For each standard reader, you can assemble the materials for the onsite reader documentation that will support each installation. Having this material available as a kit will reduce the amount of effort involved in providing this level of support for the readers. Onsite documentation contents are listed in Section 11.6.10.

12.1.4 Tag Categories

It is useful to develop tag categories and standard tag solutions as well. You might categorize packages as RF-transparent pallets, small cases, opaque small cases, pill bottles, shirts, or plasma television sets. Get a few different samples of each of these package categories for your test range. Use Protocol 9 to find the one, two, or three tags that work best for you with this category of package. Use Protocol 10 to determine the placement of these tags, along with any usage notes about how to work best with them on each package. This will give you a solution, or at least a good starting point, when you need to find tags for a new package. It also enables you to test new tags in an application to see if you want to use new tags for a particular type of package.

Tag categories represent standard solutions to challenges in your product line. The challenges may come from a product's metal or liquid content; harsh conditions such as heat, cold, vibration or moisture; handling requirements that might accidentally remove the tag; or other conditions. Some installations fail to consider adhesive requirements early in the tag selection process and run into problems later on.

As your project progresses, adding ever more packages to your set of products being tagged, you do not want to proliferate a large variety of tags to keep track of. A single tag solution, or even just a few tags in your project if you can do it, would mean you will be able to negotiate larger volume discounts, and managing your inventory of tags will be simpler.

12.2 Site Analysis

Site analysis protocols do not require that you have a dedicated test facility, but you will need access to specialized equipment. The protocols described in Table 12-2 and in the following sections enable you to profile and document the RF environment where you plan to conduct operations. The site analysis protocols will help you select a location for your test range, if you decide to build one. Use them to evaluate a site for a reader and locate and control sources of RF interference before they impact production. The decision whether to perform these site analysis tests yourself or outsource them is independent of the decision to build your own test range. You can use them without having a test range of your own. These protocols have been optimized for the installation of a UHF system.

12.2.1 Equipment for Site Analysis

To conduct the site analysis protocols, you will need the following equipment. Descriptions of each equipment type are in Section 12.1.2.

Table 12-2

The Test
Protocols

Protocol Number	Test
Protocol 1	Area Evaluation Test
Protocol 2	Area Mobile Test
Protocol 3	Introduce Interference
Protocol 4	Reader Site Survey: Horizontal
Protocol 5	Reader Site Survey: Vertical
Protocol 6	Link Margin Test
Protocol 7	Selecting Test Tags
Protocol 8	Reading Tags on a Pallet
Protocol 9	Select a Tag for Your Package
Protocol 10	SKU Testing for Tag Placement and Orientation
Protocol 11	Designing a Read Zone
Protocol 12	Reader Legal Compliance
Protocol 13	Reader Read Range Test
Protocol 14	Select a Reader for Tags in Motion
Protocol 15	Reader Read Rate Test
Protocol 16	Reader Write Range Test
Protocol 17	Reader Write Rate Test

Note that this list omits several items from the list in 12.1.2 that you will not need unless you are equipping your own test range.

- Spectrum analyzer (SA)
- Signal generator for Protocol 4
- Reader
- Antennas
- Laptop computer
- Hubs and connectors
- Tripods and rolling cart

- Software
- UPS with wattage and capacity sufficient to run your SA and reader for about an hour

You will find it helpful, particularly in Protocols 2 and 3, to have a drawing of the floor plan/blueprint of your facility. This will enable you to readily record and document locations of RF interference that you locate. If you cannot get an "official" floor plan/blueprint, you can draw one by hand. Make sure the drawing contains accurate reference points, drawing orientation, and a reasonably accurate scale.

12.2.2 Protocol 1: Area Evaluation Test

Use this area evaluation test to get a profile of the RF activity in the facility. It is important to run the test for at least 24 hours and preferably for a whole week, as certain external sources of RF may occur on certain days. One error people make is to take only a short snapshot of the day's RF activity. They then miss sources of interference that may come online at other times and interfere with their operations after they thought installation complete. The SA software should be able to automatically take and log readings every quarter hour for 24 hours. It will store the results on the computer's hard drive, and you can review them. Just seeing changes that occur over time is likely to be helpful.

If you are assessing a large warehouse, you will need to assess more than one location. Protocol 2 will tell you how to move your testing cart and test in various locations until you are satisfied that you have a good picture of the RF environment. Use Protocol 4 to test specific reader locations.

12.2.2.1 Set Up for Protocol 1

1. Put the dipole antenna on the tripod (if you're using a ¼-wave antenna, include the ground plane) and locate it in the center of your area of interest.
2. Be sure the power is turned off and attach the antenna to the SA.
3. Connect the laptop to the SA.
4. Power up the SA and tune it to the center of the local frequency range as designated in Table 12-1. Set the span to 50 MHz, the resolution bandwidth to 100 kHz, and the video bandwidth to 30 kHz. Set the amplitude attenuation to 0 dB and turn on maximum hold.
5. Verify that your laptop shows the virtual screen.

12.2.2.2 Conduct Testing

Capture a record of any RF interference and how it behaves over the course of a day. Determine also whether this single test gives you a good snapshot of the facility or you need to test other locations as well. Let the test run for at least 24 hours and preferably for one week.

12.2.2.3 Analyze Results

For any interference logged, make a judgment whether it might interfere with your RF operations. If there is any interference, you will need to find the source. If you cannot do this based on the information you have, you will have to hone your detective skills and try to correlate the interference to events and schedules. Determine what you need to do to control it (see suggestions in Section 12.2.4.3).

12.2.3 Protocol 2: Area Mobile Test

Protocol 2 enables you to capture a snapshot of all the corners and areas of your facility. It will not be run 24 hours per day for a week for each location but will involve putting your equipment on a cart and moving in a systematic way around the facility with a close eye on your SA view screen.

12.2.3.1 Set Up for Protocol 2

Load your UPS, SA, laptop, and antenna onto your cart. Set up the SA as documented in Protocol 1. Get a printed map or drawing of the layout of the facility.

12.2.3.2 Conduct Test

Travel around the facility, watching for interference in the SA's display. Pay particular attention to walls that you share with other companies or other departments because you have less visibility, influence, and control over any sources of interference they may be running.

When you find a reading, immediately document its exact location, along with the strength in decibels. To find the source, move away from the initial read point in four directions, farther and farther until one of the signals is stronger than the others. Continue moving in that direction until you locate the source of the interference.

12.2.3.3 Analyze Results

When you are satisfied that you have documented every source of interference in the facility, make a list and decide what to do about each one. If it is near a proposed read zone, it will cause you problems, and you will need to take steps to eliminate or mitigate its interference (see suggestions in Section 12.2.4.3).

12.2.4 Protocol 3: Introduce Interference

Once you have mapped the existing interference sources, you should then introduce interference of your own and document the results. The list of potential RF polluters includes machinery, equipment, devices, cleaners, mobile telephones, fluorescent lights, motors, cordless telephones, FedEx delivery people with handheld devices, and bug zappers. This protocol directs you to enumerate them all, exercise each one, and log its effects.

12.2.4.1 Set Up for Protocol 3

Set up equipment as specified in Protocol 1.

12.2.4.2 Conduct Test

Turn on every possible interfering device and log the results. If the devices move around, such as forklifts or delivery trucks outside the portal, move them around to make sure they do not generate undetected interference.

12.2.4.3 Analyze Results

Any interference that shows up represents a potential problem. In each case, determine a resolution. In some cases, merely structuring people's behavior will solve the problem. Direct the FedEx delivery person to stay away from the read zone with her bar code reader, for example (and remember to tell her replacement). In other cases, you may have to move the equipment or the read zone. In still other cases, you may have to install shielding to protect the reader's integrity.

NOTE *If you install shielding, be sure to extend it throughout the line of site. UHF waves will "bleed" around the edges of a shield. Once the shields are in place, run your tests again.*

12.2.5 Protocol 4: Reader Site Survey, Horizontal

The reader site survey is executed for every location where you plan to install a reader. This Protocol 4 is your "secret sauce," along with Protocols 10 and 11. Most other installers plug in the equipment, attach the antennas, turn it on, and tinker until it works. By following Protocol 4, along with the SKU Test in Protocol 10, you can assure yourself of the best read rates possible. Even if you have outsourced the installation, make sure your installers follow this protocol. It will save you time and money in the long run and improve the odds of success for your project.

The essence of this protocol is to test and document the reader site first, before installing any equipment. This will assure you the site is free of RF interference, and the resultant document allows you to configure the equipment and tune antenna placement to its exact RF.

12.2.5.1 Set Up for Protocol 4

1. Set up your first tripod stand in the center of your intended read zone. Mount the circular-polarized dipole antenna.

2. Position the antenna to simulate the location of a tag—usually vertically 3 feet above the ground for a dock door or 12 inches above a conveyor belt—and position the antenna there.

3. Connect the antenna to the signal generator. Set the signal generator to put out 14 dBm. This signal is simulating a tag broadcasting (only much stronger!).

NOTE   *dBm is an abbreviation for the ratio of the measured power relative to one milliwatt. It is an absolute value whereas dB is a ratio between two values. Zero dBm equals one milliwatt.*

4. Connect the spectrum analyzer to the laptop computer. Turn on the laptop.

5. Connect the UHF flat panel antenna to the spectrum analyzer. Mount the antenna on your second tripod, the same height as the other antenna, 3 feet away. Turn on the SA.

6. Tune the SA as described in Protocol 1.

12.2.5.2 Conduct Test

This test first establishes a baseline ideal read zone for this location. Then it enables you to identify where the real signal deviates from this ideal read zone so you can see where you will have problems.

The ideal read zone usually radiates from an antenna in the shape shown in Figure 12-3. Recall that the antenna produces *gain* by focusing its output in a particular direction. The large portion represents the enhanced read zone, and the small portion represents area where gain is reduced. The large portion is your ideal read zone. Next, represent the area surrounding the "tag" as a cherry pie, cut into eight pieces, as shown in Figure 12-4.

Figure 12-3
Ideal read zone

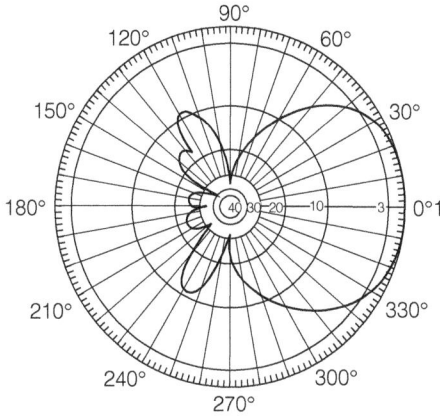

Figure 12-4
Read zone worksheet

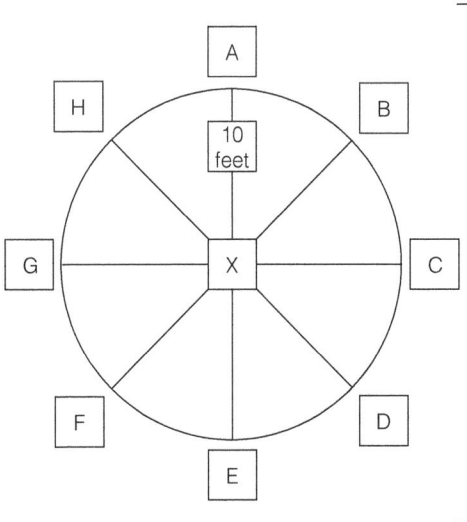

You are going to move the spectrum analyzer antenna from point to point (A, B, C, and so on) around the signal generator antenna, taking eight sets of three readings, one set at each corner of the pie. At each corner, be certain to aim the antenna so that the tag is in its read zone. The readings will take place at each of the three local frequencies from Table 12-1 (for the United States: 902, 915, and 928 MHz). The SA will give you the signal strength at each location as you move its antenna to the corner, generate a signal from the signal generator, and record the signal strength. Point A is at the 0-degree angle from the SG antenna. The distance should be the maximum distance you want to read from, shown in Figure 12-4 as 10 feet.

7. Position the SA antenna at Point A aimed at Point X.
8. Set the SG to your local frequency from Table 12-1. Generate a signal from the SG.
9. Record the strength of the signal for each frequency.
10. Perform this test at points B through H.
11. Put the 24 values into Microsoft Excel and plot it out using the radar graph option.

12.2.5.3 Analyze Results

If the results are exactly ideal, your real read zone will match the ideal. If the results deviate, that is, if some directions read weaker than the ideal, determine whether your application requires you to read in the weaker directions. If they do, you have three solution choices. One, you can work to figure out what is causing the power loss in this direction. If you can find out what it is and fix it, you may have solved your problem. In this case, rerun the test. Second, compensate for weak directions by increasing antenna power and read frequency in those directions if your reader settings permit. The third choice is to move your read zone. Decide where you will locate your read zone, relocate your Location X, and rerun your tests.

12.2.6 Protocol 5: Reader Site Survey, Vertical

Protocol 4 positioned the antennas an equal distance from the ground, but the more likely scenario is that the reader antennas will be installed higher than the tags. To capture this scenario, exercise Protocol 5.

This will enable you to test at various elevations to make sure that you maintain your read rate.

1. Position the SG antenna at Location X.
2. Elevate the SA antenna until its center is at the height you plan to locate the reader antenna.
3. Set SG for first local frequency (see Table 12-1) and generate a signal.
4. Record the strength of the signal.
5. Confirm that you have not lost signal strength.
6. Repeat for all three of your test frequencies.
7. Move the SG antenna up and down 15 degrees and test again.
8. Document the sensitivity of your performance to tag height.

Examine your use case to see the vertical range you need to cover. If any tags will be traversing the read range at heights where reading is diminished, you will have to re-engineer the read zone. This may involve any of the following steps:

- Adjust and tune your the antennas.
- Purchase and install additional antennas.
- Adjust reader settings to increase reader power or number of reads in the sequence.
- Find the cause of the diminished signal strength and remove it.

12.3 Test Facilities

The previous five protocols should be conducted at all sites where you plan to conduct RF operations. You can conduct the protocols that follow in your test range if you have decided to build or rent one, as opposed to outsourcing the testing function. Protocols 6–17 in the next sections assume you are working in your own test range.

12.3.1 How to Locate and Set Up Your Test Range

Companies locate their test range either in office space or in warehouse space. The office space option is attractive because it is more comfortable and more professional. It is satisfactory if you do not need access to heavy equipment to do your testing. However, it is difficult to test forklifts carrying pallets of goods in the office space, and conveyor systems also present a problem. If you are going to be testing pallets or cases on such conveyors, we recommend you utilize warehouse space with dock doors.

12.3.2 Site Analysis

When you have tentatively located a place for your test range, you need to conduct a thorough site analysis; see Section 12.2. for the protocols.

12.3.3 Shielding

Your test range should be completely free of RF interference in order to get consistent results. The only way to be completely certain about levels of interference in your environment is to shield the test range with materials that absorb the radio waves. A shielded facility is called an *anechoic* chamber. The internal shielding material is normally made of pyramidal urethane foam. See Figure 12-5 for a typical anechoic chamber. The external shielding is typically aluminum or copper sheets. Use aluminum or copper tape to seal the joints between the sheets.

Figure 12-5
Anechoic chamber

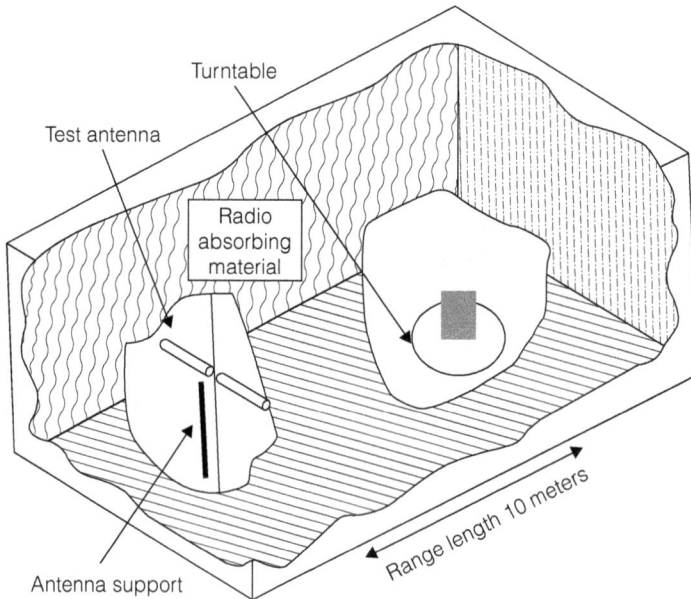

Turntable

Test antenna

Radio absorbing material

Antenna support

Range length 10 meters

12.3.4 Environmental Conditions

EPCglobal recommends that the temperature be maintained during testing at 25 degrees centigrade (77 degrees Fahrenheit), plus or minus 5 degrees. They recommend that the relative humidity be 70 percent, plus or minus 5 percent.

12.3.5 Software in the Test Range

One advantage to having your own test range is the ability to test interfaces to applications and middleware. The best solution is to have test copies of the middleware and applications software installed in your test range so you can experiment with various configurations without impacting your production systems.

12.4 The Testing Protocols

The remainder of this chapter provides testing protocols for a variety of situations you will encounter. It draws on many sources including ISO testing standards, EPCglobal testing standards, and several commercial companies that have made their methodologies available. The important point, however, is to always test and document your results before committing. The essence of radio frequency communication is this: it is subject to so many variables that only a commitment to rigorous testing can give you assurance of success and maintainability.

12.4.1 Protocol 6: Link Margin Test

The link margin measures a tag's ability to respond to a reader. It consists of establishing tag-reader communication and then *attenuating*, that is, reducing the power output of the reader until the communication performance reaches zero. That attenuation level at which communication ceases is the link margin. Use the Link Margin Test to find normal tags in a population with Protocol 7 or to calculate the reader's effective range. Use the Link Margin Test to compare readers in different circumstances and thus to support your reader selection activities.

12.4.1.1 Set Up for Protocol 6

1. Inhibit frequency hopping on the reader.
2. Connect a circular-polarized antenna to the attenuator using a coaxial cable.
3. Verify that the reader power is off and connect the attenuator to reader.
4. Set attenuation to zero. This allows the reader to deliver full power to the antenna.
5. Put the tag in center of table.
6. Position the antenna one meter from the testing table.
7. Turn on the reader.

12.4.1.2 Conduct the Test

The test must be conducted at different frequencies within the local frequency range, as described in Table 12.1, as different tags perform better or worse at different frequencies.

Increase the attenuation level and record the setting at which the read rate (number of successful reads divided by the number of attempts) goes to zero. The value when read rate equals zero is the link margin. Note that transmit power must be turned off at least 500 milliseconds between tests, in order to allow tag's stored energy level to deplete.

12.4.1.3 Analyze the Results

The Link Margin Test is sometimes used to substitute for read range tests. According to Avery-Dennison, read range tests can be difficult to measure exactly because of multipath effects, unless you have an anechoic chamber. Also, range measurements involve physically moving the tag and increasing distances from the reader, and therefore take longer to conduct than a purely electronic approach. Read range can be reconstructed from the link margin as follows:

$$R_{theoretical} = R_{measurement} \times 10^{-margin/20}$$

For example, a 12 dB link margin at 1 meter would give a theoretical read range of 4 meters. Avery-Dennison says of this test: [We have] "added calibration mechanisms that allow us to reference readings back to an absolute scale sensitivity measured in dBm, effectively making the reading independent of range...In reality, real world systems will hop in frequency....We therefore measure the link margin over the full spread of frequencies....In the past, we have attempted to make probabilistic measurements of read rate but have found inconsistencies between different types of readers and also between readers of the same type."

EPCglobal recommends a variation on the link margin test using a figure called the TurnOn value. This is useful when you are testing and comparing a number of tags. The TurnOn value is the attenuation value at which the number of successful reads divided by the number of read attempts equals 50 percent. EPCglobal also recommends computing and recording a Power Adjustment Factor correcting for antenna and cable loss.

You can evaluate readers by computing the link margin for each reader and comparing them. You can evaluate tags by computing the link margin for each tag within a lot. Select representative tags by finding a group with an average link margin and discard those with abnormally high or abnormally low values. Select a tag for a given package by comparing the link margins of various manufacturers' tags on that package. The protocols in the next sections spell out these tests in more detail.

12.4.2 Protocol 7: Selecting Test Tags

In any batch of tags, some will perform better and some worse than others. The purpose of Protocol 7 is to select a sample group of tags that will accurately reflect its population. That way, your testing can work with "normal" tags rather than tags that perform exceptionally well or poorly.

12.4.2.1 Set Up for Protocol 7

1. Choose at least 100 tags.
2. Number each tag.
3. Set up a Link Margin Test (Protocol 6).

12.4.2.2 Conduct Test

For each tag, conduct a Link Margin Test at each of the local frequencies where you will be conducting operations and record the results. Calculate the average link margin for the set of tags.

12.4.2.3 Analyze Results

Go back to your numbered set of tags and choose the set whose link margin is closest to the average. Store this as your test set. Now, when you do your testing, you can work with these and be sure you are working with tags that accurately represent their population. Now, think about your projected usage. If you plan to use the tags primarily in Europe, work with the average link margin for European frequencies. If you plan to use the tags primarily in Asia, work with the average link margin for Asia. If you plan to use the tags all over the world, use the average for the entire population.

Incidentally, if you calculate the Standard Deviation (SD) of the population, you will have a measure of the quality of this particular manufacturer and model of tag. A smaller SD generally means better quality.

For your test range, gather a test set for each standard tag category. This will make it easier to do SKU testing with normal tags for any category.

12.4.3 Protocol 8: Reading Tags on a Pallet

A pallet may contain any number of packages/cartons, and each package/carton may be tagged. Items inside the package/carton may be tagged as well. Your application may require you to record the identity of all the tags on a pallet as it moves through a portal, as it is moved on a forklift, or as a worker reads the tags with a handheld reader. The goods and the packaging may be RF-opaque, or they may be RF-transparent. Here are several considerations for managing this case:

- Your application may work well even if you only read the pallet tag and use the ASN, a database, or some other IT resource to generate the list of packages/cartons or items. This will reduce the load on your RFID system and enable it to process pallets faster. This assumes, of course, that you trust the sender or you have processes such as spot checks and penalties to ensure their accuracy.

- If you can control the packing of the pallet, have it packed with airspace between the packages/cartons to facilitate reading the tags.

- If you cannot get reliable read rates, you may wish to emulate other companies who have set aside areas where the pallet can be broken down and the tags read individually. It may not be necessary to completely break down the pallet to get 100 percent read rates.

Use this protocol to test readers and tags on a specific pallet. The test result is the length of time the reader takes to read all the tags on the pallet. Depending on your situation, you can use the same reader and test a variety of tags, or you can use the same tag and test a variety of readers. The end result is to recommend a system (that is, a workflow, a reader, antennas, placement, and a tag) to read this particular pallet configuration. Readers, of course, must generally read a variety of packages, so it may not be practical to optimize the reader selection for a particular pallet configuration.

12.4.3.1 Set Up for Protocol 8

1. Mock up the pallet using RF-neutral materials. Duplicate the packing with cardboard or Styrofoam blocks the size of your cartons. Position the tags on the blocks exactly as they will be positioned on the pallet. For Step 5, bring in an actual pallet you will be testing.

2. Set up the reader and antenna to simulate the working conditions and read zone.

12.4.3.2 Conduct the Test

3. Determine the length of time it takes for the reader to read all the tags.

4. Perform Step 3 ten times to get a valid statistical sample. Take the average.

5. Repeat Steps 3 and 4 replacing the mockup with an actual pallet.

Note that this same test can be conducted with pallets on a conveyor belt. Run the pallet through the read zone at designated speeds (for example, 200 fpm, 400 fpm, 600 fpm, and 1200 fpm) and compute the read rate. Tune the read zone and antenna configuration until you are able to read 100 percent of the tags at the speed required. Record the time.

12.4.3.3 Analyze Results

The objective of Protocol 8 is to discover the length of time it takes to read all the tags on the pallet. In the event you cannot achieve a 100 percent read rate, consider rearranging tagged cartons on the pallet so there is an unobstructed path between each tag and the antennas. You may need to add more antennas, or you may need to readjust their power levels or their sequencing. If all else fails, you will need to redesign the workflow so you partially or completely disassemble the pallet and read each carton's tags.

12.4.4 Protocol 9: Select a Tag for Your Package

Use Protocol 9 to select which tags to use for a particular package. This protocol uses the same mechanisms as Protocol 10, but the purpose is different. Protocol 9 selects a tag; Protocol 10 locates its position and orientation on the package. The two are intertwined, but you need to both select a tag and position it on the package. Protocol 9 enables you to find which tags will work best for any particular package.

It may be that there are several tags that work well. If you find this to be the case, it may enable you to reduce the total number of different tag types you will have to purchase and inventory for your project. Tag performance is a function of several variables. The variables are

- RF characteristics of the package
- Tag manufacturer and model
- Frequency
- Location and orientation on the package
- Distance between tag and reader

This protocol relies on the scientific method: hold all the variables constant except the one you are testing; vary that one and note the results. Do this test for each variable until you have a profile for successful outcomes.

12.4.4.1 Set Up for Protocol 9

Visually inspect the package to detect its size and RF characteristics.

- If there are no opaque materials, you should be able to place the tag almost anyplace with equal success. Recall that the plane of the tag antenna should be perpendicular to the plane of your radio signal.
- If you are attaching a smart label, you may be able to place it where you previously put the bar code label.
- If there are opaque materials, try to place the tag far away from them. Look for places where there is as wide an air gap as possible.
- If the package is made of metal, or metal foil, use a ¼-inch spacer made of cardboard or Styrofoam to distance the tag from the metal.

Some companies now offer "flag" tags, which occupy a plane 90 degrees from the surface of the package. These flag tags may be vulnerable to accidental removal, so they will need extra care in handling. But they improve read rates on packages of opaque materials.

Determine whether or not the package fits one of your standard tag categories (see Section 12.1.4). If so, get the test tags for that category and proceed with the test. If not, establish a new category and document its usage profile, RF characteristics, adhesive requirements, any special environmental considerations, its size, and the space available for the tag. Select a tag or small group of tags whose performance characteristics most closely match your requirements. For each tag candidate, use your test set, the group of normal tags that you assembled with Protocol 7.

12.4.4.2 Conduct the Test

Use Protocol 6 to calculate the link margin for each candidate tag. Document the results.

12.4.4.3 Analyze the Results

The result of Protocol 9 is a ranking of the tags that will work with the package. If several will work, you will need to balance cost, tag inventory management issues, and performance to select the tag you ultimately use.

12.4.5 Protocol 10: SKU Testing for Tag Placement and Orientation

SKU testing is necessary for every package you will be tagging. Even if you have a simple slap-and-ship operation, if your package contains opaque materials, you will need to perform SKU testing to determine where to place the tag to ensure its readability. The presence of even small amounts of opaque materials can detune the tag's antenna and reduce your read rates.

SKU testing is detailed and time-consuming. A number of companies have released products that automate the process, and you should consider one of these products if you believe you will have ongoing needs. Protocol 10 will show you how to do it manually.

The purpose of SKU testing is to find the "sweet spot" on the package, the location that gives the best read rates. Since tags are so sensitive to opaque materials, very small changes in tag location and orientation might make a big difference in your tag performance.

You will need an exact sample of the product you will be testing. The sample must contain the products packed exactly as they will be packed in production.

12.4.5.1 Set Up for Protocol 10

1. Make a drawing of each face of the package and number them. Divide each rectangle into nine subrectangles and number those. This drawing will enable you to document your findings.

2. Use local frequency for testing; see Table 12-1.

3. Set the reader to separate the frequency range into individual channels and report successful reads on each channel. Most readers are capable of this, but software is required to make the results meaningful. See alternative in Step 9 if you do not have access to the required software.

4. Secure at least four "normal" tags (your test tags from Protocol 7) from the candidate population.

5. Put the sample SKU on the rotating table.

6. Connect a circular-polarized antenna to the reader.

7. Put the antenna on a tripod and position it level with the tag, one meter away from it.

8. Examine the sample package and select a small number of candidate locations.

12.4.5.2 Conduct the Test

9. Run the reader for at least one minute with a tag positioned at each location. Record the number of successful reads divided by the number of tries. If you do not have access to the software to make your data meaningful, you can set the reader to record the number of successful reads and run it for three minutes.

10. Perform Step 9 at least ten times, for each of your four tags, for each of your five locations.

11. Move antenna back to two meters from the sample and perform Step 10 again.

12. Determine whether you have found any locations that give acceptable read rates. If not, try other locations and try rotating the tag 90 degrees and then testing again.

13. Once you have found an acceptable location, rotate the tag 45 degrees in each direction and test again. This will tell you how sensitive your solution is to the orientation of the tag.

14. If the results of Protocol 10 gave you no successful tags, change tags and return to Step 8.

12.4.5.3 Analyze the Results

The result of Protocol 10 is to find a place on the package where you can attach tags and get good read results. You should also document sensitivities such as orientation and placement. Note the degree of precision required.

12.4.6 Protocol 11: Designing a Read Zone

The read zone is the area in which you intend to activate and read a tag. Ideally, tags outside the read zone will not be activated, and tags within the read zone will be activated. The read zone boundaries are controlled by

- The amount of power being delivered to the antenna
- The horizontal direction of the antenna
- The vertical direction of the antenna
- Any distortions introduced by nearby opaque materials

The documentation of the antenna usually contains a contour map showing the manufacturer's intended read zone, horizontally and vertically. You can verify this chart as it applies to your location by following Protocol 4.

Use a copy of this contour map chart to document the results of Protocol 11.

12.4.6.1 Set Up for Protocol 11

1. Mount one of your test tags on an empty cardboard box.

2. Place the cardboard box on your rolling cart so it sits 3 feet above the ground for a portal (upper shelf) or 12 inches above ground (lower shelf) to simulate a conveyor belt.

3. Construct or obtain a strong, secure rack to hold the antenna.

4. Mount a patch antenna on the rack at a height and location similar to where you expect to mount the actual antenna. If the reader requires two antennas (one to transmit, one to receive), mount both on the same side.

5. Using the contour map, position the cardboard box so it is at the center of the read zone.

6. Be sure reader power is off and connect the antenna to the reader.

7. Turn on the reader.

8. Determine the use case of the reader. The use case might be similar to the following:

   - Forklift carries pallet through read zone and reader must read pallet tag.

   - Hand truck carries pallet through read zone and reader must read pallet tag or all the individual tags on the pallet.

   - Pallet travels through read zone and reader must read all the carton tags on the pallet.

   - Conveyor carries packages through the read zone 6 inches apart.

   - Person carries a single package through the read zone.

12.4.6.2 Conduct the Test

9. Verify that the read rate (number of reads divided by number of attempts) for the tag(s) is at least 50 percent. Move tag farther or closer to the antenna to find the farthest setting in a particular direction that will give you a read rate of 50 percent in one minute.

10. Mark the location of the tag on the chart with a colored X.

11. Draw a circle on the floor with its center at the antenna, passing through the tag's 50 percent read rate location.

12. Move tag 45 degrees along the circle and verify the read rate is at least 50 percent. Mark the location of the tag on the chart with a colored X.

13. Perform Step 9 at 45, 90, 135, 180, 225, and 270 degrees.

14. Raise the box to determine the height of read zone and repeat Step 10 for that height.

15. Replace the cardboard box with the actual package to be processed. Be careful to place and orient the tag on the package at the sweet spot discovered in Protocol 10, SKU Testing. Repeat Step 10.

Rotate package 90 degrees in each direction and determine how sensitive the read efficiency is to package orientation. Note that if multiple types of packages are going to pass through this read zone, you will need to run this protocol for as many of them as possible.

16. Replicate the use case(s) and record the results. This may require use of forklift or conveyor belt.

12.4.6.3 Analyze Results

If the read zone is sensitive to package orientation, examine your use case to see whether you need to add additional antennas to cover other faces, and retest. Note that most stationary reader installations have at least two pairs of antennas. Some have as many as four pairs. Some readers allow you to adjust antenna sequencing or power to individual antennas to compensate for challenging situations. See Chapter 3.

If there are areas of the read zone where read rates decrease, any or all of the following steps may help:

- Determine whether the package orientation can be assured by an appropriate workflow.
- Determine whether changes to power levels affect your read rates.
- Carefully determine whether the problematical areas are in fact necessary to achieve satisfactory read rates for all packages. It may be that these areas are outside the space where tags might go. It may also be that by changing the direction of the antennas, you can improve your read rates.
- Determine whether changes in reader settings improve your problematical read rates.
- If necessary, investigate whether specialized tags may be required.
- If necessary, investigate whether package redesign may be required.

Once you have documented your read zone, you will have an installation map showing how to install this reader and antenna at the real reader site.

12.4.7 Protocol 12: Reader Legal Compliance

In North America, UHF readers are restricted to the frequency range 902–928 MHz. The output is restricted to 1 watt maximum. Other local frequencies are listed in Table 12-1. Protocol 12 enables you to test any reader to make sure that it is in compliance with these laws. If it is not, you could be subject to a fine, and you could be interfering with other people's systems.

12.4.7.1 Set Up for Protocol 12

1. Tune your SA to the center local frequency (915 MHz for North America) with a span of 50 MHz.
2. Connect a dipole antenna to reader.
3. Set reader to maximum power.
4. Turn on reader.
5. Set Hold Lock On in the SA. This will record all signals received from 865 to 965 MHz.

12.4.7.2 Conduct the Test

Record and monitor output for at least 60 minutes.

12.4.7.3 Analyze Results

If you find any signals from your reader below the lower bound or above the upper bound of the local frequency range, or if you find any signals greater than the statutory limit, the reader is operating illegally and should be returned to the manufacturer.

Comparing Readers

The most common way to compare readers is to compare their read ranges. We provide Protocol 13 for this test, but it is not by itself a sufficient selection criterion. Portal readers must work within six to eight feet, and nearly all commercial readers will perform well at that range. Mounted readers and handheld readers, similarly, are used within a limited read range, and so finding the one that can read the farthest might not be your most productive test.

Select a reader that can best do the job you need it to perform. A portal reader may need to read a wide variety of packages. So test as many different packages as possible, preferably the same ones that will be running through the portal. If this is your use case, select the reader that does best across the widest variety. A conveyor reader may need to perform well at high speeds. If this is your use case, use Protocol 14 to select the reader that generates the best read rates at high speeds. If your application involves writing large numbers of tags, you will want to include write rate and write range in your criteria. Use Protocols 16 and 17. Reader selection criteria other than performance are discussed in Section 3.6.5.

12.4.8 Protocol 13: Reader Read Range Test

This test, and the Protocols 15, 16, and 17, are adapted from the ISO/IEC Document 18046, titled "RFID Tag and Interrogator Test Methods." It overlaps functionality with several other protocols, which gives you a choice of methodologies.

Read range is the most common measure of reader performance. It consists of the maximum distance at which the reader can excite and read a tag. As we said in the "Comparing Readers" sidebar, read range by itself is not a sufficient measure of the reader's performance for a given application. Nonetheless, it is commonly used, and it is one of the factors you will want to measure.

This protocol can be conducted for a single tag or for multiple tags. The multiple tags can be arranged in three different formations: a line, an array, or a 3-D volume. If you run the test with multiple tags, this protocol defines range as the longest distance at which 100 percent of the tag population can be read.

This protocol can be conducted for testing tags in motion. The read range is the farthest distance at which the activation signal can be sent, resulting in a 100 percent read rate.

Initially, run the test with reader factory default settings. Once you have established a read range, experiment with various reader settings to see whether they can improve it.

Note that this protocol actually tests the combination of reader and tag. You can use the same tags and use it to compare readers, or you can use the same reader and use it to compare different tags. If your objective is to find the longest read range, varying both the tags and readers in different combinations will result in identifying optimum combinations.

12.4.8.1 Set Up for Protocol 13

1. Set up reader as in Protocol 6.

2. Attach your tag to an empty cardboard box.

3. Set up the reader in the factory default setting.

4. Attach dipole circular polarized antenna to the reader using a coaxial cable.

5. Mount the antenna 3 feet above the ground and establish a read zone.

6. Position the tag at 3 feet in the center of the read zone.

12.4.8.2 Conduct the Test

7. Initiate read of tag(s) and verify 100 percent successful read.

8. Move tag(s) away from antenna or attenuate the signal. Test each tag at 1-meter intervals until the read fails. Then move tag(s) closer until you discover an exact read range to an accuracy of 10 cm.

9. Perform this test at least ten times with a minimum of ten different tags from your test set.

   If you are testing tags in motion, execute the following steps:

10. Set tags in motion at 200 fpm.

11. Initiate activation signal at read range.

12. Determine whether you achieve 100 percent read rate. If not, reposition the antenna to move the read range closer. Repeat Steps 10–12 until you achieve 100 percent read rate.

13. Increase speed by 200 fpm and repeat Steps 10–12. Repeat the test for 400, 600, 800, and 1200 fpm.

12.4.8.3 Analyze Results

Record the average of the longest distances at which the tag reads 100 percent successfully at each speed. This is the read range of the reader-tag combination. Record also the effect of any modifications of reader default settings on the results.

12.4.9 Protocol 14: Select a Reader for Tags in Motion

This protocol will enable you to select a reader for conveyor systems and other applications where the tags will be in motion. Your own circumstances will dictate the speeds at which you will be operating, and you should conduct your tests at several speeds up to double the operating speed. Wal-Mart and the DoD both have mandated 600 feet per minute as their threshold of required tag performance, so we recommend that you test and document performance at speeds up to twice that.

12.4.9.1 Set Up Test for Protocol 14

1. Select a set of tags that matches the protocol of the reader you wish to test. Choose the tags from your test set.
2. Attach the tags to several cardboard boxes.
3. Set the reader to the factory default setting(s).
4. Attach a UHF patch antenna to the reader using the coaxial cable.
5. Mount the antenna 3 feet above the trajectory of the tag and establish a read zone. Make sure the tag passes through the read zone.
6. Place a cardboard box on loop conveyor or cyclotron testing stand.
7. Turn on the reader.
8. Set conveyor or cyclotron to speed of 200 fpm.

12.4.9.2 Conduct Test

Turn on your conveyor or cyclotron and record the number of reads at the 200 fpm speed. Do this at least 50 times, record each read count, and compute the average. Do this for each of the speeds: 400 fpm, 600 fpm, and 1200 fpm.

Move the tag from the cardboard box to a RFID-unfriendly package. Make sure the path between tag and antenna is not obstructed by opaque materials. Perform the same test and record the results.

Move the tag from the RFID-unfriendly package to your most popular product. Again, make sure the path between tag and reader is not obstructed by opaque materials. Perform the same test and record the same values.

Note that it is important that all instances of this test be conducted at the same location on the same equipment. Different readers, tags, antennas, conveyors, and locations will produce different performance levels, and the results from one will not be directly comparable with results from another.

We specify that these tests be conducted with factory settings. You may be able to improve read rates by changing the settings, and this is a good time to familiarize yourself with the reader's capabilities and how they affect read rates with your packages.

12.4.9.3 Analyze Results

Protocol 14 will help you select a reader for your conveyor belt applications. Once you have completed these tests, you will still have to test your reader of choice at the actual location where it will be installed. Protocol 14 will help you select a reader, give you baseline information, and show how the readers will work with your materials.

12.4.10 Protocol 15: Reader Read Rate Test

Determining the read rate of a reader enables you to measure how fast a reader can read a tag. It reflects, therefore, how many tags can be read per second. This is a particularly useful measure when tags are being read on a fast-moving conveyor belt or a very active portal. It will also be important as you begin reading item-level tags. The theoretical read rate for Gen-2 tags is 1600 tags per second. This test will tell you what a particular reader and tag combination can actually achieve.

This protocol can be executed for a stationary tag population or a moving tag population. It can be conducted for a single tag or multiple tags. The multiple tags can be arranged in a line, an array, or a 3-D volume. If you run the test with multiple tags, record the time it takes to read 100 percent of the tag population.

Initially, run the test with reader factory default settings. Once you have established a read range, experiment with various reader settings to see whether they can improve it.

Note that this protocol actually tests the combination of reader and tag. You can use the same tags and use it to compare readers, or you can use the same reader and use it to compare tags. If your objective is to find the fastest read rate, you may vary both. You can use the same reader and tag combination and use it to optimize antenna and reader settings.

12.4.10.1 Set Up for Protocol 15

1. Set up your reader as in Protocol 1.
2. Position a tag in a responsive part of the read zone as established by Protocol 11.

12.4.10.2 Conduct the Test

3. Excite the tag and record the time from initiation of the activation signal to completion of the read process (for example, successfully retrieve all requested tag data).
4. Perform this test at least ten times with a minimum of ten different tags from your test set.

   If testing tags in motion:
5. Set tags in motion at 200 fpm.
6. Record time to read 100 percent of tags going through the read zone.
7. Increase speed by 200 fpm and repeat Steps 5, 6, and 7. Repeat test for 400, 600, 800, and 1200 fpm.

12.4.10.3 Analyze Result

Record the average of these times. This number is the average read rate for this reader-tag combination. It is useful to express the read rate as the number of tags per second that can be read.

12.4.11 Protocol 16: Reader Write Range Test

Some applications are write-intensive, and the reader's ability to write to a tag or group of tags over a longer distance may be an important criterion. Protocol 16 describes how to determine the tag's write range, the distance at which 100 percent tag writes can be achieved.

12.4.11.1 Set Up for Protocol 16

1. Set up reader as in Protocol 6.
2. Position a tag two meters from an antenna.

12.4.11.2 Conduct the Test

3. Initiate the write and verify successful write of tag.
4. Move the tag away from reader or increase the attenuation, testing each 1-meter interval until write fails.
5. Perform this test at least ten times, with a minimum of ten tags from your test set.
6. Vary the orientation of the tag(s) 90 degrees in each direction and record any differences.

12.4.11.3 Analyze Results

Record the average of the longest distances at which the tag writes succeed. If you are using multiple tags, define success as 100 percent of tags written. This is the write range of the reader-tag combination. Record also the effect of any adjustments of reader settings on the results.

12.4.12 Protocol 17: Reader Write Rate Test

The write rate of a reader enables you to measure how fast a reader can write a tag. It reflects, therefore, how many tags can be written per second. This is a particularly useful measure when tags are being written on a fast-moving conveyor belt or a very active portal. It is also useful when writing tags at the item level, as write speed will become an important limit for the entire system. This test will tell you what speed the system can actually achieve.

This protocol can be executed for a stationary tag population or a moving tag population. It can be conducted for a single tag or for multiple tags. The multiple tags can be arranged in a line, an array, or a 3-D volume. If you run the test with multiple tags, record the time it takes to write 100 percent of the tag population.

Initially, run the test with reader factory default settings. Once you have established a write range, experiment with various reader settings to see whether they can improve it.

Note that this protocol actually tests the performance of a complete system. You can use the same tags and use it to compare readers, or you can use the same reader and use it to compare tags. If your objective is to find the fastest write rate, you may vary both. You can use the same reader and tag combination and use it to optimize antenna settings and reader settings.

12.4.12.1 Set Up for Protocol 17

1. Connect a dipole antenna to your reader.
2. Position a tag or group of tags in a responsive part of the write zone as established by Protocol 16.

12.4.12.2 Conduct the Test

3. Excite the tag(s) and record the time from initiation of the activation signal to completion of the write process (for example, successfully write a test set of data to the tag(s)).
4. Perform this test at least ten times with a minimum of ten different tags from your test set.

12.4.12.3 Analyze Results

Record the average of these times. This number is the average write rate for this reader-tag combination. Express the write rate as the number of tags per second that can be written.

12.5 Chapter Summary

Testing has a key role to play in every RFID installation. RFID technology is reliable and predictable, but it is not "plug-and-play." Even the simplest tag-and-ship application will require SKU testing at the least. As you add fixed readers, you will need to perform site tests to determine the levels of RF interference that exist. No company will use all the protocols in this chapter, and undoubtedly you will develop your own as time goes on. But this chapter shows you how to do the basic tests you will need for a successful installation.

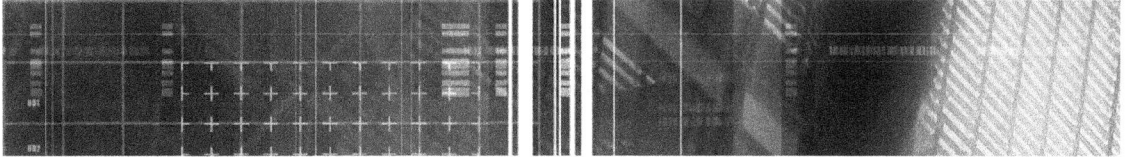

13

Data Integrity and Data Synchronization

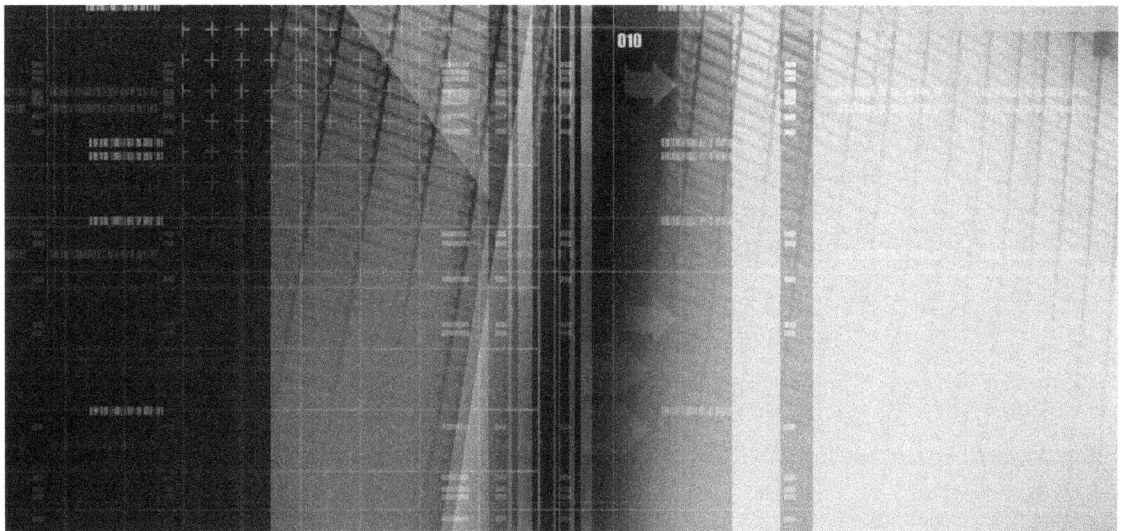

The consulting firm of A.T. Kearney listed data integrity as the most important single precursor to RFID success. A report from the company observes: "Many manufacturers and retailers still exchange inaccurate product data despite industry efforts toward standardized data formats."

According to recent research carried out by consulting company CapGemini, more than half the items in company systems contain incorrect data. Examples include wrong values and duplicate or obsolete entries. A spokesperson for Boeing reports that manual data entry produces an astonishing 1 error for every 30 keystrokes.

This is not surprising. The objective of many business processes and systems is the production of documents and messages and current information. Except for financial systems, the creation of a valuable data resource is secondary. As a result, considerable effort is spent getting the work out, but little or none is spent creating perfect data repositories.

The success of your RFID project depends heavily on the integrity of the data it consumes and produces. Nearly all of the data in an RFID application comes from an external systems. The quality of the data is paramount.

Improving the quality of a company's data should be outside the scope of your project, but you have a vested interest in its results. As you raise the issue, you may find a better reception than you might expect. Events at Worldcom, Enron, Arthur Anderson, and other high-profile business failures were, to some degree, data quality failures. The regulatory and legislative response (OFAC, HIPAA, Graham-Leach-Biley, Sarbanes-Oxley, Basel II, the U.S. Patriot Act, and International Accounting Standards (IAS)) have all served as wake-up calls to top management, highlighting the costs and dangers of poor data quality. In some cases, top management has personal liability for data quality failures, and your need for quality data so your project can deliver a return on the RFID investment may just tip management over the edge and get a separate project started in this area.

Poor data quality creates a set of problems and costs even without considering RFID. The costs include wasted materials, rework, and failed business processes. It also generates extra work such as hunting down missing data or additional reconciliations. It results in losses that can be directly linked to revenue, profit, or customer lifetime value as a result of missed opportunities, unhappy customers, and poor strategic planning. Consider the following observations:

- Up to 37 percent of invoices every year contain errors from either bad item numbers or bad prices.
- $40 billion or 3.5 percent of total retail sales is lost each year owing to supply chain information inefficiencies.

- Up to 30 percent of inbound shipments contain items whose numbers don't match with those on file.
- 30 percent of data in retail catalogs is in error, and each error costs $60 to $80 to correct.
- 25 minutes of effort per year is spent on average manually cleansing each SKU per year.
- 43 percent of all invoices result in deductions.
- Each invoice effort costs $40 to $400 to reconcile.

Businesses are using more data from more sources in more systems today than ever before. The RFID implementation will draw from, repurpose, and try to correlate customer files, order entry files, warehouse management systems, inventory systems, product master files, bill of materials files, and human resource systems. But data collected for use in the operational systems that support these functions may not be sufficiently accurate to support automated applications. Fragmented and distributed IT systems can lead to data duplication and inconsistency, erroneous entries, lack of conformity, and other discrepancies.

This chapter deals with the data that the RFID system must access in order to make use of the RFID tags' data itself. Many of the prescriptives throughout this book contemplate the data being provided by external systems or entities outside the RFID system. The RFID implementer will not "own" these systems, but the success of the project depends on having quality data to work with. This chapter provides an explanation of each of the components of data quality. In Section 13.1.6, we provide a questionnaire you can use to assess the data quality of any data source.

Chapter 10 described how to map your data flows. One step of that phase consisted of listing the data sources you will be accessing. As you complete the Business Infrastructure Document (BID), you will need to assess the data quality of each of those sources.

In order to document and evaluate the quality of your data, there needs to be a framework for evaluating and measuring it. This chapter provides such a framework and then goes on to propose a methodology for data management.

Many companies do not understand that data must be managed and maintained just like any other asset. Otherwise it deteriorates. Customers move. Products are created or discontinued or changed. Mergers, acquisitions, IT upgrades, international growth, and expansion all challenge the task of maintaining a consistent database of company, product, and customer information. Even within normal processes, the same product may have one identification number when it is in R&D, a second in engineering, and a third when marketing takes over. Proactive steps are required if the resultant data will be useful to automated systems such as RFID.

13.1 Data Quality Assessment

The Canadian Institute for Health Information (CIHI) has published a framework for data quality that is complete and very comprehensive. We have adapted this model for use in a business context. In the CIHI framework, data quality has five dimensions:

- Accuracy
- Timeliness
- Comparability
- Usability
- Relevance

13.1.1 Accuracy

Accuracy is what most people think of when they hear the term "data quality." And accuracy of the data is very important. Accuracy refers to how well the data reflects the reality it was designed to measure. Data collected for one purpose, such as order processing, may not be accurate for another purpose, such as forecasting sales.

It is very difficult for an outsider to directly assess the accuracy of a database, but there are four criteria you can assess that will enable you to evaluate its accuracy indirectly. Consider the database accurate if these five criteria are met:

- Coverage
- Quality control (QC)
- Validity
- Documentation
- Retention of raw data

Let's look at each one of these in detail.

13.1.1.1 Coverage

Coverage deals with the sources of the data. Coverage is complete if the database captures data from all the sources that should be submitting, each source is submitting all its records, and all the records are complete.

For example, one of the data sources for your RFID system might be the Warehouse Management System (WMS) for a distribution center in Missouri. Do you know the population this database should cover? The population of interest for a given application might be "all shipments from Supplier X to Warehouse Y in January." The coverage issue might arise in any of the following circumstances:

- Some shipments sent to Warehouse Y were diverted to Warehouse Z last Saturday because of a snowstorm. They were transshipped to Warehouse Y the following Monday, but the shipping records were never updated.

- Some shipments sent to Warehouse Y were recorded manually and never updated in the database.

- Some shipments sent to Warehouse Y had discrepancies between the ASN and the receiver, which are not yet resolved.

- One forklift went through the portal twice, resulting in duplicate entries.

These odd transactions might not ever be entered in the database, so a user or an application like RFID accessing it would be working with faulty data. While the shipments ultimately were stocked, distributed to stores, and the payments all got made, the intermediate data records may have not been updated. Database users, then, would be acting on erroneous input and would either have to spend time and effort tracking down the correct information (cursing "the computer" along the way!), or perhaps never become aware of the discrepancy. You can inquire whether there are ever any circumstances where the system might be circumvented and what procedures exist to update the database *ex post facto* and what problems arise.

13.1.1.2 Quality Control Measures Exist

One key aspect of data quality is whether there are procedures in place to assure its accuracy. You should determine whether each of your data sources has quality control (QC) measures in place. At a minimum, the database should be audited once a year to make sure that it is accurate.

13.1.1.3 Validity Checks

Validity of data is a subset of its accuracy. A date rendered as 12, March 2007 might be accurate but invalid because of the format. March 32, 2007 is invalid because there is no such date. A part number with two digits transposed might be valid but inaccurate. A state code entered as XY is both invalid and inaccurate. Validity deals with the format of the entry and whether it is within the allowable range. A valid entry is free from these defects, whether or not it is accurate.

In your assessment, you should determine whether all possible processes are implemented to ensure that data elements contain valid entries. If part numbers, location codes, zip codes, or any other codes are entered, do the processes exist to ensure that they are valid? If dates and times are entered, does the system make sure they are valid? If employees enter their identification, does the system check that for validity, and are there processes to inspect it?

There are two types of processes that assure the validity of the data: edit processes and referential integrity checks.

Edit checks look at the data item itself and apply business rules to determine its validity. An edit check can look at a date and make sure it is in the right format or that it falls within some specified range. It can look at a zip code and make sure it is of the form XXXXX-YYYY. It could look at a part number and make sure the third character was a dash and the first two characters were numeric.

Referential integrity (RI) checks reference an outside data source to determine the validity of the data. A system could check a state against a table of states with which you do business to make sure the state entry is valid. It could check a part number against the product master to make sure the entry is valid. RI checks are very powerful and can be helpful in assessing, monitoring, and assuring data validity.

13.1.1.4 Documentation

It is astonishing how many billions of dollars are transacted on systems that are undocumented. This is particularly true of homegrown systems, systems that were custom developed for the company by outside or inhouse programmers. The original programmers may have left the company. Many small commercial software companies have gone out of business over the years. Even the ones still in business are not all scrupulous about maintaining documentation for older systems.

As a result, although the systems do their job in terms of getting the goods to their destinations and the bills paid, the exact definition of each field's meaning or the coverage or validity processes the system employs may be obscure.

You should determine whether documentation for each source system is available and whether it has been updated since the last revision.

13.1.1.5 Retention of Raw Data

You should examine the raw data from which the application obtains its data and see if it is retained so any later project can audit the database.

13.1.2 Timeliness

Evaluate the timeliness of the data source. Many systems rely on batches of source data. This introduces delays that may impact the processing of the RFID data, which arrives in real-time. Most timeliness issues arise from frequency-of-update discrepancies, and different data sources may have different schedules, even for the same application. Besides batch input issues, you should look at manual data entry and correction processes. Sometimes changes accumulate on someone's desk for days or weeks until they have time to enter them. This raises another type of timeliness issue that may impact your project.

13.1.3 Consistency and Comparability

We tend to assume that databases are consistent over time, and they use the standard conventions such as data element definitions or reporting periods. These assumptions are not necessarily valid. This particular investigation looks first at individual databases to ensure that they are internally consistent and have had no changes in their conventions over their period of usage.

It then looks at the extent to which different databases are comparable with one another. Databases owned by different business units will tend to present comparability problems, as they may use different nomenclatures for the same items, and may use different time periods, units of measure, customer reference schemes, and even article and part identifiers. Examine your data sources for these kinds of discrepancies and document them. Data should be available to you at the finest level of detail possible, and all data elements definitions should be maintained in a master data dictionary. Codes should be standardized across all applications and used wherever possible.

13.1.4 Usability

Assess the ease with which each database may be accessed. Document any issues with making the data available to your middleware or other applications you will be installing. Several modern usability solutions are increasingly available. These include the ability to export data in XML. Many older systems offered the ability to export data as comma-delimited ASCII files. To enable system-to-system communications, many modern systems implement a SOAP interface and can respond to XML queries with XML-formatted responses. Ability to participate in generally used architectures such as J2EE (Java 2 Enterprise Edition) or Microsoft BizTalk also enhances usability, if your systems can interface with these technologies.

13.1.5 Relevance

Document the degree to which the database meets the needs of your application. If there are discrepancies, note them here.

13.1.6 Questionnaire

The following questionnaire summarizes the preceding material into a set of questions you can ask each data manager to assess the quality of the data for your purposes.

Data Source Assessment Questionnaire

1. Name the data source.
2. Name the business function for which the data is collected and maintained.
3. Name the population the data source covers.
4. Name the time period the data source covers.
5. Describe quality control measures in place:
   a. On data entry.
   b. On stored data.

    c. Are they documented?

    d. Who is responsible?

    e. Are the processes followed?

    f. Which of them are automated?

6. Describe all edit processes that are applied and which fields they are applied to:

    a. Forms of the data are valid (that is, dates, zip codes, and so on).

    b. Referential integrity is checked (that is, part numbers are valid, and so on).

7. Is documentation available and current for the processing system?

8. Is raw data maintained?

9. Is the database updated in a timely fashion?

    a. Are there instances of errors resulting from data that is not current?

    b. Will the level of timeliness support your RFID system's needs?

10. Is the database consistent in its definitions over its time period?

11. Is the database inconsistent with any other databases you will be accessing?

12. Is the database easily usable?

13. Is the database relevant and sufficient for your application?

13.2 Data Management

If you have determined that data quality problems exist, you will need to provide a recommendation for how to deal with them. One answer is to launch a huge manual effort to "clean up the databases once and for all." Follow this with installation of automated tools to support ongoing data quality processes. That way, once the data is cleansed, it stays clean.

Even if you are able to clean it, the data won't stay clean on its own. The very next day, you will begin to have new data entry errors, discrepancies, and changes in one data set that may not propagate to others. What is needed is a serious clean-up effort and initiation of processes to keep the databases clean. The first five steps tell you how to clean a data source. The sixth tells how to keep it clean.

1. Parse each database.
2. Correct the data.
3. Standardize the data.
4. Validate and verify the data.
5. Integrate the data.
6. Initiate processes to keep the data clean.

13.2.1 Parse Each Database

Parsing is the process of breaking up the data into its discrete components. Working with discrete components will enable you to simplify and enhance the data correction process and all subsequent data quality processes. It is easier to figure out that Dennis E. Brown and Dennis Brown are the same person when you have first and last names in separate fields. It is also easier to figure out that extraneous information has been included, such as supplier information in a part description field or special instructions stored in Address Line 3.

13.2.2 Correct the Data

Check all reference numbers (parts, products, locations, customers, vendors, packages, containers, invoice numbers, document numbers) to make sure they are valid. There should be no transpositions or duplicates. Similarly check all rosters, such as lists of employees, locations and equipment items in a warehouse, and lists of vehicles and other equipment that will be accessed. Correct erroneous entries, duplicates, spelling errors, and other problems in the data. Then perform the following steps:

1. Verify existence.
2. Track usage.
3. Assign attributes.
4. Locate ownership.

Each of these steps is described in the following sections.

13.2.2.1 Verify Existence

First, make sure the item listed in the database actually exists. It may have been entered in error, or it may have been replaced by a newer item and never deleted.

13.2.2.2 Track the Item's Usage

Determine whether the item is being used. It is likely that there are items in the database that are rarely or never used. Determine whether the usage level justifies the cost and effort of carrying the item.

13.2.2.3 Review Which Attributes You Are Carrying About Each Item

A detailed discussion of metadata modeling and design is outside the scope of this chapter, but one of the most difficult decisions to make is which information you should track about each item. Some of the issues around this difficult question include:

- Standards are incomplete and evolving. Schemes that meet the needs of a specific industry or product category are emerging. If one product serves several different markets, you may need to manage multiple attribute sets for that product.

- Your trading partners may have their own product information requirements and demand that you to conform. If you have multiple trading partners, you may have to manage multiple attribute sets or attributes with identical names but different definitions.

- Destination systems may or may not tolerate extra data.

- Some required attributes may not be tracked by your internal systems.

- Attributes may be needed by your internal processes and needs, even though they are not shared by trading partners.

Decide what attributes to track based on a detailed review of the application's data needs. It is easy for databases to become bloated with extraneous information based on the accurate observation that you might need to know some piece of data about the item in question. But bloated databases take up more storage; run more slowly; take more computer resources to run; take more time and effort to update, manage, validate, and maintain; and cost more to operate. Try to input, store, and maintain only the data values you need and no more.

13.2.2.4 Locate Ownership of Each Attribute

While it may first appear that some data items have multiple owners, a primary owner can usually be identified. Once ownership is established, it must be enforced. This person should, by nature of their job, have intimate knowledge of the attribute and control over whether it changes. In the event the data attribute changes, this person must notify the system owner so the change can be published.

13.2.3 Standardize the Data

Determine the correct variant for all cases where the same entity is represented differently. For example, decide whether to represent Verizon as *Verizon* or *Verizon Inc*. Propagate this standard representation throughout the entire data population. Or better yet, if your software will allow it, create a central table for each element and use codes in the database.

13.2.4 Validate and Verify the Data

This is the process of comparing the standardized data with external sources. For example, you might be storing part numbers from a supplier's catalog. You might be storing geographical descriptions from some available third party. The validation step involves checking your now-standardized entries against that third party's entries. Ideally, you will have systems that referentially check data as it is entered to make sure it is valid.

13.2.5 Integrate the Data

This step is less about data itself and more about technology and processes. Identify areas where changes to data made in one place need to be propagated to one or more other places. Examine what it will take to make that happen and automate it.

Where possible, harmonize your data usage with your trading partners. If they have a table of part numbers and you have your own part numbers, either adopt theirs or automate the interface so you work with a particular trading partner with their numbers. Similarly, if you can use their reference numbers for their locations, that will remove one source of error. Using EPCglobal standards for GTINs (Global Trade Item Numbers),

GLNs (Global Location Numbers), and so on, will facilitate this if your trading partners use them also.

13.2.6 Initiate Processes to Keep Data Clean

- Re-engineer key processes to capture data as soon as it is available.
- Assign responsibility for data integrity.
- Enable all possible edits and validations automatically.
- Set up data monitoring applications.
- Review the database at least once per year.

Keeping data clean requires re-engineering of the processes by which it is created, sent forward, and maintained. For example, consider a record in a product master database. Different parts of the company may participate in defining its attributes. Research and Development, Engineering, and various marketing groups will each assign different attributes to the same item. Other departments (with other identification systems) may assign other attributes such as logistics, user documentation, liability issues, and so forth. When the product is ready to go into the catalog, many companies assemble the information in an ad hoc approach, missing vital information resources. As attribute values change, the processes to capture those changes may be ad hoc. To keep your database clean, all these processes must be reviewed, documented, and harmonized. Determine the attributes of the items that must be carried forward from each department and establish processes for doing this, automating where possible.

Each item released for sale should be supported by a standard document or diagram detailing how it was created and who is responsible for each of its attributes. Then create a process that ensures that accurate data is available and communicated for each item in the catalog. You might consider setting up a system where individuals can "subscribe" to an item or a set of attributes and receive an e-mail notification when it changes.

13.3 Global Data Synchronization

Once you have cleaned up the internal data and streamlined the process of creating and maintaining item data internally, you can synchronize the data externally. The term used to describe this process is *Global Data Synchronization* (GDS). Data is *synchronized* when it is internally consistent and all relevant parties ascribe the same meanings to the same data items.

Industry experts point out the numerous benefits created by closer collaboration of trading partners. The ability to reduce inventory costs and transfer costs and to respond more quickly to market cues have emerged as critical success factors, and companies ignore them at their peril. These benefits accrue to companies that have documented or, better still, automated, processes for communicating vital data and collaborating. Low-hanging fruit in this arena is automated order replenishment and exception notification, real-time inventory location, track and trace applications, product recall management, and reduced order errors. Most companies are just starting to approach this end state. Collaboration in terms of production planning, sales forecasting, and product design and development, with automated support, is highly productive, but further out on the horizon. The automated support is possible only when the data is clean and synchronized.

EPCglobal makes available two separate network services. The first is the EPCglobal Network we discussed in Chapter 2. The second is the Global Data Synchronization Network. Section 13.3.1 describes the GDSN, and Section 13.3.2 recaps the discussion of the EPCglobal Network, highlighting features of comparison with GDSN. Section 13.3.3 attempts to resolve the confusion engendered in the marketplace by the existence of overlapping functionality in two network services from the same vendor.

13.3.1 Global Data Synchronization Network (GDSN)

The GDSN organization, like EPCglobal, is owned and operated by GS1, the company described in Chapter 2. GS1 was formed by the merger of the Uniform Code Council (UCC) and European Article Numbering (EAN) organization. GDSN began as UCCNet and was formed in 1999 to define standards for data compliance and synchronization. GDSN provides the GLOBALRegistry, where products are registered according to their UPC numbers. The GLOBALRegistry stores GTINs and GLNs (see Chapter 2).

GDSN provides mechanisms to synchronize product, organization, and pricing data. Manufacturers publish data describing a product to *data pools* connected to the registry. Data pools are repositories owned by independent companies under license from GSDN. Retailers subscribe to a data pool, and the data is continually synchronized by vendors using standard messages. An example message is *catalog item notification,* which notifies retailers a new product has been added to a company's catalog. Product attribute and price data are constantly updated. Using GDSN, companies can build a foundation for higher-level collaborative activities.

GDSN uses XML, and it can render messages into a number of different formats including Electronic Data Interchange (EDI) and Microsoft's BizTalk.

GDSN has a well-defined process for synchronizing your data. Once internal data integrity has been achieved and the processes to ensure its continuation are in place, the external connections that support global data synchronization can be established. The GDS process is an add/modify/delete manager for product-related information exchanged between supplier and retailer.

Each retailer and trading partner establishes a relationship with a data pool service provider who links with GDSN. Data pool service providers include UCCnet, Transora, WWRE, and QNX. A typical subscription costs a minimum $500 per year for a company with less than $25 million in annual sales to a maximum of $400,000 per year for a company exceeding $50 billion in sales.

Once a service provider has been selected, you will need to set up an on-site database with a services layer that links to the data pool. The external synchronization process generally follows six steps:

1. A subscriber creates a new product. Subscriber creates a GTIN for it following GTIN specifications and notifies the data pool of the new GTIN.

2. The data pool requests base product information to add to the catalog.

3. The data pool synchronizes the new information with the master registry. The master registry synchronizes with all other data pools.

4. Trading partners send automated invitations through the network to retailers asking them to carry the new product.

5. The retailer's system pulls master product data from their local data pool.

6. The retailer is now able to obtain data about the new product entry. They might access information such as pricing, availability, packaging, weight, and so on.

13.3.2 EPCglobal Network

The EPCglobal Network is described in Chapter 2. Briefly, it is an open, standards-based system to share item-level product and tracking information among partners in the value chain. The EPCglobal Network provides secure storage and retrieval from other sources and networks of the following:

- **Core product information**   This is a set of attributes for all products with the same GTIN.
- **Manufacturing time information**   This contains information about *this particular pallet, lot, case, or item* at the time of manufacture. Data elements such as "lot number" and "expiration date" are available.
- **Lifecycle history information**   These are the distributed details of the lifecycle of a product and is publicly available to qualified subscribers.

The EPCglobal Network is based on the following process:

1. The EPC, a unique identification number, is assigned to every instance of a product in the supply chain.
2. The EPC is held in an RFID tag that is attached to the object.
3. As the object moves through the supply chain, it is detected by RFID readers at different locations, and the information is passed to RFID middleware where it is filtered and collected.
4. The middleware aggregates information, removes duplicates, applies filters, and passes filtered information to enterprise systems.
5. When IT systems require more information about an object, they send the EPC to the Object Naming Service (ONS), and it returns a pointer to the item's Discovery Service.
6. The ONS returns the Internet address of an EPC Discovery Services (EPC DS) server, which can provide pointers to any number of EPC IS servers that can provide the item's pedigree.

13.3.3 Comparing EPCglobal Network and GDS Network

There is confusion over the relationship between the EPCglobal Network and the GDS network service. Both are offered by the same company: GS1. On the surface, they overlap, as both are worldwide, globally-accessible

Figure 13-1
Comparing
EPCglobal Network
and GDS Network

	EPCIS	GDSN	comments
Ownership	EPCglobal	GDSN	Both owned by GS1
Coding Scheme	GTIN, GLN, SSSC, GRAI,	GTIN, GLN only	GTIN for GDSN is only used for general product information. For EPCglobal, it is used for numerous special items such as books, individual luggage bags, particular items of medical equipment, etc.
Content	Transaction and movement details at the item level	Product Description Information	
Level	Individual Items	Classes of products (SKUs)	

repositories of product information. But the two networks actually store and distribute different types of information, and they are similar enough to make companies question why they would need to subscribe to both. The company insists they play separate roles, and neither by itself is sufficient. The two networks differ in significant ways, although there is some overlap. See Figure 13-1 for a comparison of the two networks.

13.3.3.1 Different Coding Schemes

EPCglobal Network works with Electronic Product Codes (EPCs), while GDSN works with the GTIN codes (EAN.UCC-13, EAN.UCC-12, EAN.UCC-8, and EAN.UCC-14) and GLN codes only. Recall that GTIN codes identify only a unique *class of object*.

EPCs can express numerous coding schemes including GTIN (which can be serialized), and it can render codes such as those adopted by various industries such as defense, aerospace, medical equipment, books, consumer electronics, airlines, logistics, and paper. So the coding scheme that underlies the EPCglobal Network is more flexible and thus the network can address a wider scope of applications than the GDSN. GSN works only with GTINs as product identifiers and Global Location Numbers.

Because the GTIN can be expressed as an EPC, both networks can use the GTIN as an index to access the product information.

13.3.3.2 Information Content

Other than the core product data, the two networks store and transmit very different data sets. The GDSN contains data such as core product data, category-specific data, and target-market–specific data, as well as data that depends on the identity and relationship of the parties in the supply chain.

The EPCglobal Network contains manufacturing time data and lifecycle history data. Both networks can access core product information.

13.3.3.3 Level of Information

EPCglobal Network is designed to distribute information about individual pallets, cases, packages, and items that are in trade (such as a case of Huggies). Its purpose is to track individual items as they move through the supply chain, providing date of manufacture, pedigree, and movement data.

The GDS Network distributes information about products (such as Huggies diapers as a product line). It stores information such as availability, pricing, size, ordering instructions, marketing support available, return policies, and so on. GLOBALRegistry is a global gateway for companies to locate a source or customer.

13.3.3.4 Network Structure

The two networks are structured very differently. The EPCglobal Network consists of systems and servers that make available transaction data that resides on a company's internal data processing systems to qualified subscribers. The GDSN, on the other hand, has licensed several data pools, specialized companies that store the data that the manufacturer has specifically uploaded for publication. Qualified subscribers find the data they need through their own data pool.

13.3.3.5 Challenges to Global Data Synchronization

The existence of two complementary networks offered by a single company creates obvious opportunities for synergy, as well as confusion and challenges. The fact that companies who wish to use both networks face separate fees for each network is a major issue. A unified charging structure would help grow the usage of both networks.

EPCglobal also needs to make a clearer case that GDSN and EPCglobal Network are both necessary for a complete solution. Until that happens, we will see a continuation of today's situation in which many incompatible business-to-business solutions exist, and higher-level collaboration along the supply chain will continue to be difficult. Many companies experimenting with EPC technology are focused on getting the tags to talk to the readers. They are not yet convinced about the need for a standards-based EPC network. Even if they were, they are not ready to address it.

The challenges to GDSN lie in three areas. First, although companies recognize that imperfect data elevates costs and elongates cycle times, the task of data cleansing is daunting. Second, the priority of data cleansing is often lower than more urgent needs. Third, the confusion surrounding the two networks stops many companies from moving forward.

13.4 Chapter Summary

Solving the problem of data quality is likely to be outside the scope of your project. The problem is sufficiently important, however, to merit your serious attention. As an intense consumer of this data, you will need to assess its quality and make the case that it must be accurate, timely, comparable, usable, and relevant. Data management is the business process that enables your company to assess, clean, and maintain the data you will need in your project. Global Data Synchronization (GDS) is the propagation of your company's clean data to other companies. It ensures that you are all working with the same data, paving the way for higher levels of collaboration. GS1, the parent company of EPCglobal, offers a product for GDS, a secure worldwide network where companies can post and retrieve each other's product descriptive information. This product, the GDSN, is often confused with EPCglobal's EPCglobal Network, which was described in Chapter 2. Until the confusion is cleared up, companies are unlikely to make widespread use of the service.

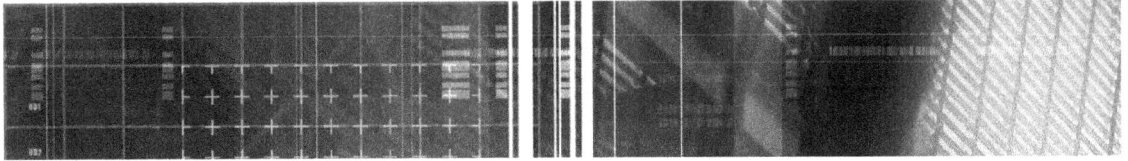

CHAPTER **14**

Privacy

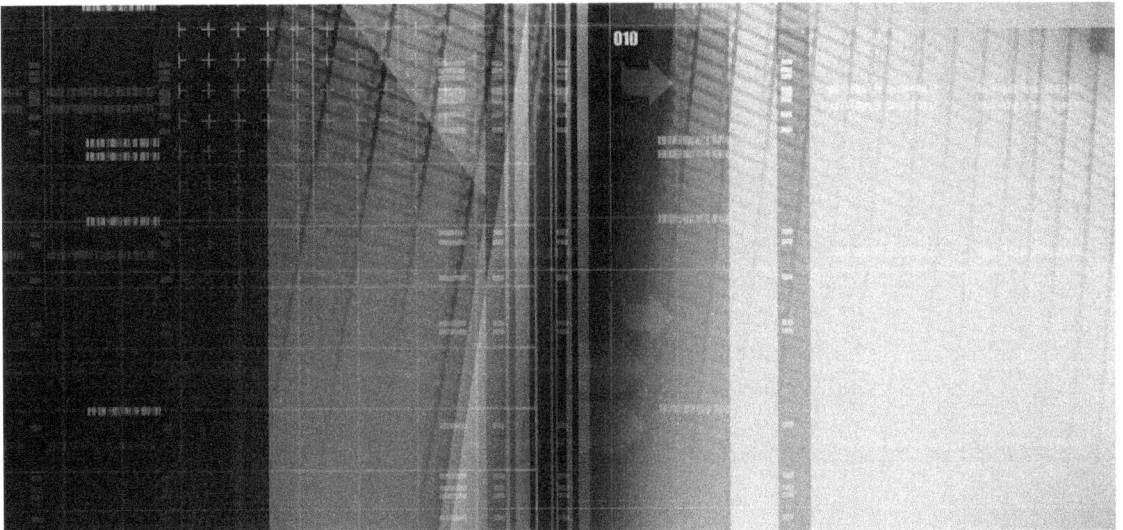

Citizens are becoming wary of technology. While they appreciate the convenience of cell phones and ATM cards, they are not so sure about video surveillance, POS terminals, digital television, and the Internet. The Internet is the defining technological achievement of our time, but citizens notice that it has brought pornography into their lives; it has enabled spam, identity theft, and viruses; and their children spend hours every day with incomprehensible mayhem, violence, and destruction.

Christians think of RFID professionals when they read in *Revelations:* "And he shall make all, both little and great, rich and poor, freemen and bondmen, to have a character in their right hand or on their foreheads: And that no man might buy or sell, but he that hath the character, or the name of the beast, or the number of his name." (Revelations 13:16–17). The citizens of free countries shift uncomfortably in their chairs as they read about the State Department putting chips in passports, the various states tagging their drivers' licenses, and the former Secretary of Health and Human Services getting a microchip implanted under his skin. It is no wonder they explode when their children come home wearing new name tags and a notice that says they're using RFID to take attendance.

RFID, while a powerful solution to intractable business problems, appears to the consumer to be another trick of the technologist, a loss of control over her own personal information, another tool for the terrorist, the brigand, the careless, or the incompetent to do us harm. Eight organizations have declared opposition to the spread of RFID, and *Spychips,* the best-selling book on this technology, is a poorly researched, fear-mongering diatribe against RFID.

The more thoughtful critics of RFID draw this nightmare scenario: *You purchase a T-shirt and put it on. The shirt has a UHF passive RFID chip embedded in its label that uniquely identifies the shirt. At checkout, you present your credit card. The store's software records that you bought this particular shirt and forever associates your name with its unique identification number. As you walk out of the store, you stop to visit an old friend from high school, a bankruptcy lawyer. Readers capture the shirt's unique ID number and relate it to you, and by the time you get home, three of your credit cards have been cancelled, and two others have raised the rates. Your application for a loan refinance is denied, and your final interview for that great job suddenly gets cancelled.*

Defenders dismiss this scenario as impossible with today's technology and infrastructure. The reader, they say, must be within a few feet of the chip in order to read it, so readers would have to be located in the lawyer's doorway and connected to the database in order to detect who comes in and goes out. The retailer's database, they say, would have to be available to credit card companies and to companies hiring new employees. But citizens are not persuaded. They know that technology becomes more powerful as time passes. In a time when a movie screen can show a spy satellite tracking Harrison Ford entering a car and read its license plate (we saw it happen in *Clear and Present Danger*!) from 50 miles away, how can we believe that they won't be able to track our private comings and goings? In a time when consumers' personal data was stolen from Check-Point, three banks, and a finance company, how are our fellow citizens to put faith in any published safeguards?

Let's take a close look at the details of this argument, for its outcome may very well determine whether the promise of accurate information to speed up and slim down our supply chain is ever realized. Let's first take a look at the values we are to defend. What is *privacy*?

14.1 Privacy in Various Countries

Privacy is difficult to define precisely. Louis Brandeis called it "the right to be left alone." He argued that it was the most cherished of all freedoms. According to Ruth Gavison, there are three elements to privacy: "secrecy, anonymity, and solitude." Robert Ellis Smith, editor of the *Privacy Journal*, called it "the desire by each of us for physical space where we can be free of interruption, intrusion, embarrassment, or accountability and the attempt to control the time and manner of disclosures of personal information about ourselves."

Smith argues that privacy is applied only to individuals. "Privacy is a human right," he says. "Organizations may want to keep secrets, but that is different from having a right to privacy. Security is important for protecting data and assets of all kinds, whether or not they pertain to identifiable persons. By contrast, privacy pertains only to data about individuals."

Numerous threats to privacy exist in modern society, and citizens are uneasy about them. Scott McNealy, the president of Sun Computers has been quoted as saying, "You have no privacy. Get over it." This section will look at how laws and customs in the United States, England, Australia, Germany, and Japan deal with the issue of privacy.

14.1.1 Privacy in the United States

In the United States, the right to privacy is not enshrined in the constitution, as many people believe. It is derived from several notions inherited from the English common law. Privacy was authoritatively defined by Samuel Warren and Louis Brandeis in the article "The Right to Privacy," which appeared in the *Harvard Law Review* on December 15, 1890. The article says:

That the individual shall have full protection in person and in property is a principle as old as the common law; but it has been found necessary from time to time to define anew the exact nature and extent of such protection. Political, social, and economic changes entail the recognition of new rights, and the common law, in its eternal youth, grows to meet the demands of society. Thus, in very early times, the law gave a remedy only for physical interference with life and property, for trespasses *vi et armis*. Then the "right to life" served only to protect the subject from battery in its various forms; liberty meant freedom from actual restraint; and the right to property secured to the individual his lands and his cattle. Later, there came a recognition of man's spiritual nature, of his feelings and his intellect. Gradually the scope of these legal rights broadened; and now the right to life has come to mean the right to enjoy life—the right to be let alone, the right to liberty secures the exercise of extensive civil privileges; and the term "property" has grown to comprise every form of possession—intangible, as well as tangible.

The Supreme Court rejected the application of the Constitution's search and seizure provision to telephone wires in 1928 in the case of *Olmstead v. U.S.* But Brandeis, by then a Supreme Court Justice, echoed this definition of the right to privacy as the right to be left alone in a famous and influential dissent.

14.1.2 Privacy for RFID in Europe

The European Union has privacy laws in place to protect consumers' privacy. Retail stores must disclose the presence of RFID tags on products and the presence of readers. They must disclose how the retailer intends to gather and control the information, the purposes for which it will be used, and who will control the data. They must show how to remove the tag from the product and how to access the information on the tag.

As technology and applications have advanced, the European Union has seen the need to expand the scope of privacy definition to address additional issues. For example, applications that link an RFID-enabled plastic card with a consumer's bank account number to enable payment processing raise concerns. RFID technology with item-level tagging increases the potential for direct marketing in stores, since shoppers could be recognized and their movements tracked.

14.1.3 Privacy in the United Kingdom

In 1990, The Calcutt Committee in the United Kingdom defined privacy as: "The right of the individual to be protected against intrusion into his personal life or affairs, or those of his family, by direct physical means or by publication of information."

However, the current privacy picture in the U.K. is grim. The Labour government's desire to appear tough on crime and its large Parliamentary majority have enabled them to pass an unprecedented number of new laws limiting rights, including freedom of assembly, privacy, freedom of movement, the right of silence, and freedom of speech. There has also been an expansion in the use of electronic surveillance, biometrics, surveillance cameras, computer databases, DNA testing, and other intrusive technologies.

To conform to the European Union's Protection Directive, Parliament approved the Data Protection Act in July 1998. This legislation came into force on March 1, 2000 and governs records held by government agencies and private entities. It provides for limitations on the use of personal information, access to, and correction of records and requires that entities that maintain records register with the Information Commissioner.

The act is quite complex and is considered not very effective in promoting privacy and implementing the EU Directive. It is often incorrectly cited as a justification for the mishandling of data by public and private authorities. The protections it does offer are being undermined. The British government is pressing public bodies to be able to share more information and to create "electronic life records" and a complete "population register," which would further undermine the protections of the act. The Court of Appeals issued a controversial decision in December 2003 narrowing the definition of personal information protected under the act and limiting individuals' rights to access personal information held in manual files. The plaintiff in that case has filed a complaint with the European commission, saying that the decision undermines the Directive. Meanwhile, the Lord Chancellor's Department issued a consultation paper called the "Data Protection Act 1998: Subject Access" in 2002 recommending restriction of individuals' access to their own files under the Data Protection Act. This occurred after several politicians and prominent people obtained their own records and found that government officials were secretly trying to undermine their efforts to obtain information from government bodies. The U.K. has a poor culture of security for the protection of personal information. Personal information from government computers is regularly disclosed inadvertently or for profit. The Inland Revenue found in 2003 there were 226 cases of employees illegally accessing the records and selling or maliciously using information. One organization has estimated that over 200,000 illegal requests for information are made each year in Great Britain by private investigators under false pretenses.

14.1.4 Privacy in Australia

In 1994, the Australian Privacy Charter provided that, "A free and democratic society requires respect for the autonomy of individuals, and limits on the power of both state and private organizations to intrude on that autonomy. Privacy is a key value which underpins human dignity and other key values such as freedom of association and freedom of speech. Privacy is a basic human right and the reasonable expectation of every person."

While privacy issues often appear in the daily news, there are few legal safeguards for personal information. Neither the Australian Federal Constitution nor the Constitutions of any of the six states contain any express provisions for privacy. There is continued debate about passing a Bill of Rights, but there are no current proposals. The Constitution limits the legislative power of the federal government. Those areas not expressly authorized are reserved for the states. Federal laws imposing privacy rules on the private sector have been questioned on constitutional grounds but not so far challenged.

Privacy Law in Australia consists of several federal statutes covering particular sectors and activities, some state or territory laws with limited effect, and the residual common law protections.

Until recently there has been no recognition of a general tort of protection of privacy. Very occasionally the common law has been used in support of privacy rights through actions for breach of confidence, defamation, trespass, or nuisance.

In June 2003, a Queensland District Court judge took up the issue and, in *Grosse v. Purvis,* awarded the plaintiff 178,000 Australian Dollars for breach of privacy occasioned by intrusion and harassment over a sustained period of time. Whether this affirmation of a common law right is appealed, upheld if appealed, or followed in other cases remains to be seen.

The main federal statute in the area is the Privacy Act of 1988, which gives partial effect to Australia's commitment to the Organization for Economic Cooperation and Development (OECD) Guidelines and to the International Covenant on Civil and Political Rights (ICCPR), Article 17. The act governs the activities of most federal government agencies.

A separate set of rules about the handling of consumer credit information, added to the law in 1989, applies to all private and public sector organizations.

The Privacy Amendment (Private Sector) Act 2000 was passed in December 2000 and took effect in December 2001. The law puts in place National Privacy Principles (NPPs) based on the National Principles for Fair Handling of Personal Information originally developed by the Federal Privacy Commissioner in 1998 as a self-regulatory substitute for legislation. Private companies are now required to observe these principles, although they can apply to the Privacy Commissioner for approval of a Code of Practice that is an "overall equivalent" to the NPPs. The act has been widely criticized as failing to meet international standards of privacy protection.

14.1.5 Privacy in Germany

Germany has one of the strictest data protection laws in the European Union. In 1977, a Federal Data Protection Law (*Bundesdatenschutzgesetz* or BDSG) was passed.

It was revised in 2002 to be in line with the EU Data Protection Directive. The general purpose of this law is "to protect the individual against violations of his personal rights by handling person-related data." The law covers collection, processing, and use of personal data collected by public federal and state authorities (as long as there is no state regulation), and by nonpublic offices, if they process and use data for commercial or professional aims.

Germany was slow to update its law to make it consistent with the EU Data Protection Directive. Under the terms of the directive, Germany should have harmonized its law by October 1998. The European Commission announced in January 2000 that it was going to take Germany to court for failure to implement the directive. An amending bill was approved by the Government on June 14, 2000, and finally passed into law in May 2001.

14.2 Parsing the Problem

Now that we have defined, as precisely as possible what privacy is and how it is enforced in many countries around the world, this section will break through the public outcries and focus attention on the realities of the threats.

The shrill voice of the opponents would have you believe that RFID is the mark of the beast and that every application of the technology is a threat. But, as you have seen, there are numerous applications for the technology, and most of them pose no threat at all to the privacy of individuals.

Ellis Smith says no threats to personal privacy are posed by the use of RFID tags before the point of sale. When warehouses and factories use the technology, the citizen has no interest in being protected.

No threats to personal privacy are posed when nonconsumer goods are tagged. Tagging the millions of items consumed by businesses, organizations, factories, and governmental organizations creates no privacy issues. Smith argues that even if the tag is later embedded in something the consumer buys, privacy is not violated. He says this is because "the identity of the item is not tied to a particular individual, and even if it were, no harm would result." This author is not so sure.

Smith argues a third case that poses no threat to privacy: consumer items that "are temporary or disposable; any possible tracking of their use will be short-lived. Should a lot of energy be expended worrying about a technology that can track the life cycle of a disposable razor? A similar category includes products whose use is innocuous or routine, even if known to a stranger. Should a lot of energy be expended worrying about RFID tracking of baby powder or a light bulb or a teapot?" Again, this author is not so sure. If a store scans my purse and finds baby powder and thereby has its kiosk offer me infant products, has my privacy not been violated?

Applications such as the tracing of cattle to prevent mad cow disease, tracking pallets of goods through a supply chain, or tracking tractors on a construction yard do not raise privacy issues. Preventing airlines from losing bags is a definite improvement in our lives, but we'll have to make sure the tags are removed or disabled when we head for home.

Thoughtful experts see privacy issues in just two scenarios. First, tags are transferred to the unsuspecting and uninformed public. Second, when people themselves are embedded with chips or find themselves required to carry tagged documents.

14.2.1 Tags Transferred to People

For tags to threaten people's privacy, the following elements must be in place:

- The tags must be readable.
- The tags must be uniquely identifiable.
- The tags must be able to be read surreptitiously.
- The data in the tag may be personally revealing or unnecessarily explicit.
- If the data is not revealing, it must access a database that provides personally revealing data.

14.2.1.1 Readable Tags

Tags that are effectively encrypted or protected by password pose little threat to personal privacy, unless the safeguards can be cracked. While many cases exist of broken security measures, the cost of breaking them probably outweighs any advantage. Most safeguards will protect against most attackers. Most successful attacks are launched discovering the keys, rather than breaking the codes. Making sure you safeguard the keys is the vital element in restricting unwarranted reading of tags. Use passwords and encryption as extensively as possible when you design your systems.

14.2.1.2 Uniquely Identifiable

Clothing that proclaims its manufacturer's name is often a fashion statement. So consumers are not always opposed to strangers knowing the manufacturer of their merchandise. RFID tags that broadcast only the type of item they identify may be a bit unsavory in the case of one's underwear, but they do not meaningfully violate a person's privacy. Tags that are uniquely identifiable can, in theory, key into a database that relates the item back to the person. This fear is largely theoretical at this point, but the technology will get stronger and the databases will contain ever-larger amounts of data. Transferring uniquely identifiable tags to consumers should be avoided if at all possible. Gen-2 tags that can be rendered inoperative upon checkout are a good solution, as well.

14.2.1.3 Surreptitious Reading

Today, defenders of RFID take comfort in how difficult it would be to read tags surreptitiously on a routine basis. They argue that read ranges are too short, and the readers are too clumsy. They take comfort in their opponents' mistaken assumption that satellites can read passive RFID tags. Industry experts say satellite reading of passive tags is impossible today, although no one will say that it will not be possible tomorrow.

The choice for item-level tagging at the moment seems to be 13.56 MHz tags, and today, they cannot be read from more than a meter away. UHF tags, however, can be read from several meters away at legal power levels.

In fact, of course, handheld readers are available. Also, inexpensive attachments are available that will turn a Palm Pilot, a Windows CE machine, or other handheld computer into a fully functional reader. It is unlikely that you, as an implementer, will be able to guarantee your tags against surreptitious reading to the satisfaction of the privacy advocates. Your privacy protection will likely have to be based on making tags unreadable or destroying their unique identifications.

One aspect of surreptitious reading is within your control. You can and should make certain that you are not open to the charge of secrecy or stealth. This means that you should put up prominent notices wherever RFID reading is taking place, especially in public places. History has shown that, even when signage is up, some privacy advocates may claim that the reading was clandestine. Your only defense is to be thorough, forthright, and conspicuous. You should also notify the public whenever you might be transferring a tag to them.

14.2.1.4 Personally Revealing Data

Tags can contain a simple license plate, or they can contain a wealth of data. A medical records application, for example, may require that medical history be stored on the tag. If the application does not absolutely require it, you should make every effort to keep personal and specific data off the tag. If you are tagging hazardous chemicals, the tags may contain instructions for use and disposal. Evaluate carefully whether any identification information you put on the tag could aid criminals or terrorists. Design your system with a database key on the tag and any sensitive information behind your firewalls, if at all possible.

14.2.1.5 Database Access

Corporate policies probably exist today in your company to safeguard sensitive databases. The policies probably cover backup, network access control, control of physical access to the facility, passwords, password authority, and remote access control. The RFID system should be designed to open no new security vulnerabilities and to utilize the existing security apparatus to safeguard the RFID data. You should carefully review the existing policies to assure yourself of three certainties:

- Existing safeguards adequately protect the data you propose to store.
- Your system will open no new security vulnerabilities.
- Your system will be able to function within the existing security framework.

14.2.2 Tagging People

Tagging people evokes very unsavory images, but for some applications it makes a lot of sense. Putting wristbands on patients in hospitals, residents in nursing homes, prisoners in custody, or employees in high-security areas all confer benefits that are clear and compelling. We have reviewed these applications in Chapter 5. But the tagging of people is too chilling an image to be undertaken lightly. It should be done only in very limited circumstances, where the benefits to the person or to society are very clear and outweigh any privacy issues.

14.3 The Solution

The solution to the RFID privacy problem is relatively straightforward. The problem has been permitted to grow out of hand by carelessness and lack of attention. Adhering to the following rules will keep your project out of trouble.

14.3.1 Mainstream Solutions

The following guidelines will help you avoid the privacy problems that RFID tagging presents:

- Always notify consumers when there are readers operating.
- Always give consumers a choice of whether to have tags continue operating or have them removed or deactivated when they complete a purchase. Make exercise of either choice easy for the consumer to

execute, and make sure there are no reductions in service level for consumers who opt for deactivation.

- Deactivate tags by default for most consumer goods. Deactivate tags on airline bags when the consumer exits the terminal.

- Continue activation by default only when there is a strong justification. Management of returns and warranties is not strong enough justification.

14.3.2 Nonstandard Solutions

Two other solutions to the RFID privacy problems have been proposed. Neither has been accepted as a standard by any recognized standards body, but these are ideas that are being discussed in the technical community and seriously considered.

14.3.2.1 Blocker Tags

Blocker tags are tags that utilize the anticollision algorithm to render tags in their vicinity unreadable. The tags are specifically designed for this function, and any consumer who carries one will temporarily disable any readers within range. The blocker tags masquerade as "any" tag, filling the reader's anticollision queue with phony reads, immobilizing the reader.

A design for blocker tags is already complete, and two companies are beginning to manufacture them.

14.3.2.2 Frequency Change

One writer has suggested that chips be built to enable them to operate on a different frequency once the item has been purchased. The advantage of this approach is that the chip could be set to work on a frequency with a much lower read range, which would protect the consumer from surreptitious reads. The disadvantage, of course, is that the chip would have to have two antennas and that would increase the cost and, potentially, the size.

14.4 EPC Privacy Guidelines

EPCglobal recognizes the threat that privacy concerns pose to its success. The organization has issued guidelines to provide consumer protection. The guidelines contain four provisions:

- Consumer notice
- Consumer choice
- Consumer education
- Record use, retention, and security

14.4.1 Consumer Notice

The guideline calls for notice to be given by using the EPC logo on the packaging: "Consumers will be given clear notice of the presence of electronic product codes on products or their packaging."

14.4.2 Consumer Choice

The guideline calls for consumers to be given choices to discard, remove, or disable EPC tags. EPCglobal anticipates that most tags will be part of the packaging and hence disposed of when the item is put in service. The organization "is committed to finding additional efficient, cost effective, and reliable alternatives to further enable customer choice."

14.4.3 Consumer Education

Consumers will be able to "easily obtain accurate information about EPC and its applications, as well as information about advances in the technology." The guideline envisions a role for the manufacturers who use the technology in their products, as well as EPCglobal organization themselves.

14.4.4 Record Use, Retention, and Security

The guideline notes that the EPC contains no personally identifiable information and calls upon organizations to comply with applicable laws regarding data stored, maintained, and protected. In addition, again, in compliance with existing laws, EPCglobal calls upon its members to publish information on their policies regarding retention, use, and protection of any personally identifiable information.

14.5 Fair Information Practices and the RFID Bill of Rights

The United States Department of Health, Education, and Welfare published a Code of Fair Information Practice (HEW73) in 1973. Simson Garfinkle, a well-known privacy activist, has updated it to reflect RFID technology and calls it the RFID Bill of Rights. The bill has five Rights:

1. The right to know if a product has an RFID tag.
2. The right to have embedded tags removed, deactivated, or destroyed when a product is purchased.
3. The right to first-class RFID alternatives.
4. The right to know what information is stored inside their RFID tags. If the information is incorrect, there must be a means to correct or amend it.
5. The right to know when, where, and why an RFID tag is being read.

Items 1 and 5 of the bill mandate there should be no covert RFID systems. It envisions placards, logos, and notices in areas where RFID readers are active, and possibly, notifications of RFID free zones as well. Item 2 is qualified with a requirement that customers not be unduly burdened with the task of tag destruction. It should be part of the normal checkout process. Some defenders have pointed to RFID tags' value in facilitating returns of merchandise. They note that killing the tags destroys that particular element of value. An alternative might be to erase the unique serial number, leaving only the prefix that identifies the *type* of merchandise but not the individual instance.

In one trial, the company put up signs that read "Electronic Merchandise Systems are in use." Privacy advocates dismissed this effort as inadequate notification.

Item 3 seeks to avoid penalizing consumers who decline RFID-enabled services. For example, if the only way to ride a particular highway were to have a transponder, that would unfairly penalize commuters who decline. That is a violation of Item 3. Similarly, merchants who condition returns on retention of the tag would be in violation of Item 3.

Item 4 may require extra equipment at the point of sale or at the exit. You might include a reader so consumers who are interested can read their tags, view the contents, and satisfy themselves that the tags have been disabled.

14.6 The Role of the Implementer

If the technologists who created and installed the routers and servers and protocols and software of the Internet had been more careful, we could, today, have a world without spam and viruses. The people implementing the technology make recommendations and decisions that control whether or not the system is vulnerable to exploitation. You must keep these issues in mind as you design and implement your system.

14.6.1 Technology Decisions

When you make your recommendations and decisions, make sure you keep the public perception of vulnerability and privacy concerns in your mind. Determine whether the tags will go to the public. If so, here is a checklist of actions to consider in your deployments:

- Make sure the public is notified clearly, completely, and unambiguously.

- Make sure the tags and system allow no individual identification outside the design specification. This issue is more subtle than it may appear. We tend to think of attackers as outsiders, and we take comfort that the all-important databases are safe "inside." But intruders find security holes, and they may easily be insiders themselves, with ready access to the databases. Take this point of view in your design decisions. Document how you will prevent unauthorized access by corporate insiders.

- Tags should be routinely disabled at the point of transfer to the consumer. This is controversial. But, as the Electronic Freedom Federation (EFF) points out, privacy is fragile and, like free speech, needs substantial "breathing space" to ensure that the technology does not outrun the social norms. Anything less than a mandatory kill at the point of transfer will open the door to those who would misuse the technology for their own gains, says the EFF.

- As the system implementer, it is very important that you take privacy considerations seriously. You must review each decision you make from the perspective of protecting privacy, and you must articulate and document the concerns. When you are in doubt, err on the side of preserving privacy.

- Encrypt all data where possible. If there is any doubt, err on the side of encrypting your data.
- Use passwords where possible. Protect all data with passwords, as the EPCglobal specification makes available.

14.6.2 Your Role as Spokesperson

When you encounter other points of view, it is important to take them seriously. Do not be dismissive. Even if they are flawed in some ways, address them seriously, patiently and completely. People with legitimate concerns may get some of their important facts wrong in their complaints, but you must be sympathetic to their fears or else you will have an implacable foe. Important trials at Metro Group in Germany and at Benetton were cancelled because of privacy concerns that were managed poorly. Projects in the Northern California school district and the United States Department of State were delayed and derailed. Gillette was forced by privacy demonstrations to halt a project in the U.K. You cannot afford to have enemies.

Implement an effective feedback system to capture and resolve any concerns related to your project. Include a privacy issues review in all of your project briefings.

14.6.3 Your Role as Privacy Advocate

Your project may go more smoothly if you take the role of privacy advocate yourself. You should take a leadership position on individual privacy and make sure your team is fully educated on the issues. Educate all of your stakeholders regarding any privacy issues your project may raise and what you are doing about them.

Review the company's published privacy policy and share it with your team. If it does not sufficiently protect the privacy of RFID data, recommend that it be updated.

Review record retention policies and make sure they are applicable to your project. Recall that RFID projects can produce huge quantities of data, and much of it may not be useful. Retention of everything will be costly.

Above all, do not ignore privacy advocates if your project has any interaction with the public. Actively seek them out, and make them part of your team. Be very open and honest with them about your plans and take their concerns seriously.

Managing Privacy Issues: A Case Study

Marks & Spencer, one of Britain's biggest retailers, began a trial to track clothing as it moves from one of the company's distribution centers to a single store. The retailer consulted with privacy groups and took pains to address the privacy implications of the trial, setting an example for others in the retail industry. The company consulted with two groups in the U.K., and with CASPIAN (Consumers against Supermarket Privacy Invasion and Number), a group that opposes the use of RFID tags in consumer products. CASPIAN acknowledged that Marks & Spencer has taken a socially responsible position by informing consumers, enabling consumers to easily remove the tags, and not using tags at the point of checkout to record which items were purchased by which customers. CASPIAN nonetheless denounced the trial, saying it set a "dangerous precedent" by putting tags in clothes. The company has since expanded the trial to become one of the largest RFID installations in the world.

14.7 Chapter Summary

The idea of privacy is slightly different in each country. But there is widespread alarm in the public mind that RFID represents a threat to a cherished right. Commonsense rules have emerged to protect the public's rights while still gaining the benefits of RFID technology. Several organizations have published "Bills of Rights" and "Privacy Guidelines." If your project has any interaction with the public, take any privacy concerns very seriously.

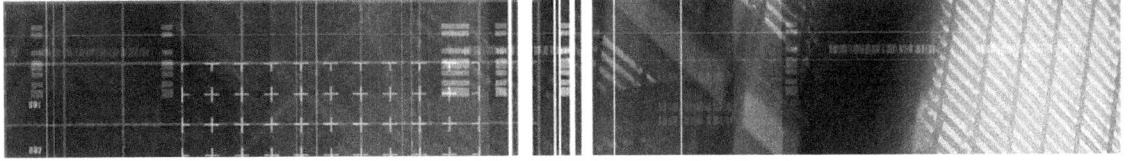

PART **IV**

References

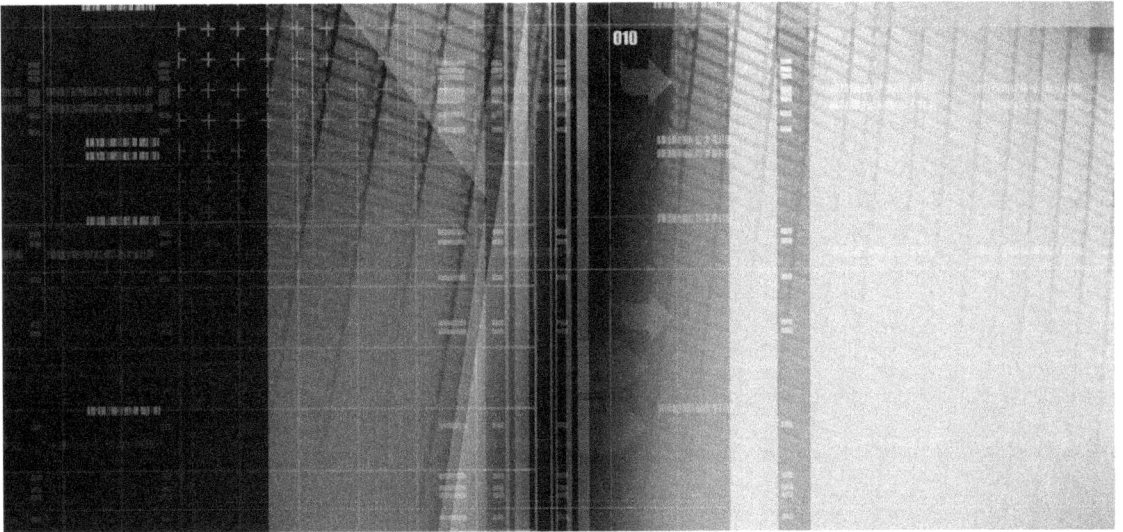

APPENDIX A

Typical Application Profiles

This table presents common technology choices for various RFID applications.

Application	Approach
Tracking and tracing of cartons and pallets in a supply chain application	UHF passive tags. Open loop system with substantial interchange of data among trading partners. EPCglobal standards. Tag stores license plate information only
Hazardous materials tracking	HF and UHF passive tags. Closed loop application. Tag can store helpful and identification data about the material as well.
Airline baggage tracking	UHF passive read/write tags. Also deployed with HF read/write tags. Closed loop application. ISO 18000-C.
Inventory monitoring and control	HF passive tags. Readers can monitor shelves. Can be an open loop application.
Fleet monitoring and management	Semi-active, read-only, and read-write tags with specialized circuits to integrate with vehicle electronics to monitor fuel level, oil pressure, and temperature. Closed loop application.
Animal tracking	LF passive tags. ISO 11784/11785.
Antitheft for high value items	Active tags with motion detector or passive tags with readers watching key portals.
Electronic article surveillance	One-bit passive tags. Some retailers are adding HF tags to report which items are being shoplifted.
Automotive antitheft immobilization	Passive LF tags work with a reader in the car, providing authentication. Newer systems use combination of passive and active tags.
Electronic payment	HF passive tags.

Application	Approach
Electronic toll payment	Semiactive tag containing customer ID and specialized electronics for displaying account status, battery level, and other information.
Access control	Passive LF and HF tags. ISO 15693 standard. Storage on tags from 64 bits to 2K bits. Closed loop application.
Smart shipping containers	Active 433 MHz tags detect and report tampering and test for explosives or radioactive contents. ISO 10374, 18185, and 23359.
Drug anticounterfeit	Passive HF, UHF, and microwave tags.
Currency anticounterfeit	Microwave tags could be used due to a very small antenna.
Tracking newborns in hospitals	Active tags at 433 MHz attaches to ankle, triggers alarms if removed. Readers at doors and in ceilings. System also tracks staff and equipment.
Tracking legal files in a law office	System uses one checkout reader and one handheld reader. Staff members can have an ID card, and the checkout process matches the person with the files they took.
Manufacturing control (automobiles)	Attach tag to vehicle carrier and track where each vehicle being manufactured is in the assembly process. System lets each station know what is coming its way so it can provision the correct parts.
Growers test effectiveness of temperature-sensing tags	Consortium of several growers put tags on pallet in Chile for transport to Florida. The tags record the temperature every five minutes. The entire log of temperature readings, up to 1500, is downloaded as pallets are unloaded.
Hospital improves access to patient data	Admitted patients are issued RFID-enabled wrist bands. Once the reader reads the tag, the medical file is displayed on the PC screen. Any new information is entered only once on the PC and becomes part of the record.

APPENDIX B

Bibliography

Accenture. "RFID/EPC Solution" (2003).

Adams Communications. "EAN Application Identifiers and the UCC/ EAN-128 Symbology Page" (July 18, 2004).

AIM/The Global Trade Association for Automatic Identification. "RFID Standards" (November 2001) http://www.aimglobal.org/standards/ rfidstds/RFIDStandard.asp.

Alien Technology. Whitepaper: "EPCglobal Class 1 Gen-2 RFID Specification" (2005).

Almyta Systems. "Inventory Management History 4" (2005).

Ascential. "Preparing for the RFID Data Explosion" (2004).

Atmel Corporation. "Understanding the Requirements of ISO/IEC 14443 for Type B Proximity Contactless Identification Cards" (August 2002).

Auto-ID Center. "Draft Protocol Specification for a 900 MHz Class 0 Radio Frequency Identification Tag" (February 23, 2003).

Avery Dennison "Selecting RFID Tags." Avery Dennison. http://www .averydennison.com/corporate.nsf/50e86d296559306188256a3b00693bd7/ 41a3657b6fd26ffe88256e6600768a97?OpenDocument.

Balanis, C. *Antenna Theory*. 2nd edition. New York: Wiley and Sons Inc., 1997.

Bass, Steve, Lisa Miller, and Bryan Nylin. "HIPAA Compliance Solutions." Microsoft and Washington (November 28, 2001).

Beizer, Doug. RFID "Embedded in Supply Chain." Washington Technology (February 7, 2005).

Bhuptani, Manish and Shahram Moradpour. *RFID Field Guide*. Upper Saddle River, NJ: Sun Microsystems Press, A Prentice Hall Title, 2005.

Boushka, Michael, Lyle Ginsurg, Jennifer Haberstroh, Thaddeus Haffey, Jason Richard, and Joseph Tobolski. "RFID on the Move: The Value of Auto-ID Technology in Freight Transportation." Accenture (2002).

Brock, David and Chris Cummins. "EPC Tag Data Specification 1.0 Last Call Working Draft Version" (September 12, 2003).

Brooke, Marlo. "RFID Best Practices Great Minds Unite." *DM Review* (September 2005).

Burt, Jeffrey. "Case Study: RFID Tags Let Purdue Pharma Monitor OxyContin's Movement through Supply Chain." *eWeek* (June 20, 2005).

CattleNetwork.com. "Would You Like Source Verification with Your Meal Today?" (October 31, 2005).

Checkpoint Systems Inc. "Using EPC/RFID to Create New Business Value" (March 2004).

Collins, Jonathan. "Automotive RFID Gets Rolling" *RFID Journal* (April 13, 2004).

Collins, Jonathan. "Gen2 Finds a Path." *RFID Journal* (December 10, 2004).

Collins, Jonathan. "Purdue Pharma Tags OxyContin." *RFID Journal* (November 16, 2004).

Contactless News. "RFID Tags and Contactless Smart Card Technology: Comparing and Contrasting Applications and Capabilities" (April 28, 2005).

Cook, Andrew, Steven Weigand, and Daniel Dobkin. "Introduction to Antenna Selection and Configuration." *RFID Solutions Online 9/22/2005.* http://www.wjcommunications.com/pdf/techpubs/ Introduction%20to%20Antenna%20Selection%20and%20Configuration .pdf.

Dekenah, Marc. "Antenna Gain Explained." *Marc's Technical Pages.* http://www.marcspages.co.uk/tech/antgain.htm.

dHondt, Susy. "The Cutting Edge of RFID Technology and Applications for Manufacturing and Distribution." Texas Instruments.

EAN-INT.ORG. "Locations" (2005).

EPCglobal, Inc. "Electronic Product Code" (2005).

EPCglobal, Inc. "The EPCglobal Network" (September 24, 2004).

EPCglobal, Inc. "The EPCglobal Network and the Global Data Synchronization Network (GDSN)" (2004).

EPCglobal, Inc. "EPC Ratifies First Software Standard for EPC/RFID" (September 21, 2005).

EPCglobal, Inc. Healthcare Life Sciences Business Action Group. http://www.epclobalinc.org/action_groups/hls_bag.html.

EPCglobal US, Inc. "Electronic Product Code" (2005). www.epcglobalus .org/Network/Electronic%20Product%20Code.html.

European Telecommunications Standards Institute (ETSI). TR 101 445 v1.1.1 (2002–2004) "Electromagnetic Compatibility and Radio Spectrum Matters (ERM); Short-Range Devices Intended for Operation in the 862 to 870 MHz B and; System Reference Document for Radio Frequency Identification (RFID) Equipment."

Federal Communications Commissions (United States). "Article 15.251 Operations within the Bands 2.9–3.26 GHz, 3.267–3.32 GHz, 3.339–3.3458 GHz, and 3.358–3.6 GHz."

Fennig, Chris. "The RFID Tagging Guide: Secrets for Handling Difficult to Read Products." Odin Technologies (February 2005).

Finkenzeller, Klaus. *RFID Handbook*. New York: John Wiley & Sons Inc., 2003.

Floerkemeier, Christian, Dipan Anarkat, Ted Osinski, and Mark Harrison. "PML Core Specification 1.0, Auto-ID Center" (September 2003).

Foley & Lardner, LLP. "New California Laws Focus on Marketing of Prescription Drugs and Use and Disclosure of Medical Information." *Law Watch* (October 21, 2004).

Garfinkel, Simson and Beth Rosenberg. *RFID Applications, Security and Privacy*. Upper Saddle River, NJ: Addison-Wesley, 2005.

Gilbert, Alorie. "U.S. Military Invests in 'Active RFID.'" *CNET News.com* (March 23, 2004).

Grocery Manufacturers of America (GMA), the Food Marketing Institute (FMI), and A.T.Kearney. "Action Plan to Accelerate Trading Partner Electronic Collaboration: Data Synchronization Proof of Concept—Case Studies from Leading Manufacturers and Retailers, 2003." http://fmi.org/evens/2004_Distribution/Data_Synchronization.pdf.

GS1 Singapore Council. "EAN Application Identifiers and the UCC/EAN-128 Symbology" (July 19, 2004). http://www.sanc.org.sg/symbol.htm.

GS1, Annual Report (2005).

Intermec. "Practical Uses for RFID Technology in Manufacturing and Distribution Applications" (2004).

The Hindu. "Mobile Commerce: Automatic Payment Systems" (July 8, 2002).

Glidden, Rob and John Schroeter. "Bringing Long Range UHF RFID Tags into Mainstream Supply Chain Applications." *RF Design Magazine* (July 1, 2005).

Geier, Jim. "EIRP Limitations for 802.11 WLANS." *Wi-Fi Planet* (July 18, 2002). http://www.wi-fiplanet.com/tutorials/article.php/1428941.

Harmon, Craig and Leslie Downey. "RFID: Will China Throw A Monkey Wrench?" *Business Week* (September 12, 2005).

Harrop, Dr. Peter. "Lessons from 1,400 Case Studies." IDTechEx.

Harrington, Tim. "Open RTLS Standards." *WhereNet* (2004).

Hou, Chua Hian. "Quick and Easy Payment." *Computer Times* (December 31, 2003).

Hugos, Michael. *Essentials of Supply Chain Management*. New York: John Wiley & Sons Inc., 2003.

Information Week. "RFID to Fight Counterfeiting of Viagra, Painkilling Drugs" (November 15, 2004).

Intelleflex. "Battery-assisted Smart Passive Tags: The Best of Both Worlds" (2005).

Intermec. "Guide to RFID Tag Selection" (2004).

Intermec. "Real World RFID." *Realtime Magazine* (2005).

Intermec. "Parts Traceability and Product Genealogy." Intermec White Paper.

Japan Ministry of Internal Affairs and Communications, SC31/WG4 Committee. "RFID Using UHF Band under Japanese Radio Law" (December 28, 2004).

Joshi, Manoj and V. Subrahamanya. "Global Data Synchronization: An End-to-End Perspective and Solution Framework" (2003). http://www.wipro.com/insights/globaldatasynchro.htm.

Inpinj. "About RFID Regulations." http://www.inpinj.com/page.cfm?ID=about RFIDRegulations.

Kleist, Robert A, Theodore A. Chapman, David A. Sakai, and Brad S. Jarvis. *RFID Labeling*. Irvine, CA: Printronix, Inc., 2004.

Laran RFID. "A Basic Introduction to RFID Technology and Its Use in the Supply Chain" (January 2004).

Laurer, George. "Development of the UPC Symbol" (October 2001). http://bellsouthpwp.net/l/a.laurergj/UPC/upc_work.html.

Lingle, Rick. "Project Jumpstart: A 'Bird's' Eye View." *Packing World Magazine* (February 2005).

Lingle, Rick. "Thomasville Rolls Out RFID 'a la Cart.'" *Packaging World Magazine* (June 5, 2005).

Lowry Computer Products. "Understanding Smart Label Technology." Brighton, MI: Lowry Computer Products (2004).

Lowry Computer Products. "RFID for the Department of Defense: The DOD Mandate." Brighton, MI: Lowry Computer Products (2005).

Macehiter, Neil. "RFID Vendor Positioning." Ovum Corporation (2005).

Machrone, Bill. "RFID: Dogs! Cats! Guitars?" *PC Magazine* (July 13, 2004).

Manias, Giles. "Smarter than 'Smart Labels'?" Ohio: Paragon Data Systems.

Manufacturing.Net. "RFID Is the Balm." Reed Business Information (May 12, 2005).

Marlin, Steven. "Visa Debuts RFID-Enabled Card-Payment System."
Information Week (February 24, 2005).

Maswelli, Jennifer. "NYK Logistics Tracks Containers." *RFID Journal*
(September 8, 2003).

Matrics. "Case Study: McCarran International Airport."

McQuivey, James and Michael Feehan. "RFID Could Enhance Pharmaceutical Marketing." *RFID Journal* (March 14, 2005).

MoreRFID. "Escort Memory Systems Moves UHF RFID into the World
of Industrial Control."

Murphy-Hoye, Mary, Hau L. Lee and James B. Rice. "A Real-World Look
at RFID." *Supply Chain Management Review* (July 1, 2005).

Napolitano, Maria Ida. "Get Ready for RFID." *Logistics Management*
(Aug 1, 2005). http://www.keepmedia.com/pubs/LogisticsManagement/
2005/08/01/972194?page=2.

O'Connor, Mary Catherine. "McCarran Airport RFID System Takes Off."
RFID Journal (Oct 25, 2005).

O'Connor, Mary Catherine. "Odin Upgrades Its Tag-Testing Tool."
RFID Journal (March 8, 2005). http://www.rfidjournal.com/arcticle/
articleview/1435/1/1/.

O'Connor, Mary Catherine. "Integral RFID Looks for RF Hot Spot."
RFID Journal (March 23, 2005). http://www.rfidjournal.com/arcticle/
articleview/1462/1/1/.

O'Connor, Mary Catherine. "Odin Benchmarks RFID EPC Tags."
RFID Journal (October 21, 2004). http://www.rfidjournal.com/
arcticle/articleview/1199/1/1/{120.

O'Connor, Mary Catherine, "Unified System for Healthcare." *RFID Journal*
(December 6, 2004).

Overby, Christine Spivey. "How RFID Improves the Order-to-Cash Process."
Forrester Best Practices (May 27, 2005).

Overby, Christine Spivey with Carrie A. Johnson and Sean Meyer. "RFID
Beyond the Supply Chain." Forrester Research (May 26, 2005).

Palmer, Roger C. *The Bar Code Book*. Helmers Publishing Inc.,
November 1995.

Petragani, Jim. "EAN.UCC System Update. Presentation to Saftermarket
Council on Electronic Commerce" (August 2, 2004).

Pfizer, Inc. "For Pharmacists: Frequently Asked Questions about
Viagra RFID" (December 15, 2005). http://www.pfizer.com/pfizer/
subsites/counterfeit_importation/mn_pharmacist_viara_rfid_faq.jsp.

Philips SemiConductors. "Item-Level Visibility in the Pharmaceutical Supply Chain" (July 2004).

Pipe, Russell. "China Launches Privacy Protection Law Project. Privacy and Business."

Polsonetti, Chantal, "The European Supply Chain Conundrum." *RFID Journal* (November 10, 2005).

Porter, Lori. "The Gen-2 Standard: What Is It, and What Does It Mean?" Paxar Corporation. http://www.hegrobelgium.be/files/RFID_gen2.pdf.

Price, Andrew. "Helping Bags Make Their Flights." *RFID Journal* (October 10, 2005).

Privacy International. "Report on Japan" (November 16, 2004).

Puleston, David J. and Ian J. Forster. "The Test Pyramid: Avery Dennison RFID Division" (October 2005).

Reca, Marcella. "Microchip Proposed as Form of Equine ID." *The Horse* (April 2005).

RFID Gazette. "A Comparison of Tag Frequencies" (October 2005).

RFID Gazette. "The RTLS Market" (October 2005).

RFID Gazette. "Frequencies for Active Tags" (September 2005).

RFID Gazette. "New Software Standard for RFID/EPC" (September 2005).

RFID Gazette. "The EPC Gen-2 System—Part 1" (October 7, 2005).

RFID Journal. "RFID Payment Systems Take Off" (June 9, 2003).

RFID Journal. "Las Vegas Airport Bets on RFID" (November 6, 2003).

RFID Journal. "RTLS Links Active and Passive RFID" (February 14, 2003).

RFID Journal. "UK Trial Addresses Privacy Issue" (January 4, 2006).

RFID Mainstreet. "RFID Know-it-all Case Studies."

RFID Wizards Inc. "Return on Investment Study for RFID Solution" (2003).

RFID Update. "Overview of 802.11 RTLS Vendors" (October 12, 2005).

RFID Update. "RTLS Market to Exceed $1.6 Billion by 2010" (September 7, 2005).

Roberti, Mark. "Active RFID Keeps Toyota Distributor Rolling" (August 1, 2005).

Roberti, Mark. "FAA to Publish Passive RFID Policy." *RFID Journal* (June 30, 2005).

Roberti, Mark. "RFID Aided Marines in Iraq." *RFID Journal* (February 21, 2005).

Roberti, Mark. "RFID Upgrade Gets Goods to Iraq." *RFID Journal* (July 23, 2004).

Roberti, Mark. "Wal-Mart Begins RFID Process Changes." *RFID Journal* (February 1, 2005).

Roberti, Mark. "What Other Retailers Can Learn from Prada." *RFID Journal* (2002).

Schuster, Edmund and David L. Brock. "Creating an Intelligent Infrastructure for ERP Systems: The Role of RFID Technology." MIT DataCenter Publications. April 14, 2004. www.mitdatacenter.org/ ERP-IntelligentInfrastructure%20Edits%20DLB%20(4-14-04).pdf.

Sensormatic. "RFID Deployment: 7 Critical Success Factors" (2004).

Shepard, Steven. *RFID*. New York: McGraw-Hill, 2005.

Smith, Brad. "RFID Takes New Turns." *Wireless Week* (July 15, 2005).

Stelzer, John L. "Data Synchronization: What Is Bad Data Costing Your Company?" (April 1, 2003). http://internetweek.bitpipe.com/detail/ RES/1051711804_812.html.

Sullivan, Laurie. "The European Union Works Out RFID Privacy Legislation." *Information Week* (February 6, 2005).

Sutherland, Ed. "Hospitals Take the Pulse of Wi-Fi Tracking." *Wi-Fi Planet* (April 12, 2005).

Sweeney, Patrick J. *RFID For Dummies*. Hoboken, NJ: Wiley Publishing Inc., 2005.

Terrill, William. "RFID: Not Just Who, but Where." The Burton Group (April 29, 2005).

Texas Instruments. "Team Tag-It News Issue No. 1" (February 2000).

ThingMagic. "Generation 2: A User Guide." http://www.thingmagic.com/ html/Generation2%20-%20A%20User%20Guide.pdf.

Traub, Ken. "ALE: A New Standard for Data Access." *RFID Journal* (April 18, 2005).

Uniform Code Council Inc. "Serial Shipping Container Code (SSCC) Implementation Guide" (May 2004).

Uniform Code Council. "Global Trade Item Number (GTIN) Implementation Guide" (2004).

Uniform Code Council. "Global Location Number (GLN) Implementation Guide" (2002).

Uniform Code Council. "Global Returnable Asset Identifier (GRAI) Implementation Guide" (2002).

Uniform Code Council. "Global Service Relation Number (GSRN) Implementation Guide" (2002).

United States Food and Drug Administration. "Combating Counterfeit Drugs" (February 2004).

United States Food and Drug Administration. "Radiofrequency Identification Feasibility Studies and Pilot Programs for Drugs Sec 400.210" (November 2004).

Wise Research. "FDA Tells How RFID Will Fight Counterfeit Drugs." *Using RFID* (February 24, 2004).

Valero, Greg. "Logistics in Reverse. Pharmaceutical Business Strategies" (March 2005).

Verisign Inc. "The EPCglobal Network: Enhancing the Supply Chain" (2005).

Voyles, Bennett. "Startup Launches Low-Cost Tracking System." *RFID Journal* (August 9, 2005).

Walker, Colonel Glenn. "Relearning Lessons Learned." *Army Logistician* (January–February 2005).

Warren, Samuel and Louis Brandeis. "The Right to Privacy." *Harvard Law Review* (December 15, 1890).

Whitaker, CDR A. Davis Jr. (U.S. Navy). "From Factory to Foxhole: In-Transit Visibility in Operation in Desert Storm. Operation Iraqi Freedom and Beyond." Naval War College (May 16, 2004).

Whiting, Rick. "Drugmakers 'Jumpstart'" RFID Tagging of Bottles. *Information Week* (July 26, 2004).

Wyld, David C. "Mad Cow Policy: The National Animal Identification System." *Contactless News* (July 21, 2005).

Wyld, David C. "RFID: The Right Frequency for Government." The IBM Center for The Business of Government (October 2005).

Zebra Technologies. "Bar Coding and RFID: The Key to Traceability and Safety in the Food Service Industry."

Zebra Technologies. "Increasing Profits and Productivity: Accurate Asset Tracking and Management with Bar Coding and RFIF" (2003).

GLOSSARY

2D Two-dimensional bar code. Conventional bar codes grow wider as more data is encoded. 2D bar codes can grow taller, as well. Thus they can store more data.

3PL Third Party Logistics company. A company that provides various logistics services such as warehousing and shipping as a service to other companies.

801.11 The IEEE classification for a family of telecommunications access methods, also referred to as *WiFi*, for wireless networks. RFID networks can join the corporate network wirelessly via a device that interfaces their readers to a WiFi base station.

active tag Active tags have a battery as their source of energy. (Compare with passive tags, which rely on power created by radio waves from the reader as their source of temporary energy.) Active tags have a read range of up to 1200 feet (approximately); passive tags' range is much less. An active tag's battery eventually wears out. Most active tags transmit continuously; passive tags transmit only when in range of a reader. Some active tags, called *beacons*, transmit at preset intervals.

actuator A mechanical device for electronic control and movement of an object.

ADC Automated Data Capture. The generic term for reading data from a bar code, magnetic card, button, or RFID tag into a computer system without manual transcription or typing.

Advance Shipping Notice See *ASN*.

agile reader A reader that can read tags operating at different frequencies or with different protocols.

AIM Association for Automatic Identification and Mobility. A global trade association representing the automatic identification and data collection industry.

air protocol The specification for communication between a tag and reader. It describes the operating frequency, the call-and-response characteristics, transmission distances allowed, and regulations. It also defines the response characteristics (AM, half-duplex, and so on) and protocols for collision avoidance and dense reader operations.

Aloha The name associated with a protocol for *collision avoidance*. When numerous tags are within range of a reader they will all try to transmit simultaneously and the reader will see "garbage" data. Aloha is a simple method whereby tags "back off" for a random amount of time and then retransmit. This "spreads" the transmissions to enable the reader to read every tag.

Amplitude Modulation (AM) AM is a method of combining a radio frequency and an information signal. It is widely used in commercial radio (the alternative being FM, or Frequency Modulation). AM is also used in RFID tags.

ANSI American National Standards Institute. ANSI administers the United States standardization and conformity assessment system.

antenna Converts a radio wave into alternating current (a *receiving* antenna), or an alternating current into a radio wave (a *transmitting* antenna). The antenna within a passive tag provides the power as well as the data received by the tag. Tag antennas can be either copper coils or made of metallic ink printed on some material as part of the tag.

antenna gain The increase in RF energy output of the antenna in a particular direction.

ASN Advance Shipping Notice. This is notification that suppliers must provide to their customers so that the customers' computers will know what pallets and cases to expect along with the RFID information. When the items arrive, they are matched by the RFID scan to the correct order, invoice, and other business information.

Auto-ID Center MIT formed the Auto-ID Center in 1999 to create standards for RFID. The center's work has been taken over by EPCglobal.

auto-ID tags RFID tags used in the supply chain.

backscatter Radio waves transmitted by a passive tag using power from the reader. Backscatter is modulated (or encoded) with data the reader can read.

bar code Automatic identification technology that encodes information as an array of parallel rectangles, spaces, and bars. The information is encoded in the width of the various bars. Bar code technology is widely used but has limited capabilities as compared with RFID.

beacon An active tag programmed to wake up and broadcast a signal at preset intervals.

beam power tag Another term for passive tag.

capture window, field, or zone The region around a reader antenna within which the tag will transmit.

case   A product package, a unit of shipment. A case contains items and may be itself contained in a larger unit such as a pallet or a shipping container.

case analysis   Analytical process to discover the profile and characteristics of a smart label or RFID tag that will work for a particular application.

certified label   Label approved by an equipment manufacturer for use with their equipment.

channel   A specific frequency within a band, used for a specific purpose. The UHF band in the United States is divided into 200 channels, and readers transmit on only one channel at a time. The regulations limit the amount of time a reader can occupy a single channel to 4 seconds, after which the reader must "hop" to another channel.

checksum   A number computed and stored with a value in tag memory. The reader recomputes the checksum as a way to verify that it has received the correct value.

circular polarization   Antenna design, where energy is broadcast in a circular pattern. Contrast with linear polarization. Circular polarized antennas are used when you cannot guarantee the orientation of the tag.

closed loop system   An RFID system used within a single business. A closed loop system can use nonstandard designators, since there is no need to share information with any other entity.

collision avoidance algorithm   A method for sequencing the broadcast of a group of tags located within range of a reader at the same time, so they do not interfere with one another. This is usually implemented in the firmware of the tag and reader.

commissioning   Initially assigning a value to a tag. Usually occurs when the tag is first put into service.

controller   A device that controls a group of readers. RFID controllers are a relatively new category, and they perform a number of functions in an RFID network including management of the readers and insulation of the application software from multiple duplicate reads.

CPG   Consumer Packaged Goods. A category of products such as diapers, detergents, razors, and soap that are packaged and sold to consumers.

CRC   Cyclic redundancy check. The checksum calculation field in an EPC Class 1 tag or EPCglobal Gen-2 tag. The CRC enables the reader to verify that it has received a valid response.

cross-docking   Moving material across the dock from receiving to shipping without putting it away first. Cross-docking eliminates at least half the time the material is handled, reduces cycle times, improves customer responsiveness, and reduces costs.

curtain The area of a packaging line where reader antennas are arranged to read tags passing between them.

cycle The complete wavelength from peak to trough to peak of a radio wave.

data transfer rate The rate at which a reader can read data from a tag. The rate is measured in bits per second.

dB Decibel. A unit of measure for antenna gain and power output.

DC Distribution Center.

dead zone An area where a radio frequency signal cannot be read.

detune The reduction of power available to the tag from the reader caused by the presence of metals, liquids, sodium, graphite, and simple distance.

dielectric loss The loss of power due to a substance's resistance to electricity.

dipole An antenna of half the wavelength of the frequency used, interrupted in the center.

discovery services Elements of the EPCglobal network, the discovery services enable a system to access information about the item. The two discovery services announced are ONS and PML Server.

DoD United States Department of Defense.

EAN European Article Number. A code, 8 or 13 numeric digits long, originally used by companies outside North America to identify themselves and their products. EAN is identical to UPC version A, except for the number of digits. EAN-13 has 10 numbers, plus two characters to identify the country that issued the number, and a checksum.

EAN International A nonprofit European organization that administers the EAN. Now part of EPCglobal.

EAS Electronic Article Surveillance, a theft detection system. EAS tags are attached to clothing and other items in a retail store. EAS tags are very inexpensive; they usually store just a single bit, turned "off" when the purchaser pays for the product. This enables the reader located at the door to pass the item without sounding the alarm. EAS is a simple system. It affords none of the management benefits of item or pallet tracking.

ECC Error Correcting Code. Supplemental bits in a transmission that enable the system to compute the value of any bits that are missing.

ECCnet A service of the Electronic Commerce Council of Canada. A data synchronization service, providing a secure, online, single source of standardized item data. It is continuously synchronized with trading partners.

EDI Electronic Data Interchange. An older format used for exchanging business data such as invoices, purchase orders, and Advanced Shipping Notices.

EEPROM Electronically Erasable Programmable Read-Only Memory. A storage method used in some tags. The distinguishing characteristic is that, while its contents are normally unchangeable, they can be over-written using special, qualified equipment.

EIRP Equivalent Isotropic Radiated Power. A measure of the antenna power used in the United States. EIRP is measured in watts. EIRP = 1.64 ERP.

electromagnetic field A field of energy produced when electrons in a wire accelerate and decelerate. Electromagnetic fields are characterized by wavelength and frequency.

electronic pedigree An electronic record of each movement of a particular item through the supply chain.

EMI Electromagnetic Interference. Electrical signals at the same frequency as the communication, caused by motors, fluorescent lights, adjacent readers, sunspots, or alternative transmitters. These signals can disrupt radio communications between reader and tag.

encoder Part of a smart label printer that writes information to tags.

encoder/printer A smart label printer that encodes the tag, prints the label, and verifies the functioning of the tag.

EPC Electronic Product Code. Created and managed by EPCglobal, backed by the UCC and EAN International, this labeling code identifies manufacturer, product category, and individual item. The code can be 8 to 32 characters, or 64 to 256 bits in length.

EPCglobal EPCglobal is a joint venture of EAN International and UCC. EPCglobal is the organization that oversees Electronic Product Codes and publishes standards for tags, readers, encoders, and reader-to-tag communications.

EPC Header The part of the EPC data format that identifies the EPC version number. It also designates the length or type of EPC. (The header is normally 8 bits, or two characters long, but it can be up to 8 characters long under certain specifications.)

EPC manager The manufacturer of the product, who is also the organization responsible for managing the information in subsequent fields. The EPC Manager field is 28 bits in length.

EPC middleware   Software that enables a network of RFID readers to exchange data with applications. Middleware also performs functions of reader monitoring and management.

EPCglobal Network   A system and method for using RFID technology in the global supply chain by using inexpensive RFID tags and readers to pass Electronic Product Code numbers and then leveraging the Internet to access associated information that can be shared among authorized users. There are five components of the EPCglobal Network: EPC (the Electronic Product Code), the ID System, EPC middleware, discovery services, and EPC information services.

EPC object class   The EPC object class identifies the type of product. The EPC object class is assigned and managed by the EPC manager. It corresponds to an SKU.

EPC serial number   The unique identification number for the item. Inclusion of this field enables EPC to track products at the item level, not just the pallet, case, or container level.

EPCIS   EPC Information Service. A computer system for maintaining, updating, and disseminating the electronic pedigree of an item.

ERP   Enterprise Resource Planning, a software application. ERP systems manage data pertaining to both finance and operations, helping companies run the business and make informed decisions. ERP systems can benefit from the real-time data that RFID provides. In most cases, the existing ERP systems will need to be upgraded to handle real time data and item-level data that RFID systems provide.

ERP   Effective Radiated Power. A measure of antenna power used in Europe. See also *EIRP*.

ETSI   European Telecommunications Standards Institute. A regulatory body in Europe that publishes standards for various communications devices. Readers and other transmitters operating in Europe must conform to ETSI standards.

factory programmed tags   Tags that have information written into them during the manufacturing process.

far field   The distance where the electromagnetic field separates and radio waves propagate away from the magnetic field. (See also *near field*.)

FCC   Federal Communications Commission. The United States government agency responsible for regulating communications by radio, wire, television, satellite, and cable.

field programming   Programming information into the tags after the tag has been shipped from the manufacturer to an end user, or in some cases to the manufacturer's distribution locations. Field programming generally occurs immediately before the tag is attached to the object to be identified. The capability for field programming permits the introduction of data relevant to the specifics of the application into the tag at any time. In some cases, programming of all data in the tag is performed in the field. In other cases, some portion may be reserved for factory programming, and the remaining portion is written later in the supply chain process.

field strength   A measure of radio signal reception.

firmware   Control software programmed chip installed in a reader or part of a tag. If the chip is an EEPROM, it can be reprogrammed by special equipment. In a reader, this is the technology that allows equipment, once purchased, to be updated with new capabilities or bugs to be fixed. Generally the manufacturer issues updates to the firmware.

fixed RFID Reader   A reader permanently installed at a location, not mobile.

frequency   Number of repetitions of a waveform in one second, measured in *hertz (Hz)*. For example, 10 kilohertz equals 10,000 waveforms per second. 13.56 megahertz equals 13.56 million waveforms per second.

frequency hopping   A protocol to minimize interference with other devices. Readers and tags hop across at least 50 channels between 902–928 MHz in the United States. They use the Aloha algorithm and a combination of Quiet, Scroll, and Talk commands to read and sort multiple simultaneous tag signals.

full duplex   A method of communication whereby transmission in both directions use the same band at the same time. Full duplex is contrasted with half duplex transmission.

GDSN   Global Data Synchronization Network. A service to help companies keep their product-related data synchronized. GDSN networks are used by trading partners to communicate new and changed product descriptions, definitions, prices, packages, and so on.

GIAI   Global Individual Asset Identifier. A code used to identify fixed assets.

GLN   Global Location Number. A code used to identify physical, functional, and legal entities. Any organization may apply for a GLN manager number and then assign numbers to reference any of its locations (offices, loading docks, ATM stations, desks, airports). The GLN is used in bar codes. Organizations are encouraged to work through UCC rather than implementing their own location identification designators to facilitate data exchange with other companies.

global data synchronization   A process whereby item data in various databases is rendered into standard formats and given consistent definitions. Data synchronization is often required even among different business units within the same enterprise. Data synchronization is necessary so the data and formats can be used among trading partners to designate and describe the item. Data synchronization is a prerequisite to successful RFID system implementation if the tags are to be generally useful among the different organizations.

good label   A label wherein the RFID data is written correctly, the correct information is printed, and the content data is verified against the source.

GRAI   Global Returnable Asset Identifier. A code used to designate returnable containers such as pallets, barrels, trays, or cargo containers.

GTIN   Global Trade Item Number. A code used to designate items that are products and services traded by companies. Can be bar coded or read by an RFID system. Adding a serial number (making it a SGTIN, a serialized GTIN) enables the code to represent individual items.

half duplex   A method of communication where the same band is used in both directions, but only one direction at a time. The sender must relinquish the frequency to the receiver in order to receive the answer. Half duplex transmission is contrasted with full duplex transmission.

handheld RFID Reader   A reader that can be carried in a person's hand to read tags and locate tagged items.

header   The part of the Electronic Product Code that identifies the EPC version number. It also designates the length and identity type of EPC.

hertz   The measure of frequency oscillation. 10 MHz, or 10 million hertz, indicates that the waveform reproduces 10 million times per second.

high frequency tag   RFID tag operating in the 13.56 MHz band.

HiTag   A Philips chip optimized for operating ranges in excess of one meter. These are passive chips operating at the frequency of 100 to 150 kHz. HiTag chips support anti-collision features, mutual authentication between reader and tag, and encrypted data transmission. HiTag chips are used for livestock identification, casino game chips, and access control. They are used for ski ticketing, gas cylinder identification, and electronic keys, as well.

host computer   Computer that runs application and/or middleware software that interacts with RFID and other devices.

iButton   A computer chip enclosed in a small stainless steel "can." The chip can be read by touching the reader to the top of the can. The iButton offers address-only, memory, real-time clock, and secure and temperature transmission capabilities.

inductive coupling   A method of creating a current in a conductor without touching it directly to a power source. A tag using near-field communications responds to a reader by inductively coupling with its signal.

inlay   The combined chip and antenna mounted on a substrate. Inlays may be encased in paper to create smart labels, or in plastic or glass to create other RFID products.

interference   Electrical noise at the same frequency as the communication signal, interfering with its successful reception. (See also *EMI*.)

interrogator   Another term for reader.

IP Number Rating   A rating system for enclosures, specifying the level of protection they give their contents, as in the following table. (See also *NEMA rating*.)

(First #)	Description
0	No Special Protection
1	Protected Against Solid Objects > 50 mm in diameter
2	Protected Against Solid Objects > 12 mm in diameter
3	Protected Against Solid Objects > 2.5 mm in diameter
4	Protected Against Solid Objects > 1 mm in diameter
5	Dust Protected
6	Dust-tight
(Second #)	
0	No Special Protection
1	Protected Against Dripping Water
2	Protected Against Dripping Water When Tilted Up to 15° C From Normal Position
3	Protected Against Spraying Water
4	Protected Against Splashing Water
5	Protected Against Water Jet Spray
6	Protected Against Heavy Jet Spray
7	Protected Against the Effects of Immersion
8	Protected Against Submersion

Example

IP54 is dust protected and protected against splashing water.

item An individual product in trade. The item is a particular instance of the product.

item-level tracking Individual identification of items. Contrasted with lot-level tracking, which only identifies shipments of lots or cases or pallets of items and does not enable capture of information regarding individual items.

linear polarization An antenna design where energy radiates mostly in a straight line pattern at right angles to the antenna's plane. A linear polarized antenna's performance is highly sensitive to tag orientation.

line of sight A requirement that the receiver be able to "see" the sender (that is, there are no obstructions between them). Bar code operations are restricted to line of sight; RFID operations are not.

low-frequency tag RFID tag operating in 125 KHz band.

manager EPCglobal term for the organization responsible for the global data regarding an item, container, location, or individual. The EPCglobal manager is generally the manufacturer.

microwave frequency Radio waves oscillating at more than one GHz.

microwave tag RFID tag operating in 2.45 GHz band or 5.8 GHz.

NEMA Rating NEMA (National Electrical Manufacturers Association) is the most commonly referenced authority for control enclosures in the U.S. Other countries may have similar agencies—for Canadian installations, it's CSA (Canadian Standards Association) or EEMAC (Electric Equipment Manufacturers Advisory Council). The following table defines the meaning of the NEMA code values.

NEMA #	Description
1	General Purpose (Indoor)
2	Water Drip Proof (Indoor)
3R	Dust-tight, Rain-tight, and Ice Resistant (Outdoor)
4	Watertight and Dust-tight (Indoor/Outdoor)
4X	Watertight, Dust-tight, and Corrosion Resistant (Indoor/Outdoor)
9	Indoor Hazardous Locations
12	Industrial Use—Dust-tight and Drip-tight (Indoor)
13	Oil-tight and Dust-tight (Indoor)

Example

NEMA-4 is a watertight and dust-tight indoor/outdoor enclosure.

near field The portion of the electromagnetic field where the magnetic field is located. The near field is utilized by encoders to encode a tag. The workflow must position the tag so it is isolated and alone within the near field to ensure that the correct tag is encoded.

ONS Object Name Service. A system for looking up unique EPCs through a computer linked to the Internet. The ONS gives access to information provided by the manufacturer about the item.

opaque Material that reflects or absorbs a radio wave. Opacity is frequency-dependent; metal and liquids are opaque to UHF and microwave frequencies.

order cycle time A measure of supply chain efficiency, order cycle time is the time it takes for a location to replenish its stock of a given item or list of items.

overstrike A printer's ability to recognize a bad label, back up, and mark it bad by printing a grid on top of it.

pallet A skid full of cases that can be handled by a forklift.

passive RFID tag A passive tag is powered and activated by the electromagnetic waves of a reader, with a read range from a few centimeters up to about 25 feet. Passive tags generally have no battery, but a few companies are building passive tags with batteries to provide power for sensors and recording devices. They are still passive, because they transmit using only the backscatter from the reader.

patch antenna An antenna consisting of a metal plate mounted on a substrate. Also called a planar antenna.

PDF417 A two-dimensional (2-D) bar code symbology patented by Symbol in the early 1990s and now in the public domain. PDF417 is a bridge technology between bar codes, which store small amounts of data (typically 10–20 characters) and full EPC, which can store thousands of characters. PDF417 can be read with a laser scanner, but other 2-D symbologies must be read by an imaging scanner.

pilot The stage of implementation where the technology, process, and methods are installed and evaluated before being committed to their use in a production environment.

PML Physical Markup Language. An Auto-ID Center–designed method of describing products readable by humans and by computers. PML is a dialect of XML (Extensible Markup Language). The PML server is part of the EPCglobal network.

polarization The orientation of flux lines in an electromagnetic field.

portal A doorway or passageway. Readers are mounted in such a way as to read all tags passing through the portal. Examples of portals include dock door, shop door, or a location on a production or conveyor line.

POS   Point of sale. Refers to equipment at the point the sale occurs, such as a cash register.

print quality   The measure of compliance of a bar code symbol to AIM requirements. Elements of compliance are reflectance, voids, spots, edge roughness, dimensional tolerance, quiet zone, and encodation.

quiet label   A smart label that cannot be read from a normal distance.

radio waves   Electromagnetic waves at the lower end of the electro-magnetic spectrum.

read after print   A step in producing smart labels, where the printer interrogates the RFID tag and reads the contents of the tag it has just written to verify the label is good.

read rate   The speed at which the tags can be interrogated by the reader. Expressed in reads-per-second.

read redundancy   The number of times a tag can be read when in the read zone.

read-only memory (ROM)   A computer chip whose contents cannot be overwritten.

read-only tag   A tag encoded during the manufacturing process so its information content cannot be changed. EPC Class 0 tags are read-only.

read/write tag   A programmable tag, one that is capable of storing incoming information as well as broadcasting it.

reader   A device that reads data from and writes data to a tag. Fixed readers are permanently installed in a location; mobile readers are mounted on a forklift or truck. Some readers are handheld and can operate while being held in a worker's hand.

RIED   Real-time in-memory event database.

replenishment   Supply chain term for ordering and receiving stock.

reprogrammable   A tag's ability to have its contents rewritten after it has been commissioned.

RF   Radio frequency.

RFID   Radio Frequency Identification. A method of tracking using radio waves that trigger a response from a device attached to an item.

RSS   Reduced-Space Symbology. A technology commonly used to bar code label items that are too small for traditional bar codes, or where there are additional data requirements, such as expiration and weight for random-weight items (such as produce and other perishables). RSS Expanded can encode up to 74 characters. RSS 14 encodes 14 characters.

savant   Auto-ID Center term for distributed network software that manages and moves data related to EPCs. Now called RFID middleware.

semi-passive tag   A passive tag that uses battery power to gain additional transmission range.

sensor   A device that can detect a physical condition such as temperature, moisture, radiation, presence of certain chemicals, vibration, or tampering.

serialization   Giving objects unique numbers for identification purposes.

shrinkage   The measure of product loss after receipt from a supplier. Shrinkage consists of employee theft, stock mismanagement, poor record keeping, waste, and loss.

signpost   Signposts activate only those active tags within their immediate vicinity, enabling precise identification of tagged items at specific locations within the business operation. Signposts can also transmit commands and configuration information to tags, enabling data read-write, adjustment of tag communication rate, and adjustment of tag frequency. In addition, signposts can activate and deactivate tags, a feature especially important for air cargo and other items that must adhere to strict FAA regulations.

skimming   Reading a tag without authorization. Opponents of RFID tags in identification documents worry that their contents could be read—skimmed—and thus reveal personal information to unauthorized persons.

SKU   Stock keeping unit. A particular class of item, such as an 8-foot 2×4, or a Heinz 24-oz. bottle of mustard. Individual items within that class have the same SKU.

slap-and-ship   Manually affixing RFID tags or smart labels to packages just before shipment. Many companies faced with a customer mandate have adopted the practice of slap-and-ship as the lowest cost means of compliance. They lose the management benefits of the system, but they do comply with the mandates.

sleep mode   The command to a tag to stop transmitting. It is used as part of the collision-avoidance protocol to reduce redundant reads.

smart label   A bar code label that also contains an RFID tag. It can store information and communicate with a reader, but it also has a bar code and text that can be read by a human. Smart labels are printed and encoded by an encoder/printer at the same time.

smart shelves   Shelves with RFID readers on them. Connected to a suitable system, they can report low stocks or empty shelves and thus trigger replenishment orders without human intervention.

SSCC   Serial Shipping Container Code. A code used to identify cases, pallets, and other containers.

strap   The section of an RFID tag surrounding the chip. The antenna is attached to the strap.

supply chain   A group of companies working together to manufacture, inventory, and supply goods and materials. The members of a supply chain are called trading partners.

symbology   A set of symbols documented to represent textual or numeric information.

tag   Generic term for RFID device. Tags are attached to items, cases and pallets, equipment, animals, documents, and persons to identify them. Tags are read by equipment called readers, which read their contents and transmit them to a computer system.

tag-it   A set of RFID protocols published by Texas Instruments and supported by numerous vendors.

telemetry   Measurement data. Examples of telemetry include temperature, radiation, location, time stamps, incidence of tampering, location, moisture, and vibration.

tote   A reusable container that can be carried by hand.

trade item   A product packaged in lot sizes for transport and sale.

transponder   Another word for *tag*.

UCC   Uniform Code Council. A membership organization that jointly manages the EAN.UCC system, including Uniform Product Code in the United States and Canada.

UCCnet   A product registry service for electronic commerce. Based on the Internet, it enables synchronization of item and location information among trading partners.

UHF   Ultra high frequency. This designates waves in 300 MHz to 3GHz range. UHF offers good range, but the waves don't penetrate opaque materials well.

UHF Generation 2 Foundation Protocol   The new EPC class standard for passive RFID tags. It replaces Class 0 and Class 1 specifications.

UID   Universal Identification Designator. A code used by the Department of Defense to mark and track assets.

UPC   Universal Product Code. The name of the bar code standard in North America. Version A consists of ten identification digits and two overhead digits. The first digit designates the type of product; the final digit is a checksum used to detect errors. Characters 2–6 designate the manufacturer. These are followed by a separator (called "guard bars"), and a 5-digit product code. Version E of the UPC is used on packages too small for Version A; it drops zeros and the middle guard bars.

wavelength Measure of one peak to the next peak of a radio wave. The wavelength is distance, measured in meters.

WiFi See *802.11*.

workflow A set of steps to accomplish a goal. A workflow is the definition of the steps people must take and actions systems must perform, in the correct order. The workflow may include alternate paths, to accommodate various contingencies and decision points. Also called a work process.

WORM tag Write once, read many tag. This tag can be written to only once, and then the information is fixed and can be read as many times as necessary.

XML Extensible Markup Language. A language for defining and sharing documents. XML can be read by computers and by humans.

INDEX

www.ingramcontent.com/pod-product-compliance
Lightning Source LLC
Chambersburg PA
CBHW062010190326

41458CB00009B/3033